STRENGTH OF MATERIALS

By

J. P. DEN HARTOG

PROFESSOR OF MECHANICAL ENGINEERING
MASSACHUSETTS INSTITUTE OF TECHNOLOGY

DOVER PUBLICATIONS, INC.
NEW YORK

Published in Canada by General Publishing Company, Ltd., 30 Lesmill Road, Don Mills, Toronto, Ontario.

Published in the United Kingdom by Constable and Company, Ltd., 3 The Lanchesters, 162-164 Fulham Palace Road, London W6 9ER.

This Dover edition, first published in 1961, is an unabridged and unaltered republication of the first edition published by McGraw-Hill Book Company, Inc., in 1949.

International Standard Book Number: 0-486-60755-0

Manufactured in the United States of America
Dover Publications, Inc., 31 East 2nd Street, Mineola, N.Y. 11501

PREFACE

No apology can logically be expected from an author who has had the temerity to add still another to the existing large collection of college textbooks on strength of materials, and hence no apology is offered. This text is intended for use in the standard one-semester course on the subject in engineering schools. The material treated in it is more extensive than can conveniently be covered in such a short time, and hence the instructor must leave out some of it. It is suggested that the material *not* indicated by asterisks in the table of contents is so fundamental that it has to be included in any course. From among the remaining paragraphs, marked with asterisks, the instructor can then make a suitable choice to round out the course, and it is pretty well immaterial which paragraphs he so selects.

The book is intended to be a sequel to my "Mechanics," published in 1948, in that it employs the same notations, but it also can be used independently, following the study of any other text on engineering statics and dynamics.

In writing a preface to "Strength of Materials," my thoughts naturally go back to the days when I was a student, learning the subject for the first time. I remember that first course with pleasure and gratitude because I took it under a truly great and warm-hearted teacher, Professor C. B. Biezeno, of the Polytechnic in Delft, Holland. It was Biezeno who introduced me to the mysteries of the celebrated Dr. Myosotis, whose formulas (page 88) he called the *vergeet-mij-nietjes*, in good Dutch. After the alphabet and the tables of multiplication, nothing has proved quite so useful in my professional life as these six little expressions, and undoubtedly some of my students will have the same experience.

Pleasant memories of the past are precious possessions of the present. If therefore a few of my readers in later years can think back to their course in strength of materials with half the pleasure with which I remember mine under Biezeno, I shall hold myself well rewarded for the labor of writing this book.

J. P. Den Hartog

Cambridge, Mass.
April, 1949

CONTENTS

*Subjects marked with an asterisk * may be omitted in a simplified course.*

CHAPTER VI—SPECIAL BEAM PROBLEMS

CHAPTER VII—CYLINDERS AND CURVED BARS

CHAPTER VIII—THE ENERGY METHOD

CHAPTER IX—BUCKLING

CHAPTER X—EXPERIMENTAL ELASTICITY

STRENGTH OF MATERIALS

CHAPTER I

TENSION

1. Introduction. By long tradition and common practice our subject is called "strength of materials," and, as is often the case with a traditional name, it does not strictly fit the facts. An uninitiated person would naturally think that under the heading "strength of materials" we would study the forces that various materials can withstand before they break or deform, but *that* subject is commonly called "properties of materials" or "materials testing." The traditional content of a course in strength of materials can more aptly be described as the "statics of deformable elastic bodies." In our subject we shall calculate the stresses and deflections or deformations in the beams, shafts, pipes, or other structures as functions of the loads imposed upon them and of the dimensions of the structure. As we shall see later, these stresses are usually independent of the material: a steel beam and a wooden beam of the same dimensions under the same loads will have the same stresses. The question of the conditions of failure of such a beam under the load, surprisingly enough, is only of secondary significance in our subject. In a typical calculation almost all of the work, say 95 per cent of it, will be "statics," independent of the material in use, and only at the very end will we substitute numerical values for an "allowable working stress" and for a "modulus of elasticity" of the material at hand.

The reason for this nomenclature is largely historical. Our subject owes much of its development to a great school of French mathematicians in the first half of the last century, of which the most outstanding names are Poisson, Lamé, Navier, Poncelet, Saint-Venant, and Boussinesq. Being mathematicians, they naturally considered their problem completely solved as soon as they had a formula relating stress to loading, and moreover they were convinced that they were working on a "practical" subject. Hence they gave to their subject the practical name *résistance de matériaux*, and their influence was so great that the name has persisted to this day among engineers in the English-speaking world.

Another distinction in nomenclature exists between "strength of materials" and "theory of elasticity." The general mathematical problem of finding the stresses for given applied loads on a body of arbitrary shape is extremely difficult and in fact has not been solved even by the mathematicians. On the other hand, we do possess solutions for simplified bodies, such as "beams" or "shafts," which are bodies in which the sidewise dimensions are negligibly small with respect to the length, or "plates," "slabs," and "curved shells," in which the thickness is negligibly small with respect to the other two dimensions. Although no practical structure exists in which one or more dimensions are mathematically negligible, they are negligible in practice, so that we "idealize" our given structure into equivalent beams, shafts, or plates. To this idealized structure we apply the theory. The more simple parts of that theory (bending, torsion, etc.) are called "strength of materials," while the more complicated parts are usually named "theory of elasticity." Hence the distinction between the two is not great. This book deals with the more elementary portions of strength of materials, sufficient to give a practically adequate solution for a great many of the commonly applied structures and machine elements. Therefore, to an engineer our subject is just about the most important of the subdivisions of mechanics that he must know.

Besides the questions of stress and deflection of structures, the subject also embraces the problem of stability. The main question in this category is the buckling of columns, which is the determination of the compressive load that can be placed lengthwise on a long, thin column before it buckles out sidewise. This problem is discussed in Chap. IX.

2. Hooke's Law. In the usual "tensile test" a bar of steel or other material is placed in a tensile-testing machine, and while it is slowly being pulled, readings are made of the pulling force and of the change in length (elongation) of the center portion of the bar. When these two quantities are plotted against each other, a diagram such as Fig. 1 results. In it we distinguish three stages, OA, AB, and BC. During the first stage OA, the diagram is substantially a straight line, and for ordinary steel the elongation OA is about one part in a thousand (0.001 in. elongation per 1 in. gage length), while the stress OA is about 30,000 lb/sq in. Moreover, if the load at A is let off again, the force-elongation diagram goes back along the same line AO to O, and in particular the bar returns to point O, which means that after release of the load there is no "permanent elongation." This first stage OA is called the "elastic" stage. The next, or second, stage AB usually has some undulations in the diagram as shown, but it is substantially a horizontal piece of curve at constant force (constant stress), and the elongation will increase from 0.001 in./in. at A to about 0.020 in./in. at B. Thus the horizontal distance AB is very much greater than the horizontal distance OA, which could not be conveniently shown in the diagram. This second stage is known as the "plastic" stage, in which the stress is substantially constant and equal to the "yield stress," about 30,000 lb/sq in. for structural steel. In the third stage BC, the deformations become very large, and the test piece "necks," or shows a locally diminished diameter, and finally breaks in the center of the necking at point C in the diagram. If the load is let off at some point between A and C, before the final failure, the return curve is substantially a straight line parallel to OA, as shown by the dotted lines, and in that case the bar shows a "permanent elongation" after release of the load to zero.

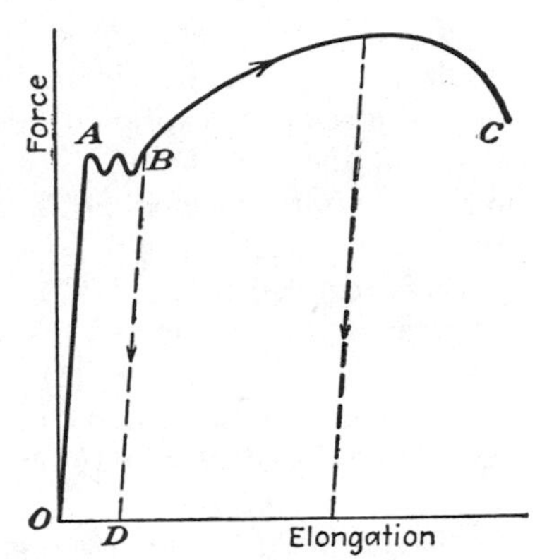

FIG. 1. Stress-strain diagram, shown distorted for purposes of illustration. In reality the slope OA is so steep that point A practically lies on the ordinate axis.

Our subject of strength of materials deals only with the material in the first, or elastic, stage OA. Another branch of mechanics, named "plasticity," dealing with the second stage A to B, was started about twenty-five years ago. It is not only much younger but also much more difficult and complicated than our present subject and hence altogether outside the scope of this text.

The elastic, linear behavior of a material in the region OA of Fig. 1 is known as "Hooke's law."[1] It states that the elongation

[1] The law was first enunciated by Robert Hooke (1635–1703) in Elizabethan England in connection with his invention of applying a hairspring to a watch or clock. Before his day no portable clocks existed, since they all depended on pendulums or on even more primitive devices. Hooke, after the manner of his age, published his invention in the shape of a riddle, or "anagram,"

is proportional to the force, or expressed in a formula:

$$s = \frac{P}{A} = \frac{E\,\Delta l}{l},$$

or

$$\frac{s}{E} = \frac{\Delta l}{l} = \epsilon. \tag{1}$$

The quantity $\epsilon = \Delta l/l$ is the "unit elongation," or "strain," expressed in inches per inch and hence dimensionless. The quantity s is the stress or the force per unit area, expressed in pounds per square inch. Finally, E is a proportionality constant, which also must be expressed in pounds per square inch. It is known as the "modulus of elasticity," or also as "Young's modulus," after its inventor Thomas Young (1773–1829). The formula states that E has the dimension of a stress. In fact, it is the stress required to make $\Delta l/l = 1$, or to elongate the test piece to double its original length. Now this is an artificial statement, because steel obeys the law (1) only up to elongations of about one part in a thousand (Fig. 1), and the test piece will have broken long before it has doubled its length, but still the statement is useful for the visualization it conveys of the size of E. For steel, E = 30,000,000 lb/sq in.; the yield stress s_y is about 30,000 lb/sq in.; and hence the yield unit elongation, or yield strain, is 0.001.

Within the limit of elasticity, then, the elongation is very small compared with the original length, and it will be assumed infinitesimally small in the sense of the calculus. This assumption leads to conclusions that are in good agreement with experimental facts, except for rubber or similar materials, which can double their lengths and in which the strain consequently can be large. The behavior of rubber under stress is not treated in strength of materials. It forms a subject in itself called "elasticity of large deformations," which is extremely complicated and still in the beginning of its development.

Problems 1 *to* 21.

3. Simple and Compound Bars. If a bar of a cross section A, constant along its length, is subjected to a tensile or compressive

---

ceiiinosssttuv, which, when unscrambled, was supposed to read *Ut tensio sic vis*, Latin for "As the extension, so is the force."

force P, that is, to a pair of forces directed along the length, then the elongation of that bar is determined by Eq. (1):

$$\Delta l = \frac{Pl}{AE}. \tag{1a}$$

If the cross section varies along the length x, that is, if A is a function of x, then the same relation still holds for a small element of length dx (Fig. 2). The elongation $\Delta\, dx$ of an element originally of length dx is:

$$\Delta\, dx = \frac{P\, dx}{AE},$$

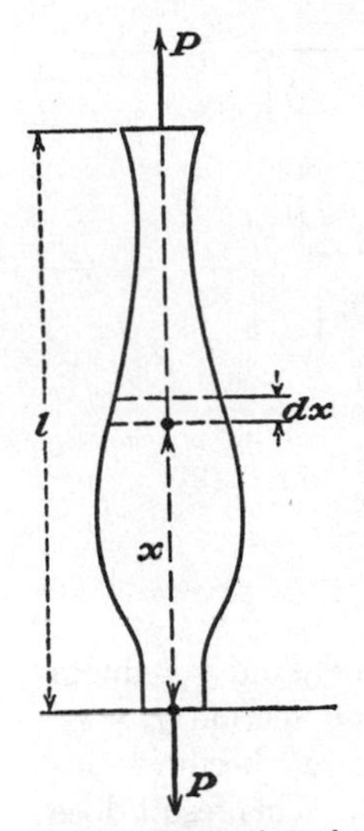

FIG. 2. Elongation of a bar of variable cross section $A(x)$ loaded by a tensile force P.

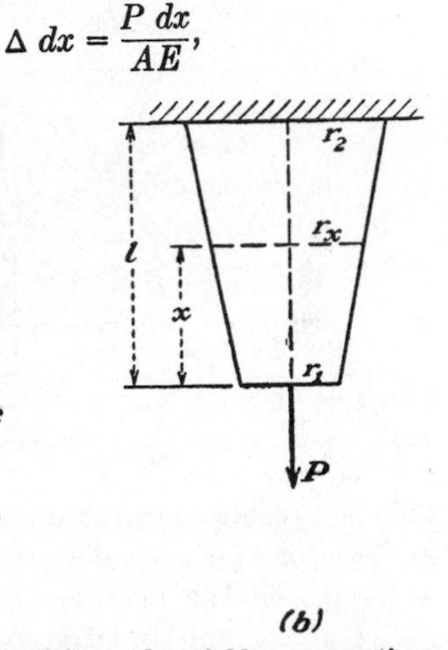

FIG. 3. Examples of bars of variable cross section. Equation (2) applies with good accuracy to case b, but not so well to case a. See also Fig. 167 on page 204.

and the total elongation of the entire bar is:

$$\Delta l = \int_0^l \Delta\, dx = \int_0^l \frac{P\, dx}{AE} = \frac{P}{E}\int_0^l \frac{dx}{A} \tag{2}$$

As we shall see later, on page 204, this relation is strictly valid only if the bar changes its cross section gradually, *i.e.*, if the curve of A plotted against x does not show steep slopes. In particular, it does not apply strictly to the practical case of Fig. 3*a*, where two pieces, each of constant cross section, are attached to each other, forming an abrupt change in that section. For that case Eq. (2) becomes:

$$\Delta l = \frac{P}{E}\left(\frac{l_1}{A_1} + \frac{l_2}{A_2}\right),$$

and it is only approximately true. However, for the truncated cone of Fig. 3*b* the cross section changes sufficiently gradually so that the expression is perfectly good. In this case, the diameter of the bar changes linearly with the length, so that we can write:

$$r_x = r_1 + \frac{x}{l}(r_2 - r_1),$$
$$A_x = \pi r_x^2.$$

By Eq. (2):

$$\Delta l = \frac{P}{\pi E}\int_0^l \frac{dx}{r_x^2} = \frac{P}{\pi E}\int_0^l \frac{dx}{\left[r_1 + \frac{x}{l}(r_2 - r_1)\right]^2}$$

$$= \frac{P}{\pi E}\frac{l}{r_2 - r_1}\int_{x=0}^{x=l} \frac{d\left[r_1 + \frac{x}{l}(r_2 - r_1)\right]}{\left[r_1 + \frac{x}{l}(r_2 - r_1)\right]^2}$$

$$= \frac{P}{\pi E}\frac{l}{r_2 - r_1}\left[\frac{-1}{r_1 + \frac{x}{l}(r_2 - r_1)}\right]_{x=0}^{x=l}$$

$$= \frac{P}{\pi E}\frac{l}{r_2 - r_1}\left(\frac{1}{r_1} - \frac{1}{r_2}\right) = \frac{Pl}{\pi E r_1 r_2}.$$

This is a result of the same dimension as Eq. (1*a*), and it reduces to (1*a*) for the special case of a constant cross section $r_1 = r_2$. We also see that the elongation Δl becomes infinitely large for a full cone $r_1 = 0$. This result does not appear strange any more when we reason that the stress at the apex of the cone is infinite, so that the problem then reduces to a practically impossible mathematical abstraction.

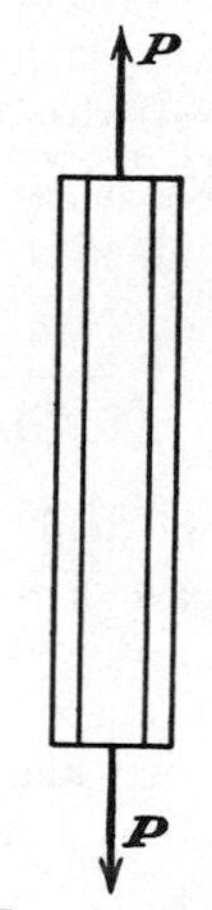

FIG. 4. Steel bolt with a shrunk-on bronze bushing, in tension.

Next we consider the case of Fig. 4, a steel bolt with a bronze bushing shrunk hard around it, so that the two pieces form a solid unit. The modulus of steel E_s is different from that of bronze E_b; also the cross sections A_s and A_b are different. The length l and a possible elongation Δl must be the same for the steel and the bronze, but the total force $P = P_s + P_b$ divides itself into unequal parts between the two. Thus we can write:

$$\frac{\Delta l}{l} = \frac{P_s}{A_s E_s} = \frac{P_b}{A_b E_b},$$
$$P_s + P_b = P.$$

From these two equations we solve for P_s and P_b with the result:

$$P_s = P \frac{A_s E_s}{A_s E_s + A_b E_b},$$

$$P_b = P \frac{A_b E_b}{A_s E_s + A_b E_b}.$$

The reader should now verify that, for the case that $A_b = A_s/2$ and for $E_s = 30 \times 10^6$ lb/sq in. and $E_b = 12 \times 10^6$ lb/sq in., one-sixth of the load is carried by the bronze and five-sixths by the steel. How does the stress in the bronze then compare with the stress in the steel? Could the fact that these stresses come out in ratio of the two E's have been seen immediately without calculation?

Another problem is when the compound bolt of Fig. 4 is not subjected to a tensile load P as shown but instead is heated to a temperature $T°$ above room temperature. We continue to assume that the shrink fit is so strong that no slippage occurs. The bronze will tend to expand longitudinally more than the steel but cannot do it. As a result, the bronze will not quite expand to its full length freely and hence will find itself in compression, while the steel is pulled by the bronze to expand more than it likes to and hence is in tension. Before bringing this story into formulas, we remark that the free thermal expansion of bronze is $\alpha_b Tl$, where α_b is the expansion coefficient expressed in inches expansion per inch length per degree temperature rise. If the bronze so expands, there is no stress in it; but if it is elongated by a different amount, that difference must be caused by a pull which is a stress. Hence:

$$\Delta l - \alpha_b Tl = ls_b/E_b,$$

$$\Delta l - \alpha_s Tl = ls_s/E_s, \quad (-)$$

$$(\alpha_b - \alpha_s)T = \frac{s_s}{E_s} - \frac{s_b}{E_b}.$$

By statics:

$$0 = P = P_s + P_b = s_s A_s + s_b A_b = 0.$$

From the above two equations we solve for the stresses s_s and s_b or for the corresponding forces, with the result:

$$P_s = A_s s_s = -P_b = -A_b s_b = (\alpha_b - \alpha_s)T \frac{A_s E_s \times A_b E_b}{A_s E_s + A_b E_b}.$$

Before numerical values are substituted (Problem 30), the formula should be checked for dimensions and for special cases. It can be seen, for example, that the stresses become zero for $\alpha_b = \alpha_s$, or for $T = 0$, or for one of the two cross-sectional areas becoming zero. Also it is noted that the two forces P_s and P_b are equal with opposite signs; the negative force is interpreted as compression and the positive one as tension.

In the first statement of this thermal-stress problem it was said that the temperature was raised T degrees above "room temperature." We tacitly assumed at that time that there were no internal tensions or compressions along the bar at that room temperature, and with that assumption the above calculation is correct. However, when we ask what these internal longitudinal stresses are at room temperature, the question is unanswerable, because any amount of internal longitudinal stress (up to the yield stress) may be locked up in the bar, depending on how it is assembled. In fact, it is extremely difficult, and in practice just about impossible, to assemble the bolt of Fig. 4 so that it has no internal locked-up longitudinal stresses at room temperature. Then the above calculation only gives us the change in these stresses caused by the change in temperature, and the total longitudinal stress in the bolt when hot is the sum of the stresses just calculated and the locked-up stresses in the cold state.

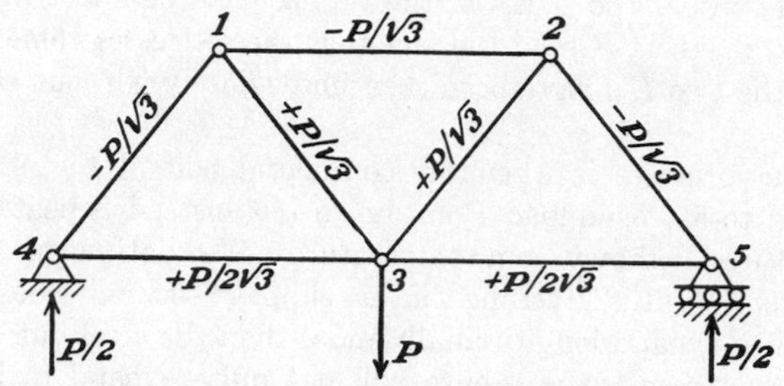

FIG. 5. Forces in the bars of a simple truss.

Problems 22 to 32.

4. Trusses. Consider the truss of Fig. 5, made up of seven bars, all of length l, including angles of 60 deg with each other, loaded with a single central load P and supported at its two ends with reactions $P/2$. By the usual methods of statics the tensile or compressive forces in the bars can be determined. The reader is asked to perform these calculations and to verify that the answers are as shown in the figure, where positive quantities denote tensile forces and negative signs indicate compressive ones, as usual.

To find the stresses in the bars we simply divide the forces by the cross-sectional areas of the bars, and if these cross sections are all alike, as is often the case in practice, then the stresses are proportional to the forces shown in Fig. 5. This completes the problem for most practical purposes, because as a rule we are not much interested in the deflection of a statically determined truss such as the one we are considering. However, for statically indeterminate ones the bar forces depend on the deflections, and hence it is

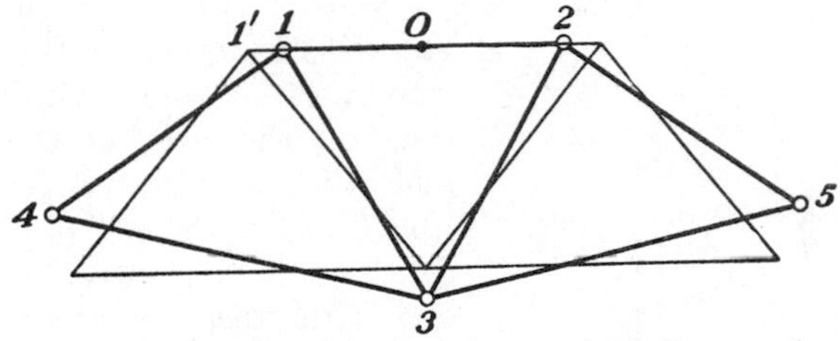

FIG. 6. Distorted shape of the truss of Fig. 5. The thin lines are the unloaded, undistorted shape; the heavy lines show the distorted shape.

important to have a method for determining them. The first step in this process is the calculation of the elongations of all individual bars by Hooke's law, Eq. (1*a*). Then we know the distorted lengths of all the bars, and in principle we could construct the truss graphically in its distorted form. This has been done in Fig. 6 in an exaggerated manner. The actual elongations of the bars are 1 part in 1,000 of their lengths at the most, so that the difference between the distorted and unstressed bars cannot be seen in an actual drawing. Figure 6 represents what the figure would look like if the strains were of the order of 10 per cent, or 100 times as large as they really can be. Since the loading is symmetrical, we start from the middle O of the top bar and find the joints 1 and 2, the bar 12 being shortened by an amount $Pl/AE\sqrt{3}$. Then we take into our compass the new length of bars 13 and 23, which is $l + Pl/AE\sqrt{3}$, and with this length as radius we describe arcs about the new positions 1 and 2 as centers. These arcs intersect at the new joint location 3. From there we find joint 4 by making 14 shorter than l and 34 longer than l by the proper amounts. From the figure we can now find the vertical sag of the center joint 3 with respect to the two supports 4 and 5. Only, as has been said before, if Fig. 6 had been drawn true to scale, it would have been indistinguishable from the original truss, and hence the accuracy of the construction, as described, is totally inadequate.

Here we are helped out of the difficulty by a construction published in 1877 by the French engineer Williot. Instead of laying off the total lengths of the bars as in Fig. 6, he proposed to draw only the elongations themselves, as shown in Fig. 7. Again we start from the center of the top bar, which we call O. Now joint 1 moves to the right by an amount $Pl/AE2\sqrt{3}$, which we shall call one-half unit of length for further convenience. We lay off that distance to the right of O and so find point 1. Thus the stretch from O to 1 in the Williot diagram (Fig. 7) is the distance 1′-1 of Fig. 6, *i.e.*, the distance that joint 1 has moved between the unloaded and the loaded conditions of the frame. Similarly we find point 2. Next we consider bar 13, which stretches by a full unit. We think of that bar as loosened from its connection with the others at joint 3, but still attached at 1, so that it can turn about joint 1. First we allow the bar to stretch but keep it parallel to the old position. This stretch is parallel to the bar at 60 deg and leads us to point $3a$. Now we allow the bar to turn slightly about point 1 of Fig. 5 so that point 3 describes a circular arc. In the Williot diagram, Fig. 7, the radius of the circular arc is more than a thousand times the length 1 to $3a$ (Why?) and hence "infinite," so that the circular arc reduces to the dotted straight line through $3a$, perpendicular to $1\text{-}3a$. This dotted line is one locus of point 3. Now we consider bar 23 of Fig. 5, allowing it first to stretch appropriately, leading to point $3b$ of Fig. 7, and then allowing it to turn, leading to the dotted line through $3b$ as a second locus for point 3. Where the two loci intersect we find point 3 in Fig. 7. Again, as before, the stretch $O3$ represents the displacement of point 3 when point O is held steady. Now we proceed to construct two loci for joint 4, the first one by considering bar 34, which stretches by half

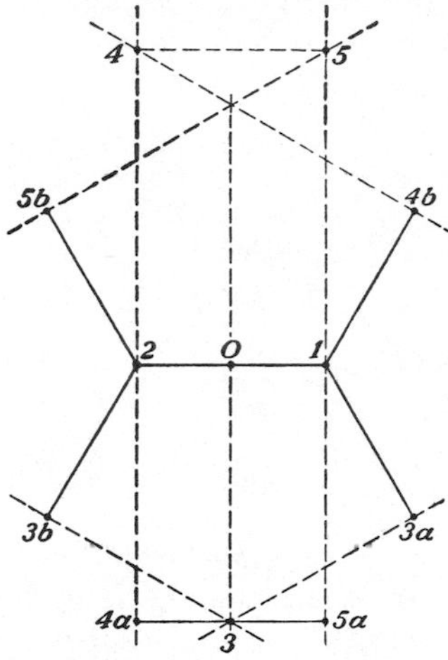

FIG. 7. Williot diagram for the truss of Fig. 5.

a unit (point $4a$ in Fig. 7) and then turns about point 3 in Fig. 5 or about a point very far to the left of point 3 in Fig. 7. The locus thus is the dotted line $4a$-2. The second locus is found by considering bar 14, which contracts by one unit (point $4b$) and then turns. The two loci intersect at 4, and then point 5 is found similarly from the bars 25 and 35. Now the Williot diagram is complete, and we find the downward displacement of the center joint 3 (Fig. 5) with respect to the fixed joints 4 and 5 as the vertical distance between 3 and 4 or 5 in Fig. 7. Since in Fig. 7 all angles are 60, 30, and 90 deg, the reader should check that that distance is $11/2\sqrt{3}$ units. One unit is the stretch of one bar under force $P\sqrt{3}$ or $Pl\sqrt{3}/AE$, so that the sag of the center of the truss comes out to be

$$\delta_P = \frac{11}{6} \frac{Pl}{AE},$$

or the sag of a solid bar of length $11l/6$, loaded by a force P.

The diagram of Fig. 7 is quite simple because we have chosen a simple and symmetrical example. In a general case the Williot diagram is more complex, but the steps are all the same. We start by considering one point of the structure and the direction of one bar through that point as fixed and proceed from that base logically from point to point.

Problems 33 *to* 38.

5. Statically Indeterminate Truss. A truss built up of consecutive triangles is said to be statically determinate, because for any loading the bar forces can be determined by the rules of statics only, and in particular the result is not dependent on Hooke's law. In other words the bar forces in a truss of which the material would have a quadratic elastic law (s = constant ϵ^2) would be exactly the same as when Hooke's (linear) law would hold, provided, of course, that the deformations in both cases are small in comparison with the bar lengths.

If in such a rigid, statically determined truss we insert an additional bar between two joints, that additional bar has to be made of just the right length in order to fit. If the new bar is a few thousandths of an inch long, we have to compress it elastically to insert it and we shall find our truss with internal stresses without any load. Such internal stresses without load are impossible in a statically determined truss.

If an indeterminate truss, *i.e.*, a truss with one or more "redundant" bars, is loaded, the distribution of the bar forces does depend on the elastic law involved and on the cross-sectional areas of the bars. If such a truss has internal stresses without load, the total forces will be the sum of those internal-no-load forces and the forces caused by the load.

As an example of all this, consider the truss of Fig. 8, consisting of two 45-deg triangles with a redundant diagonal bar. The two diagonals are not attached to each other in the center; they can freely slide over each other there. The lengths of the sides are l and of the diagonals, $l\sqrt{2}$. Assuming no internal stresses in the unloaded state, what are the various bar stresses when loaded with two forces P as shown? To answer a question of this sort we always follow the same procedure:

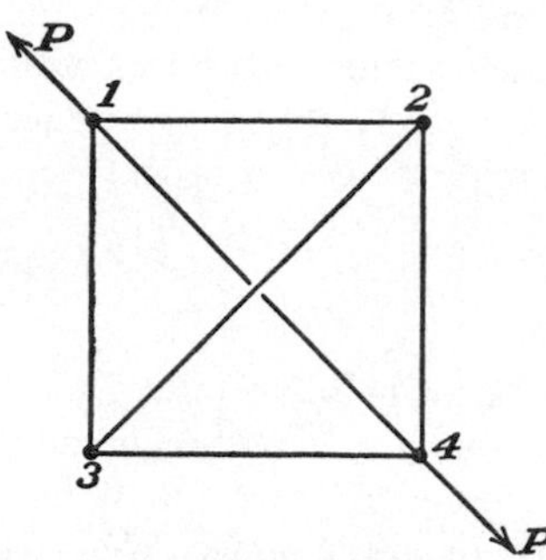

FIG. 8. A statically indeterminate truss with one redundant bar.

a. We replace the redundant bar (or bars) by the unknown force X (and Y, Z, etc.), thus making the truss statically determinate.

b. We calculate the bar stresses by statics in terms of the loadings P, X (and Y, Z, etc.).

c. We calculate such deformations of the determinate truss by Williot's construction as are necessary to find the elongations of what would have been the redundant bar (or bars).

d. We set those elongations equal to Xl/AE (and to Yl/AE, etc.), thus obtaining as many equations as there are ununknowns X (Y, Z, etc.)

e. With the known forces P, X (Y, Z, etc.) we calculate all the bar stresses.

Applying this general procedure to the example in hand, we replace the diagonal 14 by tensile forces X, so that the loading on the (determinate) truss becomes $P - X$. By statics we find for the tensile force in any one of the four side bars $(P - X)/\sqrt{2}$, and in the diagonal bar 23 we find $-(P - X)$. The shortening of the diagonal then is $(P - X)l/AE$, which, for convenience, we designate as the "unit" shortening. The Williot diagram of Fig. 9 is started from the point O, taken in the middle of the diagonal 23. Since that diagonal shortens, point 3 moves up and to the right with respect to O by one-half unit. Thus the points 2 and 3 are found in Fig. 9. Now we proceed to joint 1 via the bars 21 and 31. Bar 21 elongates $1/\sqrt{2} = 0.707$ unit, and we find point $1a$ in Fig. 9 by laying off 0.707 in a direction parallel to bar 21. Then we turn that bar about point 2 (Fig. 8), and the circular arc that point 1 (Fig. 8) describes becomes a dotted straight line through $1a$ in Fig. 9, being the first locus of point 1. We perform similarly on bar 31, finding point $1b$ in Fig. 9 and the second locus. The point of intersection is point 1. We see from the figure that the distance $O1$ is $1\frac{1}{2}$ units, so that, by symmetry, the complete diagonal 14 is 3 units. Hence the distance 14 increases by $3(P - X)l/AE$. This expression is now equated to the extension of the diagonal 14, of length $l\sqrt{2}$, pulled by a force X, and the equation is solved for X:

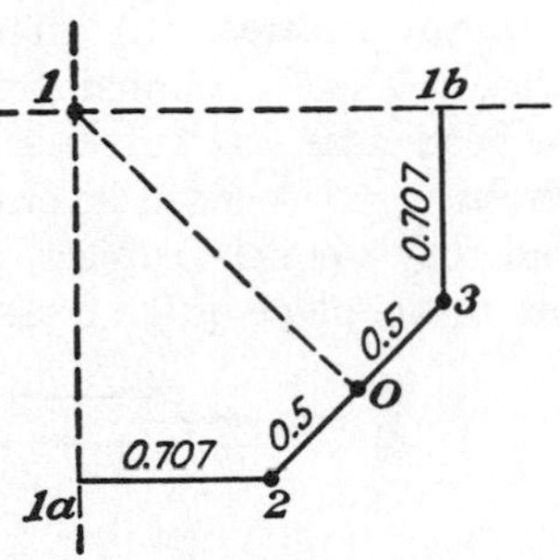

FIG. 9. Williot diagram for the truss of Fig. 8.

$$X = P\frac{3}{3 + \sqrt{2}} = 0.68P.$$

This then is the tension in the diagonal 14 of the statically indeterminate truss Fig. 8. The diagonal 23 is in compression with force $P - X = 0.32P$, and the four side bars are in tension with force $0.23P$.

From this derivation it should be clear that the result depends on the cross sections of the various bars. For example, if the diagonal 23 had a cross-sectional area small in comparison with all the others, then 14 would have carried the entire load P and all other bars would have been stressless. What would have been the force distribution if diagonal 14 had a cross section small in comparison with all others?

Problems 39 *to* 44.

CHAPTER II

TORSION

6. Shear Stress. A stress, or force per unit cross-sectional area, is called a "normal stress" when the direction of the force is perpendicular to the plane of the cross section. Such normal stresses are either tensile or compressive, depending on their sign, and they were the subject of the first chapter. When the stress lies in the plane of the cross section, it is called a "shear stress."

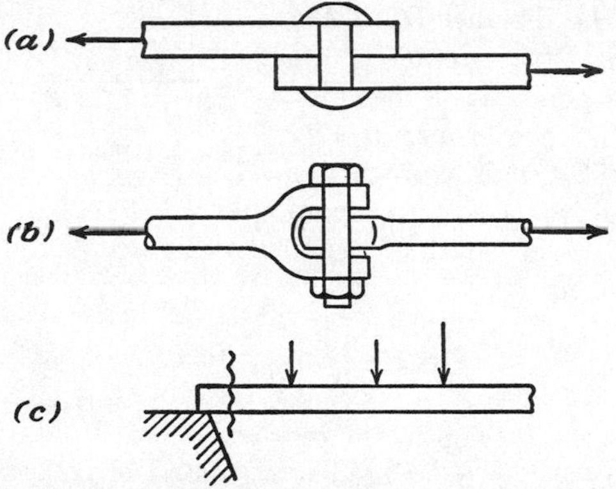

FIG. 10. Three frequently occurring cases in which shear stress is important.

Simple construction elements in which shear stresses occur are shown in Fig. 10. The first of these shows two boiler plates connected by a riveted lap joint. If the rivet is not in tension and hence the two plates do not press against each other, there is no friction between them and a tensile force between the plates must be transmitted as a shear stress across the mid-section of the shank of the rivet.

The second case shows two tension bars connected together through a fork-eye-bolt construction. The tensile force is transmitted from the eye to the two prongs of the fork in the form of a shear stress across two sections in the shank of the bolt. Finally, Fig. 10*c* shows the left end of a beam on its support. The section of the beam just to the right of the support must transmit the support reaction force from left to right, and hence this cross section must be subjected to a shear stress. In all these cases we know the total shear force and also the cross-sectional area that must take the shear force, but as yet we do not know how this shear force will be distributed over the area. The simplest assumption, of course, is that of uniform distribution, where the shear stress is the same all over the cross section. In most practical cases this is not quite true; the shear force distributes itself unevenly, showing stress concentrations in some parts of the area and leaving other

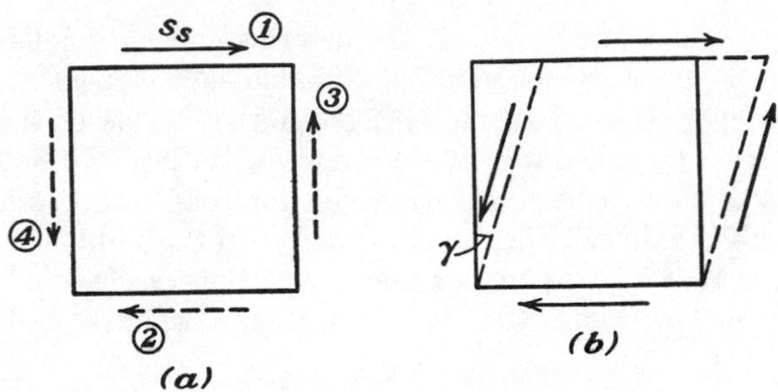

FIG. 11. An element subjected to pure shear.

parts with a small stress. The details of this distribution can be calculated only in a few rather simple cases (page 44), while mostly we can only guess. Therefore, in practice we assume uniform distribution and multiply the "average shear stress" thus found by a factor, that we politely call "factor of safety" because the expression "factor of ignorance" sounds too cynical. The stress thus multiplied is then designed equal to the yield stress, if we do not want yielding, or equal to the ultimate stress, if we are willing to allow yielding but do not want failure. We shall return to this general question on page 221.

Now we are ready to examine the details of shear stress, the deformation associated with it, and the relation between stress and deformation.

Consider a small cube of material of side length dx, Fig. 11, and let a shear stress s_s be acting on the top face as indicated by the number 1. The force on that face then is stress times area or $s_x\ (dx)^2$. For equilibrium in the left-right direction, it is necessary that an equal shear stress act on the bottom face in the opposite direction, shown by arrow 2. But now there is a clockwise moment acting on the particle, and in order to prevent it from spinning we must apply another pair of equal and opposite shear stresses 3 and 4 to it. A shear stress on a single face of a particle thus cannot exist by itself; it always appears in combination with three other equal shear stresses, as shown in Fig. 11*a*. This is a very important fact, of which we shall make use repeatedly in what follows. The reader is advised to repeat the proof of this relationship on a parallelepiped of three different sides, dx, dy, dz, instead of on the cube of Fig. 11*a*. He will find that the shear *forces* on the various faces will be different but that the four shear *stresses* must all be equal for equilibrium.

Figure 11*b* shows how the cube deforms under the influence of the shear stress by becoming a diamond-shaped parallelepiped. If the shear stress s_s is plotted against the "angle of shear" γ, sometimes also called the "shear strain" γ, a diagram results very much like the tensile stress-strain diagram of Fig. 1. Again the first portion is linear, and for structural steel the yield shear stress is about 15,000 lb/sq in., or about half the tensile yield stress. The linear law from O to A in Fig. 1 is expressed as:

$$s_s = G\gamma \qquad \text{or} \qquad \frac{s_s}{G} = \gamma. \tag{3}$$

Here γ is the angle of shear, expressed in radians and hence dimensionless. Then the proportionality constant G, known as the *shear modulus,* must have the same dimension as the shear stress. For structural steel its value is:

$$G_{\text{steel}} = 12 \times 10^6 \text{ lb/sq in.}$$

For a yield shear stress of 15,000 lb/sq in. the yield strain thus is:

$$\gamma_{\text{yield}} = 0.00125 \text{ radian} = 0.07 \text{ deg,}$$

too small to be visible to the naked eye. An important practical difference between the tensile strain-stress diagram of Fig. 1, and its shear stress counterpart $[s_s = f(\gamma)]$ is that the tensile diagram can be easily obtained by a simple test in a testing machine, which is not the case for the shear diagram. The experimental determination of Young's modulus E consists in measuring a tensile force (with some form of dynamometer) and an elongation (usually optically with a microscope) of a tensile test piece. On the other hand the shear modulus is measured in a much more indirect manner, usually by a torsion, or twisting, test, in which the state of stress and deformation is deduced from Eq. (3) and Fig. 11 only by a process of reasoning, including an integration. The most important occurrences of shear stress are not in cases that we commonly call shear, such as illustrated in Fig. 10, but rather in the torsion of circular bars.

Problems 45 *to* 49.

7. Solid Circular Shafts. Consider a long, straight shaft of length l, having a solid circular cross section of radius r, subjected

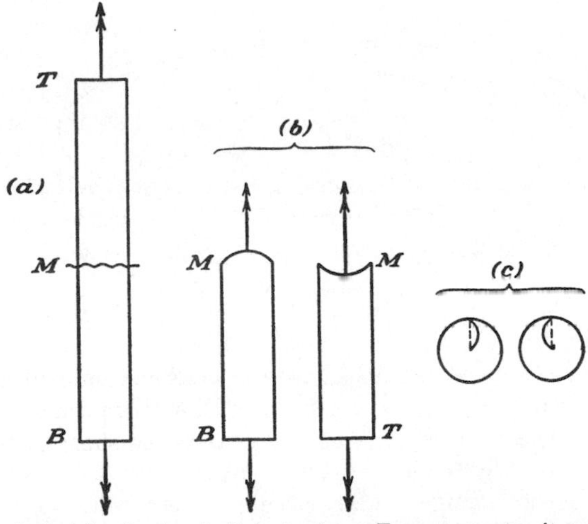

FIG. 12. A straight circular shaft in torsion. From symmetry it can be concluded that plane cross sections remain plane, that straight radii remain straight, and that the angle of twist is proportional to the length locally.

to equal and opposite twisting couples at its two ends. These couples can be represented by curved arrows and also by straight, double-headed arrows directed perpendicularly to the plane of the couple, as shown in Fig. 12*a*. This shaft is quite symmetrical in a geometric sense, and it is also loaded symmetrically. From this symmetry we shall be able to deduce considerable information regarding the deformations caused by the couples. First imagine the bar cut in two parts across the middle M. For static equilibrium of each half shaft the mid-sections must be subjected to twisting couples like the end sections. Then the two halves of the shaft are identical, and they are identically loaded. Hence they must deform identically. From this we can conclude (*a*) that plane cross sections across the shaft remain plane after deformation and (*b*) that straight radii in such cross sections remain straight and radial after the deformation. To understand conclusion *a*, assume first that it is not true. The plane mid-section M assumes the shape of a body of revolution; it becomes cup-shaped after the deformation takes place. We now take the top half of the shaft and turn it upside down, placing it next to the bottom half, as in Fig. 12*b*. We notice that of the two identical half shafts, identically loaded, one has a convex top and the other a concave one. This is impossible, and hence plane sections remain plane. To understand

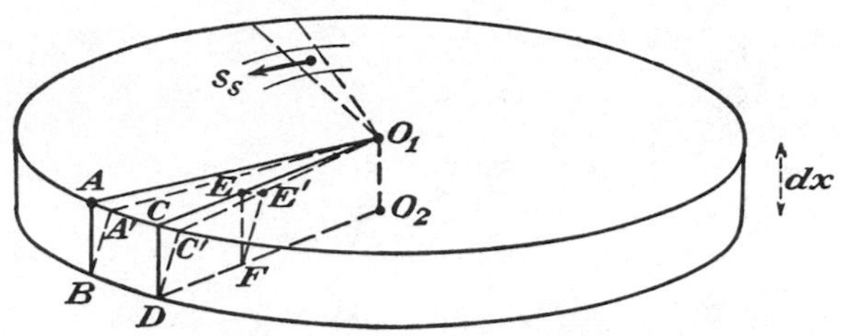

FIG. 13. An element dx of the twisted shaft.

conclusion *b*, again assume it not to be true. Let a radius of the mid-section M become curved by the deformation. Placing the two half shafts side by side as in Fig. 12*b*, the two sections M will appear as shown in Fig. 12*c*. Again these two figures must be identical, because they depict two identical half shafts, identically loaded. Hence straight radii must remain straight.

Then the deformation of the shaft must consist of a turning of the various cross sections, each as rigid figures, with respect to each other. Let, for example in Fig. 12*a*, the bottom section B be held steady and the top section T turn through 1 deg. Then the mid-section must turn through 0.5 deg, and all sections must turn through angles proportional to their distance from the base. This proportionality also follows from symmetry, because, if it were not so, we could isolate two different quarter lengths of the shaft and find different angles of twist along them. Since the two quarter lengths are identical geometrically and identically loaded, they must deform in the same way as well.

If we draw a set of straight longitudinal lines on the cylindrical outside surface of the shaft and then twist the shaft, these lines become spirals.

In Fig. 13 we consider a slice of the shaft of length dx. The deformation, as we have seen, consists of a turning of the top circle as a rigid figure with respect to the bottom circle. Let the bottom circle BO_2 be the basis of operations, and let points A and C of the upper circle go to A' and C' owing to the deformation. The line BA then becomes the spiral BA', and the straight radii O_1A and O_1C become the straight radii O_1A' and O_1C'. The little quadrangle $ABCD$ becomes the parallelogram $A'BDC'$ and hence, by Fig. 11*b*, is subjected to a shear stress $s_s = G\gamma$, where γ is the angle of twist $\angle ABA'$ or $\angle CDC'$. If we consider a point E not on the periphery, we see that EE' is less than CC', while the height $EF = CD = dx$ is the same. Hence the angle of *shear* γ is variable; it diminishes when moving from A to O_1, and at the center O_1 there is no shear at all.

The angle of *twist* is the angle by which the entire top circle turns with respect to the bottom circle. It is $\angle AO_1A'$, denoted by $d\varphi$, and is of course independent of the distance from O_1. We now denote $O_1E = r$, a variable radius, growing from 0 at O_1 to R at C. Then:

$$EE' = r\,d\varphi = \gamma\,dx = \frac{s_s}{G}\,dx.$$

Hence:

$$s_s = G\frac{d\varphi}{dx}r, \tag{a}$$

or the shear stress is proportional to the radius r, because the ratio $d\varphi/dx$ is a constant: the angle of twist per unit length of the shaft. This shear stress acts in the plane of the circle of Fig. 13, in a tangential direction, furnishing an elemental contribution to the torque M_t.

Taking an element dA of area of the circle, the tangential force is $s_s\,dA$, and its contribution toward the moment about O_1 is $s_sr\,dA$. Hence:

$$dM_t = s_sr\,dA = G\frac{d\varphi}{dx}r^2\,dA,$$

and

$$M_t = G\frac{d\varphi}{dx}\int\int r^2\,dA = GI_p\frac{d\varphi}{dx}.$$

The integral is known as the polar area moment of inertia; it is denoted by I_p, and we shall calculate it a little later. Now $d\varphi/dx = \varphi/l$, where φ is the total twist of the shaft and l its total length. Hence:

$$\varphi = \frac{M_tl}{GI_p}. \tag{4}$$

This is an important formula. It states that the angle of twist is proportional to the twisting couple M_t and to the length of the shaft l. It is inversely proportional to the product GI_p, which is called the *torsional stiffness* of the shaft. Of the two factors in this stiffness, one, G, expresses the property of the material of the shaft, and the other one, I_p, expresses the influence of the size of its cross section. Equation (*a*) expresses the stress in terms of the angle of twist, which is not convenient, because as a rule we know only the twisting couple, and the angle of twist must be calculated, *e.g.*, by Eq. (4). Therefore we eliminate algebraically the angle of twist $d\varphi/dx$ from between (*a*) and (4) and find:

$$s_s = \frac{M_t r}{I_p}. \tag{5}$$

This is another important equation. We note that the letter G does not appear, so that the stress is independent of the material. A steel shaft and a brass shaft of equal dimensions and subjected to equal torques will have the same stresses in them. But, by Eq. (4), they will not twist through the same angle!

Now we must return to the calculation of the integral I_p, which is a question of simple calculus. Taking polar or r, θ coordinates, we have:

$$\begin{aligned} I_p &= \iint_A r^2\,dA = \iint_A r^2 \cdot r\,d\theta\,dr = \int^{2\pi} d\theta \int_0^R r^3\,dr \\ &= 2\pi\frac{R^4}{4} = \frac{\pi R^4}{2}. \end{aligned}$$

$$I_p = \frac{\pi R^4}{2} = \frac{\pi D^4}{32}, \tag{6}$$

where D is the diameter of the shaft. We notice that I_p has the dimension of D^4 and hence is measured in inches^4.

This completes the discussion of the twist of straight shafts with a solid circular cross section. This theory will be extended on page 112 to shafts consisting of two materials, such as a steel core with a bronze sleeve around it. Before leaving the subject, it is well to consider the order of magnitude of the various quantities involved. First we look at the angle $\gamma = \angle ABA'$ of Fig. 13. This angle must be less than the yield angle, or less than 0.07 deg for mild steel, because otherwise the whole above theory, including Eqs. (4) and (5), would not apply. Hence the "spiral" effect of the lines BA' of Fig. 13 will be too small to be visible to the naked eye. This brings us to a question that is often asked in connection with Fig. 13 or Fig. 11*b*. In Fig. 11*b* the dotted line on the top side is drawn to coincide with the solid line. We have thus assumed that the height of the element has not changed. But then the length of the sides has increased in ratio $\sqrt{1+\gamma^2}$ by Pythagoras' theorem. If then a shaft is twisted and plane cross sections remain plane, the outside longitudinal fibers become longer and the center-line fiber keeps its length. Then there must be a longitudinal tensile stress in the outside fiber, which we have neglected to mention. The argument appears important as long as we do not calculate how *much* these stresses are. For an ordinary steel shaft, twisted just to yield at the periphery, the unit elongation of the outside fiber (being the difference in length of the skew and vertical lines of Fig. 11*b*) is:

$$\epsilon = \sqrt{1+\gamma^2} - 1 = 1 + \frac{\gamma^2}{2} + \cdots - 1 = \frac{\gamma^2}{2} + \cdots$$

$$= \frac{0.00125^2}{2} = 0.78 \times 10^{-6},$$

and the tensile stress would be:

$$s = E\epsilon = 30 \times 10^6 \times 0.78 \times 10^{-6} = 23 \text{ lb/sq in.}$$

This stress is so small as to be completely negligible. In its derivation we notice the *square* of γ; the deformation is said to be small of the second order, and such second-order quantities are not considered in our subject. Of course, if we take a rubber shaft and wind it up 360 deg in, say, 10 diameters length, the effect does become very considerable. But then it becomes a question in the "theory of large deformations," which we cannot consider here.

Problems 50 *to* 52.

8. Examples; Hollow Shafts. *a*. As a first example of the foregoing, we consider a crank of a gasoline engine (Fig. 14). The single-engine cylinder has a 3½- by 4-in. bore and stroke; its connecting rod length $l = 3r$, where $r = 2$ in. is the crank radius; the main shaft diameter is 1¼ in.; the gas pressure at 90 deg crank angle is 100 lb/sq in., and at 45 deg crank angle (from top dead center) it is 150 lb/sq in. What is the torsion stress in the shaft at these two positions?

The principal point of this problem is the calculation of the torque. Since the bore is $3\frac{1}{2}$ in., the gas-pressure force on the piston in the first case is: $100\pi(3\frac{1}{2})^2/4 = 960$ lb. As shown in Fig. 14*b*, to this force must be added a small sidewise force from the cylinder walls on the piston to combine into the connecting-rod compression. It is not necessary to calculate this sidewise force, because when transmitted to the crankpin through the connecting rod and re-

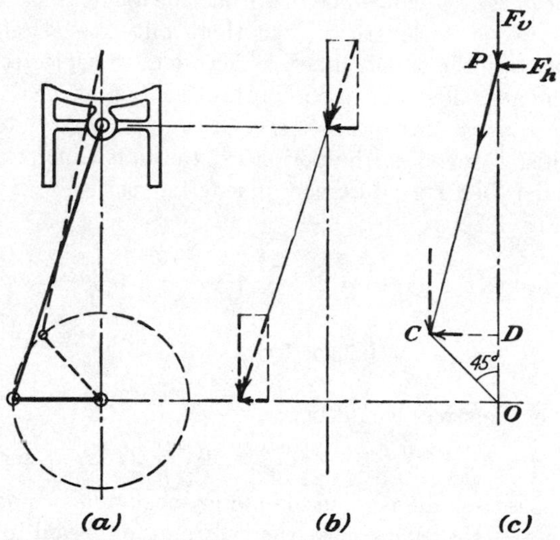

FIG. 14. Example of torsion in the crankshaft of an internal-combustion engine.

solved again into components we see that only the 960-lb force has a moment arm of 2 in., while the small force is radial. Hence the moment on the crankshaft is 1,920 in.-lb. The shaft constant by Eq. (6) is:

$$I_p = \frac{\pi}{32}\left(1\frac{1}{4}\right)^4 = 0.24 \text{ in.}^4$$

and the stress, by Eq. (5):

$$s_s = \frac{1{,}920 \times \frac{5}{8}}{0.24} = 5{,}000 \text{ lb/sq in.}$$

For the 45-deg crank position the gas force is 50 per cent greater, or 1,440 lb, and as is shown in Fig. 14*c* we now must calculate the small sidewise force. We know that $PC = 3r$ and $CD = r/\sqrt{2}$, so that $PD = r\sqrt{9 - \frac{1}{2}} = r\sqrt{17/2}$. The ratio of the two forces is $F_v/F_h = \sqrt{17/2} : \sqrt{\frac{1}{2}} = \sqrt{17}$. These two forces are transmitted to the crankpin C and there have the same moment arm $r/\sqrt{2}$ about O. Hence the moment is:

$$(F_v + F_h)\frac{r}{\sqrt{2}} = 1{,}440\left(1 + \frac{1}{\sqrt{17}}\right)\sqrt{2} = 2{,}540 \text{ lb-in.},$$

with a resultant stress of:

$$s_s = \frac{2{,}540 \times 5/8}{0.24} = 6{,}600 \text{ lb/sq in.}$$

b. The second example is the determinaton of the main lineshaft diameter of a large ocean liner having a 50,000-hp drive at 185 propeller rpm. The shaft is to be designed to a 5,000 lb/sq in shear stress in torsion.

The shaft torque is:

$$M_t = \frac{33{,}000 \text{ hp}}{2\pi N} = \frac{33{,}000 \times 50{,}000}{2\pi \times 185}$$
$$= 1{,}420{,}000 \text{ ft-lb} = 17{,}100{,}000 \text{ in.-lb.}$$

By Eqs. (5) and (6):

$$s_s = 5{,}000 = \frac{17.1 \times 10^6 \times r}{\pi r^4/2},$$

or

$$r^3 = \frac{17.1 \times 10^6 \times 2}{5{,}000\pi} = 2{,}170 \text{ in.}^3,$$

and

$$r = 12.93 \text{ in.} \approx 13 \text{ in.} \qquad \text{or} \qquad D = 26 \text{ in.}$$

This is a good example of conservative marine design. Another extreme is found in the main propeller driving shaft of a Navy PT boat, which transmits 2,000 hp at 2,000 rpm from an aircraft-type in-line gasoline engine. The shaft is made of special high-grade steel; it is 100 in. long with a diameter of 1½ in. We find:

$$M_t = \frac{2{,}000 \times 33{,}000}{2{,}000 \times 2\pi} = 5{,}250 \text{ ft-lb} = 63{,}000 \text{ in.-lb},$$

$$I_p = \frac{\pi}{32}\left(\frac{3}{2}\right)^4 = 0.497 \text{ in.}^4,$$

$$s_s = 63{,}000 \times 0.75/0.497 = 95{,}000 \text{ lb/sq in.},$$

a stress that can be withstood only by the best of special steels and by a shaft without any "stress concentrations" (page 202) and without any stress variations of an alternating nature. It is interesting to calculate the wind up in the shaft at full power by Eq. (4):

$$\varphi = \frac{63{,}000 \times 100}{12 \times 10^6 \times 0.497} = 1.06 \text{ radians} = 60°.$$

Thus the propeller of this PT boat runs 60 deg elastically behind the engine. How large is the corresponding angle for the case of the ocean liner, where the shaft length is 200 ft?

c. Hollow Shaft. If we once more look at Fig. 13 and think over the development of the formulae for the solid shaft, we notice that no stresses exist on the curved outside surface of the shaft. On a cut perpendicular to the shaft center line, as in Fig. 13, there are shear stresses directed tangentially, *i.e.*, perpendicular to the radii. On a plane cut passing through the center line (as in Problem 50, page 240) there are shear stresses parallel to the center line. If we cut up the shaft into concentric straight thin-walled tubes, all fitting into each other like a sort of linearized onion, no shear stresses or any other stresses exist between these individual tubes. Therefore, if a shaft were so cut, it would not be weakened in torsion and the assembly of loosely fitting tubes would carry the same stresses and have the same deformations as the solid shaft. Then, if we were to throw away some of the inside tubes, the remaining ones outside would carry their part of the torque in the same manner as before. Therefore, the formulae (4) and (5) are still valid for a circular shaft with a concentric hole in it, while Eq. (6) remains practically the same, becoming:

$$I_p = \frac{\pi D_o^4}{32} - \frac{\pi D_i^4}{32} = \frac{\pi}{32}(D_o^4 - D_i^4), \tag{6a}$$

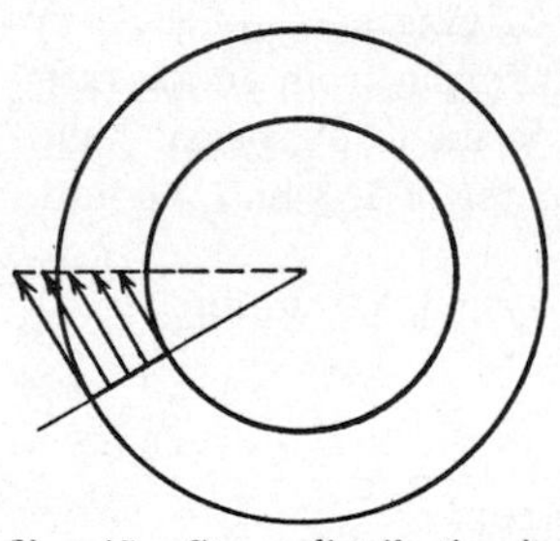

FIG. 15. Stress distribution in a concentric hollow shaft in torsion.

where D_o and D_i are the outside and inside diameters of the hollow shaft.

If weight is to be saved, it is of advantage to transmit a given torque through a hollow shaft of somewhat larger outside diameter than through a solid shaft. This is because the stress distribution in a hollow shaft (Fig. 15) is more uniform than in a solid one. In a solid shaft the material near the core contributes to the weight of the shaft, but not to its torque-carrying capacity, because it carries very little stress. As a numerical example let us calculate for the ocean-liner shaft of page 23 what bore diameter is permitted if the outside diameter is stepped up from 26 in. to 30 in.

and the maximum stress is held at the same value of 5,000 lb/sq in. The last three lines of the calculation then modify to:

$$s_s = 5{,}000 = \frac{17.1 \times 10^6 \times 15}{\pi(15^4 - r_i^4)/2},$$

or

$$15^4 - r_i^4 = \frac{17.1 \times 10^6 \times 15 \times 2}{5{,}000\pi} = 32{,}700;$$

$$r_i^4 = 50{,}600 - 32{,}700 = 17{,}900;$$

$$r_i = 11.6 \text{ in.} \qquad \text{and} \qquad D_i = 23.2 \text{ in.}$$

With steel weighing 0.28 lb/cu in., the reader should check that the weight of the hollow shaft per foot length is 956 lb, while that of the equally strong solid shaft is 1,780 lb.

d. Shaft of Non-circular Cross Section. The theory thus far given holds for hollow or solid circular shafts. Suppose we now consider the twisting of a bar of non-circular cross section, for example, of a rectangular bar or of an I beam or a channel. When we look over the argument with Fig. 12, leading to the result that plane cross sections remain plane, we realize that the reasoning there depended on rotational symmetry about the shaft center line. No such symmetry exists for a non-circular cross section so that we cannot prove in this manner that plane sections remain plane. In fact, they don't: for all cross sections other than the solid or hollow circular one a plane section will bulge perpendicular to its plane when the shaft is twisted. The complete theory of this case is complicated and outside the scope of this book. Let it suffice to say that Eq. (4) and (5) are not even decent approximations for sections far different from the circular one and should, therefore, never be applied to any case but the circular one. This can be understood without much theory when we consider the twist of a flat rectangular bar. If we make the cross section quite flat and thin, like a sheet of paper, our intuition or experience tells us that the torsional stiffness is small; it is easy to twist a piece of paper. But Eq. (4) tells us that the stiffness is proportional to I_p, which is fairly large for a narrow rectangular section, because most of the material of the section is rather far away from its center. If, in a rectangular cross section, we leave the area bh constant but gradually make the width b greater and hence the thickness h smaller, Eq. (4) tells us that the stiffness (proportional to I_p) goes up, while experiments show that the torsional stiffness goes down. A complete theory of all this was worked out by

Saint-Venant, about 1850, and in order to express the behavior of thin rectangular sections Saint-Venant proposed an empirical formula as follows:

$$I_{p\ \text{equiv}} = \frac{A^4}{4\pi^2 I_p} \tag{7}$$

Here A is the cross-sectional area and I_p is the actual polar moment of inertia. $I_{p\ \text{equiv}}$ is *not* a polar moment of inertia but a quantity of the same dimensions that, when substituted into Eq. (4), gives an experimentally correct, decent approximation of the torsional stiffness. The quantity I_p, as it appears in the denominator of the right-hand member, is the true moment of inertia. The reader should verify that Eq. (7) becomes an identity when applied to circular sections, as it should be, and he also should check that the dimensions are correct.

Furthermore, it is important to note that the approximation of Eq. (7) does *not* apply to the stress equation (5). That equation states that the stress is proportional to r, that is, increases with the distance from the center. In a narrow rectangular section the maximum stress appears at a point of the periphery that is closest to the center. No empirical formula, good for all sections, has ever been proposed to take care of these stress relations, and the reader is referred to more advanced books for further information.

Problems 53 *to* 70.

9. Closely Coiled Helical Springs. For the purpose of finding the force and moment transmitted by an arbitrary section A of a spring (Fig. 16) loaded by a pair of tensile forces P, we isolate the part of the spring above A and set up its conditions of static equilibrium. In order to satisfy the up-and-down equilibrium, the section A has to transmit a shear force P. Then, this shear force and the load P at the top end of the spring form a moment in the plane passing through A and the center line of the coil. This moment can be represented by a double-headed arrow perpendicular to that plane, which is in a direction almost that of the spring wire at A. If the wire on account of its helical shape includes a small angle β with the perpendicular to the plane APP, then the moment of the shear force and load force P can be resolved into a component along the wire and one across the wire, as indicated in Fig. 16*b*. Now a spring is commonly called "closely coiled" if the effect of the latter, small component $PR \sin \beta$ is considered negligible and if the main component $PR \cos \beta$ is considered to be equal to PR itself. This is a very good approximation, which simplifies the analysis considerably. Then the cross section A carries a shear force P and a twisting moment PR. The stress in the wire caused by this twisting moment, by Eqs. (5) and (6), is:

$$s_s = \frac{PRr}{\pi r^4/2} = \frac{2}{\pi}\frac{PR}{r^3}. \tag{8}$$

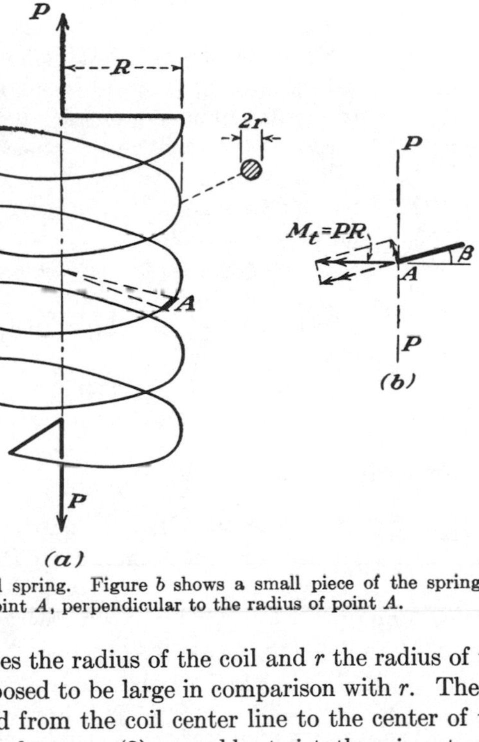

FIG. 16. Helical spring. Figure *b* shows a small piece of the spring wire in a plane through point A, perpendicular to the radius of point A.

Here R denotes the radius of the coil and r the radius of the wire, and R is supposed to be large in comparison with r. The value of R is measured from the coil center line to the center of the wire. In addition to the stress (8) caused by twist, there is a stress caused by the shear force. At this stage of our knowledge the best we can say is that the shear force is distributed uniformly over the cross section so that the stress is $P/\pi r^2$. The stress (8), due to twist, is $2R/r$ times as large as this. Thus we find that for a spring of the usual dimensions ($r \ll R$) the stress due to twist is very much larger than the stress due to direct shear, and we are accustomed to neglect the latter and to consider Eq. (8) an acceptable expression for the total stress. However, it is of interest to note that the two components of stress are additive at the inside of the coil and that they subtract from each other at the outside of the coil, for springs in both tension and compression. Broken springs in which the wire diameter r is a considerable fraction of R always show their initial failure at an inside point of the coil.

Now we are ready to tackle the more difficult job of finding the elongation of the spring under the influence of the loads P. This is done by integration. We consider a short piece of wire near A of length $dl = R\,d\theta$ and of cross-sectional area πr^2 and allow this piece to deform, while the rest of the spring is thought of as rigid, or undeformable. If the top of the spring is held in place, the whole part below A will move as a rigid body, and the bottom point P will displace a short distance. This distance is calculated and then integrated along l, so that the total displacement of the bottom P is thought of as the sum of the individual displacements it would get from each element dl consecutively.

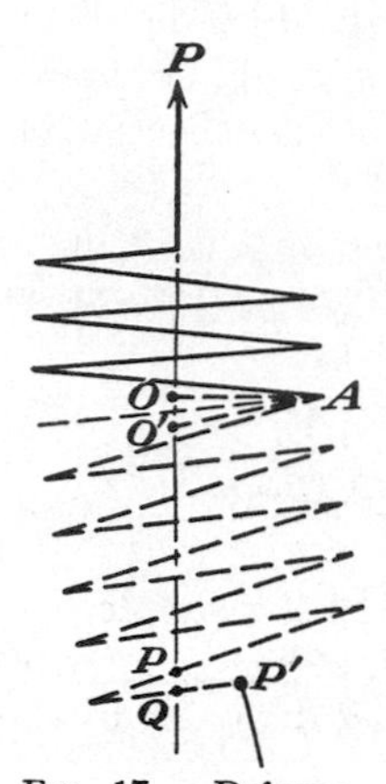

FIG. 17. Deformation of a coil spring if only a small element dl at point A is allowed to twist while the rest of the spring is thought of as rigid.

By Eq. (4) the bottom part of the spring turns about point A through an angle

$$d\varphi = \frac{PR\,dl}{G\pi d^4/32}.$$

In Fig. 17 the plane of the paper is through point A, the plane being that of PAP of Fig. 16. The bottom point P moves to P', which is down and to the right. The amount downward is $PQ = OO' = R\,d\varphi$; the amount to the right is greater as a rule, but we will not calculate it. Now, if we consider various points A along one full coil of the spring, the downward deflections of P are all downward but the sidewise deflections are evenly distributed around the circle and cancel each other. Thus, for a number of full coils of the spring, the bottom point will not go sidewise, but only downward by an amount

$$\delta = \int R\,d\varphi = \frac{PR^2}{G\pi d^4/32}\int dl = \frac{PR^2}{G\pi d^4/32}\,2\pi RN = \frac{64PR^3N}{Gd^4},$$

$$\delta = P\cdot\frac{8D^3N}{Gd^4}, \tag{9}$$

where N is the number of coils, D is the coil diameter, and d the wire diameter. If the number N is fractional, say $5\frac{1}{2}$, then the first five full coils integrate into a downward displacement only, while the last half coil will give a mixed downward sidewise movement of the end. However, a glance at Fig. 17 shows that the sidewise displacement is proportional to the distance OP, which is zero or very small for the *last* half coil. Thus Eq. (9) describes the deformation of the spring caused by the twisting couple PR in the wire. Another part of the deformation, caused by the shear force P in the wire, is very much smaller than (9), as will be shown later on page 108, so that Eq. (9) is used for the total deflection of the spring, including the effect of all causes.

The quantity P/δ, usually denoted by k, expressed in pounds per inch, is called the *stiffness* of the spring and is used often in specifications.

From Eq. (9) we write immediately:

$$k = \frac{Gd^4}{8D^3N}. \tag{9a}$$

Thus the stiffness of a spring can be doubled, for example, by taking only half its number of coils, or by increasing its wire diameter by 19 per cent, or by decreasing its coil diameter by 26 per cent.

Problems 71 *to* 75.

CHAPTER III

BENDING

10. Bending-moment Diagrams. The determination of shear-force and bending-moment diagrams for beams is a problem that should properly be classified under statics of rigid bodies and not under strength of materials. The bending moment in a beam depends only on the loads on the beam and on its consequent support reactions; it is the same whether the beam is rigid or deformable, elastic or plastic. However, most textbooks on statics do not make much mention of the subject, so that we shall now discuss it briefly as a preliminary to the real problem in strength of materials of finding the stresses and deformations in the beam as a result of the existing bending moment and shear force.

We start with a "statically determined" beam, *i.e.*, a structure which is straight and long, of which the cross-sectional dimensions are small compared with the length, supported on such supports as are just necessary to prevent the beam from falling down. Then the support reactions can be determined by writing the static equilibrium equations of the complete structure. The "shear force" and "bending moment" in an arbitrary section of the beam then can be calculated by isolating one part of the beam (either to the right or to the left of the section) and by writing the static-equilibrium conditions of this isolated part. In this process no mention at all is made of Hooke's law or of any other stress-strain law.

As an example, consider the "simply supported" beam of Fig. 18, of which the loads P and Q are given and of which the two support reactions are as yet unknown. Writing the static-moment equations of the whole beam about the two ends consecutively allows us to solve for the support reactions, with the results shown in the figure. In order to find the shear force and bending moment in a section at an arbitrary distance a from the right support we write the equilibrium equations of the isolated right part of the beam. For vertical equilibrium the shear force must be $P/2 + Q/4$, and for moment equilibrium about any point the bending moment at the cut must be $(P/2 + Q/4)a$. We isolated the *right* part of the beam because it happens to be the simpler one; the same result must follow from isolating the left part of the beam. The vertical-equilibrium equation then is:

$$S = P + Q - \left(\frac{P}{2} + \frac{3}{4}Q\right) = \frac{P}{2} + \frac{Q}{4},$$

and the moment equation about the cut is:

$$M_b = \left(\frac{P}{2} + \frac{3}{4}Q\right)(l - a) - Q\left(\frac{3}{4}l - a\right) - P\left(\frac{l}{2} - a\right),$$

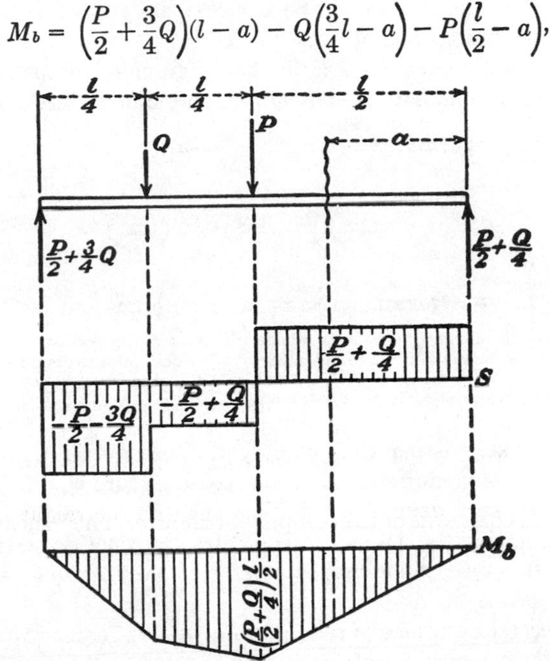

Fig. 18. Shear-force and bending-moment diagram of a beam on two supports at the ends, loaded by two forces P and Q.

which, when worked out, leads to the same previous result. If in these expressions a is now considered variable, between $a = 0$ and $a = l/2$, we have functional expressions for S and M_b in any section of the right half of the beam, which we can plot in appropriate diagrams, as shown in Fig. 18. To find these quantities in the other parts of the beam, again static-equilibrium equations of an isolated portion of the beam have to be written. The reader should do this for Fig. 18 and check all portions of both diagrams.

Before proceeding to a next example, it is useful to agree once and for all (for the rest of this book at least) on the signs, on what we shall consider to be a positive or negative bending moment and shear force. The definitions are shown in Fig. 19. We shall always count distance along the beam from left to right; deflection will be considered positive downward; a bending moment will be positive when it tends to bend the beam convex upward; and a shear force is positive when it shears the beam upward at the right. In working problems frequent reference must be made to Fig. 19, and the reader's first reference should be a check that Fig. 18 is in accord with the definitions of Fig. 19.

There is nothing compelling in these sign conventions; any one or all of them could have been reversed, and in fact there are hardly

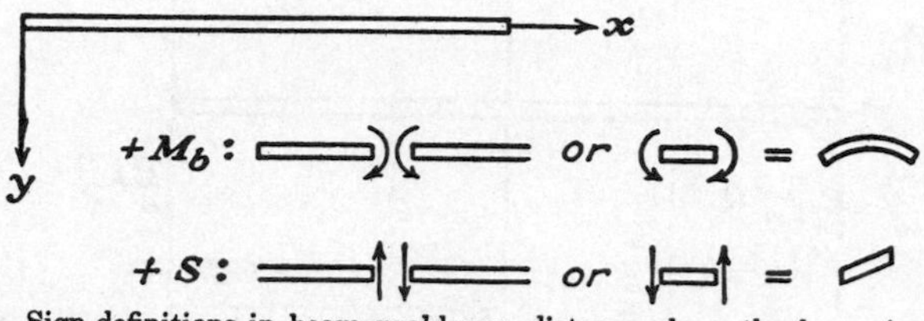

FIG. 19. Sign definitions in beam problems: distance along the beam is positive to the right; deflection is positive downward; a bending moment is positive when it makes the beam shed water; and a shear force is positive when it turns an element counterclockwise.

two textbooks existing that employ the same definitions. The matter is not of great importance; one could go through life without ever making such definitions and just get things straight for each problem as it comes up. But for future reference to past work, time can be saved if one makes a habit of always using the same sign convention.

As a second example we take a beam with different supports, a "cantilever," which is loaded partially with a "uniformly distributed" loading (Fig. 20). In a section at a distance a to the left from the free end of the cantilever, we have by vertical equilibrium of the right-hand portion:

$$S = -wa = -\frac{2Pa}{l},$$

negative by the convention of Fig. 19. By taking moments of the free right-hand portion about the cut we find:

$$M_b = +\frac{wa^2}{2} = +\frac{Pa^2}{l},$$

positive by Fig. 19. For variable a these expressions are valid until the middle of the beam, and, in particular, the bending moment in the middle of the beam ($a = l/2$) is $Pl/4$. The rest of the diagram should be calculated and checked by the reader by repeated

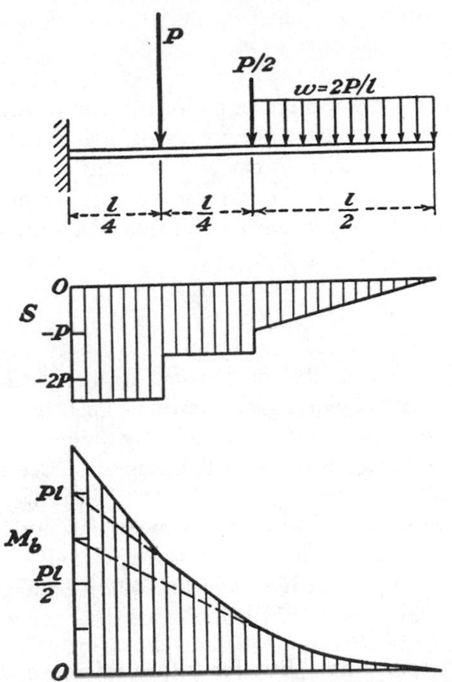

FIG. 20. Shear-force and bending-moment diagrams of a cantilever beam.

equilibrium calculations. However, there is a quicker and somewhat handier way to proceed in a case like this by using a pair of general theorems that will now be derived with the aid of Fig. 21.

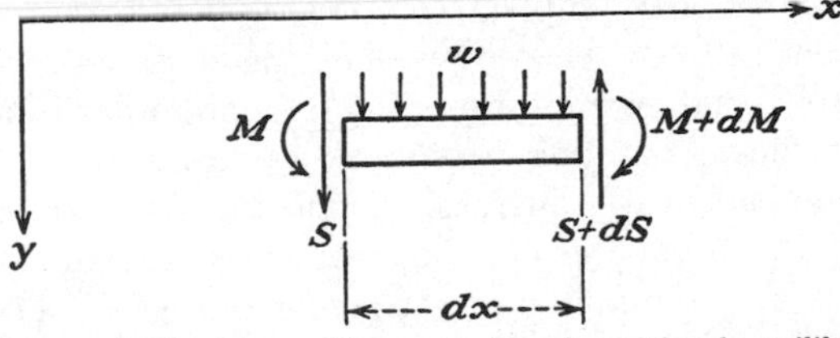

FIG. 21. Element dx of a beam. The vertical and rotational equilibrium conditions are expressed by Eqs. (10a) and (10b).

That figure shows a short piece dx of the beam on which are acting a distributed load w and the shear forces and bending moments. These are not quite the same on the two sides of the element but differ from each other by differential amounts as indicated. Then, vertical equilibrium requires that $w\,dx = dS$, or

$$w = \frac{dS}{dx}. \tag{10a}$$

The moment-equilibrium equation of the element about the left-hand end of it contains many terms, which the reader should write down for himself. Most of the terms are small of second order, being proportional to expressions like dx^2 or $dx\,dS$. Neglecting these second-order terms and retaining only the first-order ones, we have $dM = S\,dx$, or

$$S = \frac{dM}{dx}. \tag{10b}$$

The pair of equations (10) express the fact that the slope of the shear-force diagram equals the ordinate of the loading or w diagram and that the slope of the bending-moment diagram equals the ordinate of the shear-force diagram or, conversely, that the shear is the integral of the loading and the bending moment is the integral of the shear. With these properties in mind we return to the cantilever beam of Fig. 20. In the right half of the beam the loading is $w = 2P/l$, which should be equal to the slope of the S diagram. At the free end of the beam the shear force is zero; hence in the middle it is $2P/l$ multiplied by $l/2 = P$. At this point there appears a concentrated load $P/2$, which means a distributed loading of infinite intensity on an infinitely short length of beam. Hence the slope of the S diagram is infinite (vertical), the jump in height being $\int w\,dx$, which is the concentrated load $P/2$. Beyond that to the left there is no load w, hence no slope in the S diagram, until the jump P.

Now we turn to the bending-moment diagram. At the free end there is zero bending moment. The S diagram has a linearly increasing ordinate, starting from zero; therefore, the slope of the M diagram is linearly increasing, starting from zero. In the middle of the beam the bending moment is the integral of the shear force, *i.e.*, the area of the triangle or $P \cdot l/4$. The slope of the M diagram in the middle is $-P$, so that the tangent to the curve there rises from $Pl/4$ to $3Pl/4$ over a base distance $l/2$, as indicated by the dotted line. In the next $l/4$ section, between two concentrated loads, the S diagram has an ordinate $-3P/2$, and hence the slope of the M diagram is 50 per cent greater than before, as again indicated by the dotted line. In general a concentrated load in the w diagram corresponds to a jump in ordinate in the S diagram and to a jump in slope of the M diagram.

The reader is now advised to return to Fig. 18 and derive all the details of both diagrams by repeated applications of Eqs. (10). A possible difficulty may occur in the end conditions. It is advisable to imagine the beam extended slightly beyond the two

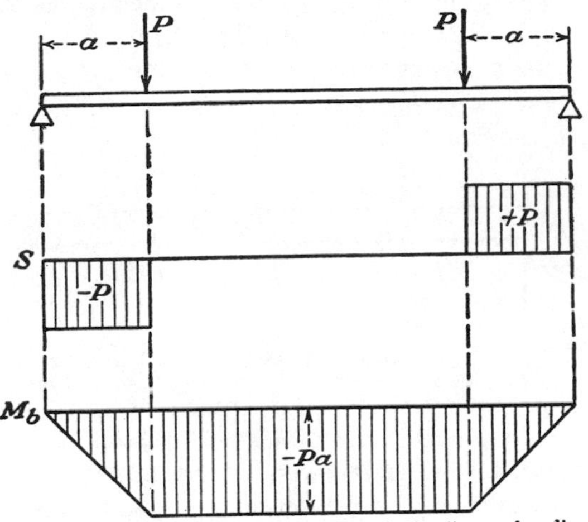

FIG. 22. A beam of which the center portion is in "pure bending," without shear force.

supports and to reason that in those free extensions there can be no S and no M. From one of these ends, then, the integration can start, and upon arriving at the other end a check is obtained.

Problems 76 *to* 88.

11. Pure Bending Stress. Consider the beam on two supports of Fig. 22, symmetrically loaded with two loads P. Both support reactions will be P upward, and in the central portion of the beam the shear force S will be zero and the bending moment constant, equal to $-Pa$. Then the center portion is said to be in a state of "pure" bending, *i.e.*, without shear, and it is for this simple case that we will now calculate the stress in a cross section of the beam. From a reasoning of static equilibrium it follows that the top fibers of the beam must be in compression and the bottom fibers in tension. This is clear when, as in Fig. 23, we look at an exaggerated picture of the left portion of the beam of Fig. 22, isolated. The load P and the support reaction P tend to rotate that portion clockwise; it is in equilibrium; hence the forces transmitted by the right half of the beam to the left half must form a counterclockwise couple Pa, which we call the bending moment. This can be done only by a push-pull situation, as indicated in the sketch. The details of the distribution of this push-pull arrangement are the subject of this article. Statics alone cannot give us the answer. If we were to assume, for example, that the top 10 per cent of the cross-sectional area were in compression, the bottom 10 per cent

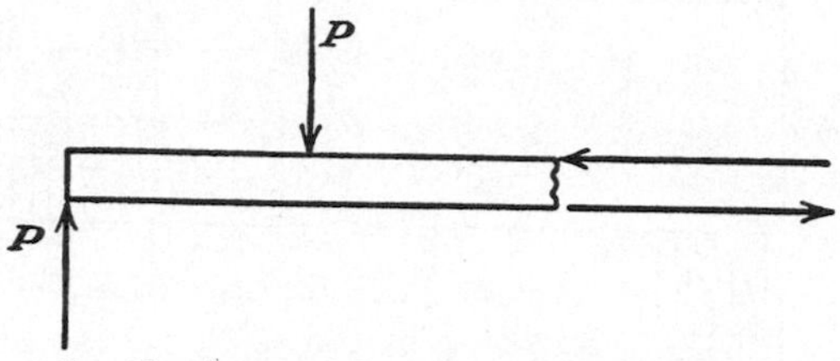

FIG. 23. A part of the beam of Fig. 22, isolated.

in tension, and the remaining 80 per cent stressless, we should get an answer that is perfectly correct from the standpoint of statics. Moreover, many different answers, all equally correct statically, could be easily obtained. The real answer, *i.e.*, the stress distribution that occurs in practice, not only satisfies the static equilibrium condition but also gives us continuity of deformation. In the arbitrary assumption of the 10-10-80 per cent, for example, the top fibers would shrink elastically, the bottom fibers would elongate, and the middle portion would retain its original length. After this deformation these three pieces of the beam could not be fitted together; in other words, the proposed deformation leaves the beam "discontinuous," with tears and cracks.

We therefore now turn the problem around and first assume a continuous deformation, judiciously chosen so that the top fibers become shorter and the bottom fibers longer. From this state of deformation we calculate the stresses by means of Hooke's law, and then we calculate the bending moment these stresses cause. Before we do this, we restrict our problem once more: to beams of cross sections that are symmetrical about a vertical center line, such as are shown in Figs. 24*a*, *b*, and *c*, but excluding sections like Fig. 24*d*. Furthermore, we now consider only cases where the loads on the beam act in that vertical plane of symmetry and not off side it. The more general cases of Fig. 24*d*, *e*, or *f* will be discussed later, in Chap. VI, page 114.

With this limitation the assumption about the deformation is that plane cross sections of the beam perpendicular to its axis re-

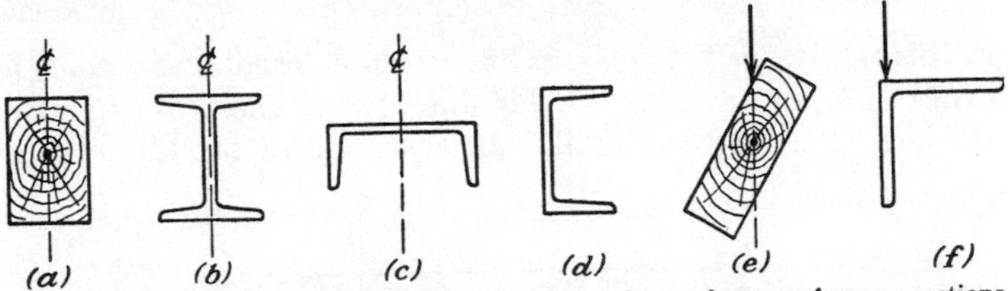

FIG. 24. The theory of this chapter applies only to beams of cross sections having a vertical axis of symmetry like (*a*), (*b*), and (*c*); the more complicated case of vertically non-symmetrical sections, like (*d*), (*e*), or (*f*), will be postponed to page 114.

main plane. Figure 25 shows a short piece of the beam, of length $2\,dx$, between two such sections, the full lines *abcd* in the unstressed state and the dotted lines $a'b'c'd'$ in the stressed or deformed state.

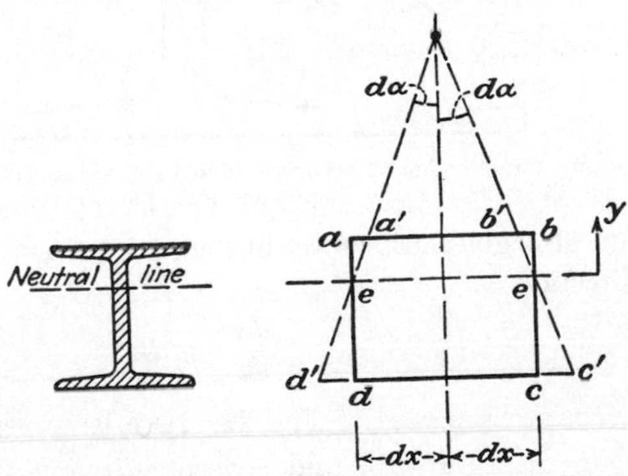

FIG. 25. An element of beam of length $2\,dx$, in which the original plane cross sections *ad* and *bc* deform to $a'd'$ and $b'c'$, remaining plane.

We see that the bottom fibers are elongated and the top fibers are shortened, in line with Fig. 23. There are "neutral fibers" *ee* between top and bottom that retain their length, and as yet we do not know just where these neutral fibers are located. Wherever *e* may be located, for our subsequent analysis we measure y upward from these neutral fibers. Then, for small angles $d\alpha$ the lengths aa', bb', cc', and dd' are expressed by $-y\,d\alpha$, where y is a variable, positive above the neutral line, negative below it. The elongation of a piece dx thus is $-y\,d\alpha$, and the unit elongation or strain is $\epsilon = -y\,d\alpha/dx$.

By Hooke's Law, Eq. (1), the stress is:

$$s = E\epsilon = -Ey\frac{d\alpha}{dx}. \qquad (a)$$

In this expression E and $d\alpha/dx$ are constant for all fibers, while y varies from fiber to fiber. Below the neutral line, y is negative, and hence the stress s is tensile; above the neutral line the stress is

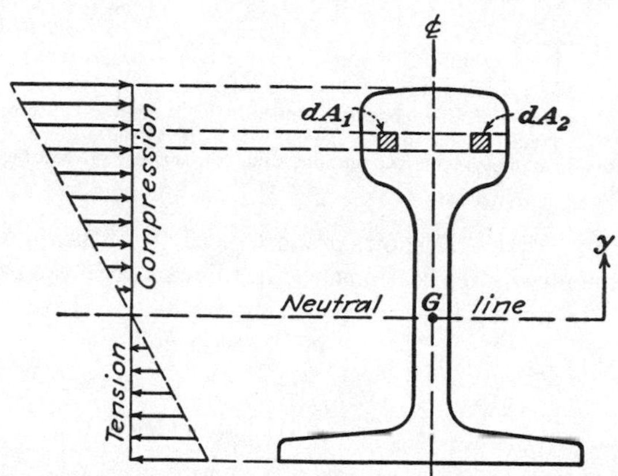

Fig. 26. The elastic forces acting on a horizontal strip dy across the cross section have a resultant in the vertical center line on account of symmetry.

compressive. The total tensile force of the entire cross section in a horizontal direction is:

$$P = \int_A s\,dA = -E\frac{d\alpha}{dx}\int_A y\,dA.$$

Now, if we consider the case of Fig. 22, there is no tensile force along the beam; hence $P = 0$, and consequently $\int_A y\,dA$ is zero. If we remember our definition of center of gravity, we recognize that this means that the origin of y, that is, the neutral line, passes through the center of gravity of the cross section. (In case we had a beam subjected to bending *and* simultaneous tension or compression, P would not be zero, and the neutral line would not pass through the center of gravity. We shall return to this case on page 55.)

Assuming that the origin of y passes through the center of gravity of the cross section, the tensile *force* on a small element dA_1 is $s\,dA_1$. As is indicated in Fig. 26 (which satisfies the limitation imposed by

Fig. 24), we can find another element dA_2 symmetrically situated, with the same force, and the resultant of the two elemental forces lies on the vertical center line of Fig. 26. The moment of this small resultant about the center of gravity is $ys(dA_1 + dA_2) = ys\, dA$, and the total moment of all stresses, being the bending moment M, is:

$$M = \int ys\, dA = -E\frac{d\alpha}{dx}\int y^2\, dA = -EI\frac{d\alpha}{dx}. \qquad (b)$$

The integral, denoted by I, is the *area moment of inertia*, sometimes also called the "second moment" of the cross section. It can be calculated for each cross section by integration and is expressed in inches^4.

Now we have Eqs. (a) and (b) for the stress and bending moment, each expressed in terms of the quantity $d\alpha/dx$ of Fig. 25. We can now eliminate this quantity from between the two and express the stress directly in terms of bending moment:

$$s = \frac{My}{I}. \qquad (11)$$

In this important result M is the bending moment in the cross section, measured in inch-pounds; I is the area moment of inertia $\int y^2\, dA$, measured in inches^4; and y is the vertical distance from the "neutral line," which is a horizontal line across the cross section through its center of gravity. Equation (11) was derived for the case of pure bending, *i.e.*, for the center portion of the beam of Fig. 22, but in the derivation we mentioned only the properties of one cross section; hence, the equation is also valid for the more general case of variable bending moment (Fig. 18 or 20), but then M is the value of the local bending moment. The maximum stress occurs, of course, in the fiber that is farthest away from the center of gravity, so that we can write:

$$s_{max} = \frac{My_{max}}{I} = \frac{M}{I/y_{max}} = \frac{M}{Z}. \qquad (11a)$$

The combination I/y_{max}, measured in inches^3 and denoted by Z, is called the *section modulus*. It is the geometrical property of the cross section determining the ratio of stress and bending moment. The moment of inertia and section modulus can be calculated for each cross section by integration. For a rectangular one, as shown in Fig. 27, we have:

$$I = \int y^2\,dA = \int_{y=-\frac{h}{2}}^{y=\frac{h}{2}} y^2(b\,dy) = b\,\frac{y^3}{3}\bigg|_{-\frac{h}{2}}^{\frac{h}{2}} = \frac{bh^3}{12}. \tag{12a}$$

The longest distance y_{max} from the neutral line is $h/2$, so that the section modulus becomes:

$$Z = \frac{I}{y_{max}} = \frac{bh^3/12}{h/2} = \frac{bh^2}{6}. \tag{12b}$$

For a circular cross section, as shown in Fig. 28, the moment of inertia can be calculated in the same manner by integrating hori-

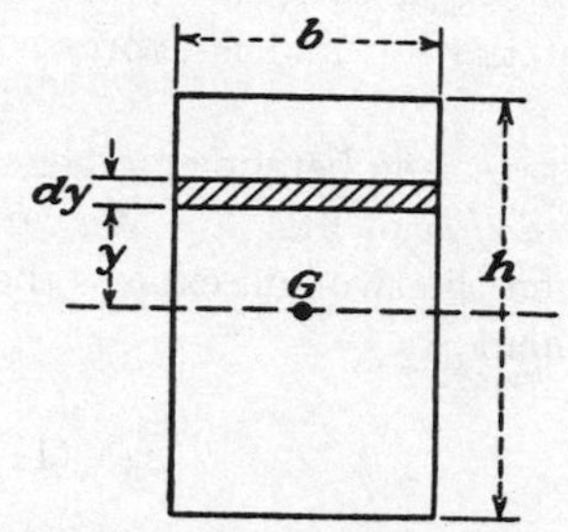

Fig. 27. Moment of inertia $bh^3/12$ and section modulus $bh^2/6$ of a rectangular cross section.

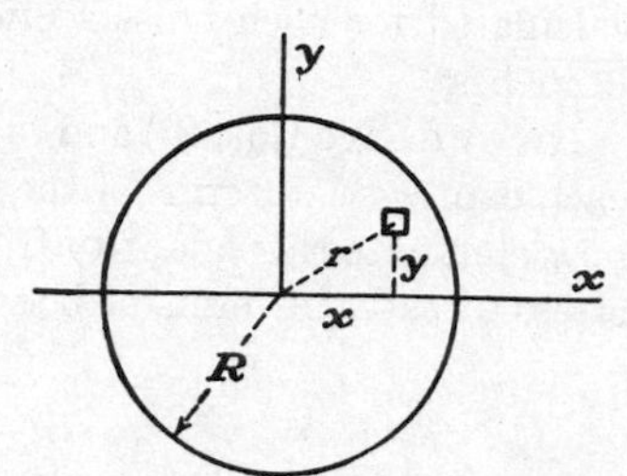

Fig. 28. For a circular section the diametral moment of inertia $\pi D^4/64$ is half the polar moment of inertia $\pi D^4/32$.

zontal strips dy, but it can be done more simply by observing that the polar moment of inertia I_p [Eq. (6), page 20] is:

$$I_p = \int r^2\,dA = \int (x^2 + y^2)\,dA = \int x^2\,dA + \int y^2\,dA.$$

The two axial or diametral moments of inertia about the x and y axes must be the same for reasons of symmetry. Hence:

$$I = \int y^2\,dA = \frac{1}{2}\,I_p = \frac{\pi R^4}{4} = \frac{\pi D^4}{64}, \tag{13a}$$

and the section modulus is:

$$Z = \frac{I}{R} = \frac{\pi R^3}{4} = \frac{\pi D^3}{32}. \tag{13b}$$

Beams of rectangular section occur frequently, usually in timber; circular beams in bending are hardly ever used, and in most practical cases the beams used are steel I, T, or Z sections, channels, angles, or built-up box sections, such as are shown in Fig. 29. The reason for giving a beam an I-shaped cross section is eminently practical: by Eq. (11a) a small stress occurs for a given bending moment and a given beam height when the moment of inertia I is large, and by the definition of I it becomes large for a given weight when as much of the material as possible is far away from the neutral line, so that that material can carry stress. Just as in torsion the core of a round shaft is almost dead and useless, so is the material near the neutral line in bending. The only reason for having material there at all, *i.e.*, for having a "web" in an I beam, is to keep the two useful flanges apart and in their place.

The values of moment of inertia I, section modulus Z, area A, and weight per foot length for various structural sections (Fig. 29) have been calculated, and their tabulations can be found in any

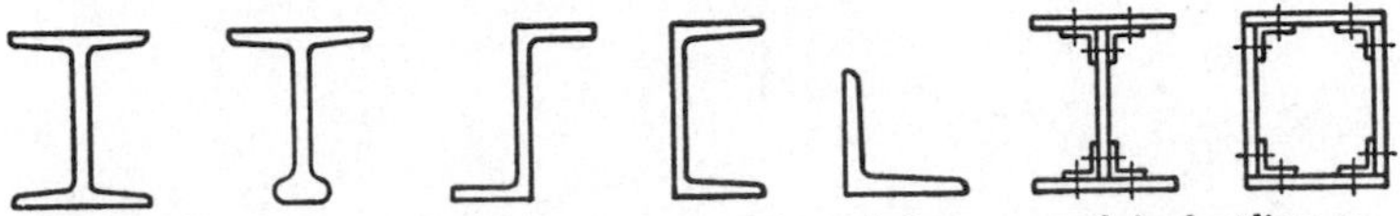

Fig. 29. Various cross sections of structural steel beams, used in bending applications.

civil or mechanical engineering handbook or in special publications by the steel companies, such as the "Carnegie Pocket Companion," the "American Institute of Steel Construction Handbook," and others. The reader is advised to look up such tables where he can find them most conveniently, see that he understands them, and roughly calculate a few cases to check for dimensions and orders of magnitude.

Problems 89 *to* 100.

12. Shear-stress Distribution. In a case where a cross section of a beam is not in "pure" bending, it must transmit a shear force in addition to the bending moment. The existence of such a shear force is always coupled with a variation of the bending moment along the beam, as can be seen from Eq. (10*b*) or from the examples in Figs. 18 and 20. From this it will be possible to calculate the distribution of the shear force S over the cross section by a reasoning of static equilibrium, making one assumption during the process. We now start doing this for a beam of rectangular cross section (Fig. 30) that is subjected to a positive bending moment (and hence, by Fig. 19, has tensile stress in its upper fibers). The bending moment increases with increasing distance x along the beam, to the right.

For the purpose of this analysis we now isolate from the beam a rectangular piece of dimensions dx, b, and $h/2 - y_0$ and examine the equilibrium of that piece. For clarity the piece is shown once more in projection (Fig. 31). By Eq. (11) the tensile stress on the left face at any point y is:

$$s_{\text{left}} = \frac{My}{I}.$$

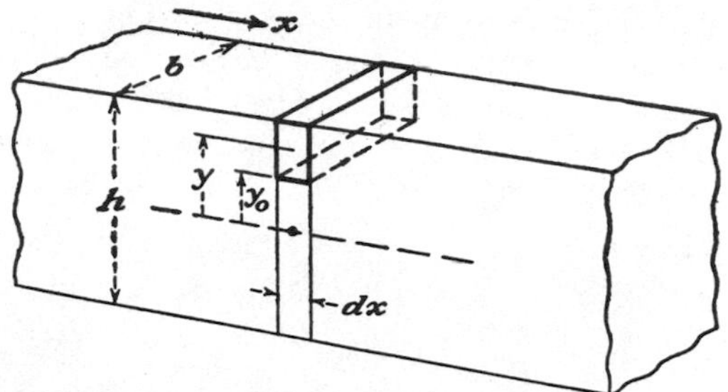

FIG. 30. Small parallelepiped isolated from a rectangular beam in bending and shear, for the purpose of calculating the shear-stress distribution.

On the other face, distance dx farther to the right, the bending moment is somewhat greater, so that:

$$s_{\text{right}} = \frac{(M + dM)y}{I} = \frac{(M + S\,dx)y}{I},$$

by Eq. (10b). Hence the excess stress (right less left) is $S\,dx\,y/I$, and the excess force to the right ΔF is this stress integrated over the area:

$$\Delta F = \int \frac{S\,dx\,y}{I}\,dA = \frac{S}{I}\,dx \int_{y=y_0}^{y=\frac{h}{2}} y\,dA.$$

In the last step the constant quantities have been brought before the integral sign, and the integration extends over the entire face of our isolated piece. Now that piece is being pulled to the right by the force ΔF, just calculated, and for equilibrium there must be an equal force pulling to the left. Our piece has six faces, three of which are on the outside periphery of the beam and hence have no stress or force. Of the three inside faces we have investigated two and found an

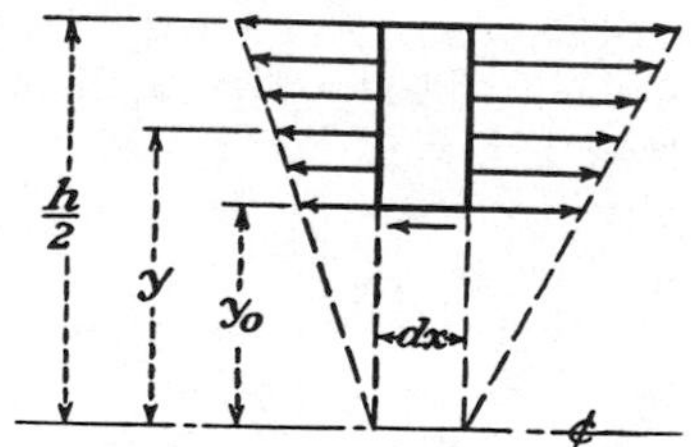

FIG. 31. The excess of tensile force to the right over that to the left must be compensated for by a shear stress on the bottom of the element.

excess force ΔF in the $+x$ direction. Shear stresses on those two faces have no x component. We are thus left with the sixth, or bottom, face only and conclude that there must be shear stresses acting on it to the left, as indicated in Fig. 31. How these stresses are distributed over the width b we have no means of knowing, but we make the reasonable *assumption* that they are distributed uniformly, *i.e.*, that the shear stress s_s is the same over the entire width b. Then we write:

$$s_s \cdot b\, dx = \Delta F = \frac{S}{I}\, dx \int_{y_0}^{\frac{h}{2}} y\, dA,$$

and consequently:

$$s_s = \frac{S}{bI} \int_{y_0}^{\frac{h}{2}} y\, dA. \tag{14}$$

In this fundamental formula S is the total shear force of the section; I is the moment of inertia of the section; b is the width at the spot where the stress is calculated, while the integral, to be extended from the spot of the stress s_s to the top, is the "static moment of the area above the cut with respect to the neutral line."

Now we return to Fig. 31, and we call on Fig. 11 (page 15) for reference. If in Fig. 31 there is a shear stress s_s on the bottom face of the isolated piece, then there must be equal shear stresses on the right face upward and on the left face downward. Or, to be more precise, these shear stresses must act on those faces near the bottom edge only, because, by Eq. (14), s_s varies with y_0, and our isolated piece is of small length dx but of finite height $h/2 - y_0$. These shear stresses when integrated over the entire right or left cross sections of the beam must furnish the shear force S, which should constitute a check on our calculation. We now proceed to do this, first by evaluating the integral in Eq. (14). It is:

$$\int y\, dA = \int yb\, dy = b \int_{y_0}^{\frac{h}{2}} y\, dy = \frac{by^2}{2}\Bigg|_{y_0}^{\frac{h}{2}} = b\left(\frac{h^2}{8} - \frac{y_0^2}{2}\right).$$

Remembering Eq. (12*a*) we find for the stress:

$$s_s = \frac{S}{bI}\, b\left(\frac{h^2}{8} - \frac{y_0^2}{2}\right) = \frac{3}{2}\frac{S}{bh}\left[1 - \left(\frac{y_0}{h/2}\right)^2\right].$$

We see that the shear stress depends on y_0, that is, on the location in the cross section. It is zero for $y_0 = +h/2$, as it should be. It attains its maximum value at $y_0 = 0$, where it becomes 3/2 times

S/bh, or, in words, fifty per cent larger than the average shear stress. The distribution of the stress, given by the above formula,

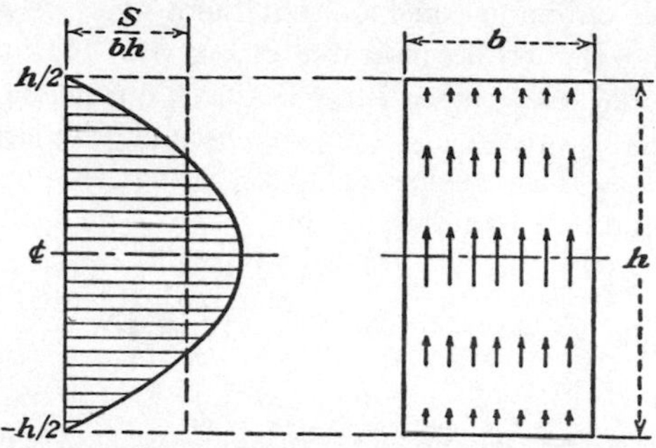

FIG. 32. Shear-stress distribution across a beam of rectangular cross section.

is shown in Fig. 32 and is represented by a parabola. The total shear force across the entire face then is:

$$\int s_s\,dA = \int_{-\frac{h}{2}}^{\frac{h}{2}} s_s b\,dy_0 = \frac{3}{2}\frac{S}{h}\int_{-\frac{h}{2}}^{\frac{h}{2}}\left[1 - \left(\frac{2y_0}{h}\right)^2\right]dy_0$$

$$= \frac{3}{2}\frac{S}{h}\left[y_0 - \frac{4}{h^2}\frac{y_0^3}{3}\right]_{-\frac{h}{2}}^{\frac{h}{2}} = \frac{3}{2}\frac{S}{h}\left(h - \frac{h}{3}\right) = S.$$

This comes out to the answer S, which constitutes a check on our derivation and completes the analysis for the rectangular cross section.

We shall now apply the general result, Eq. (14), first to the case of a circular cross section, and after that to an I beam.

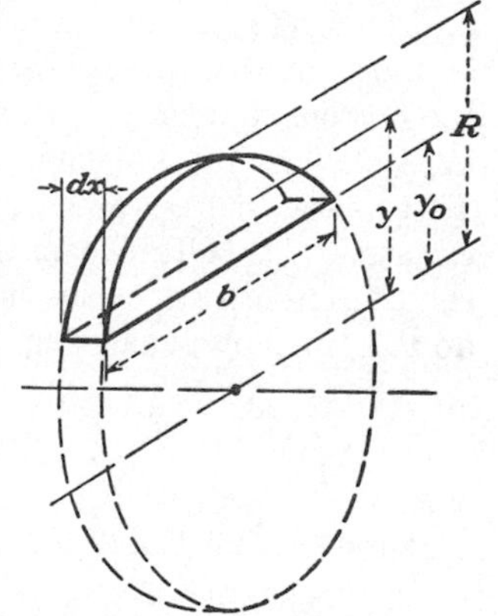

FIG. 33. Element of beam isolated for calculating the shear stress in a circular cross section.

These sections differ from the rectangular one mainly in the fact that the width b is no longer constant but depends on the distance y_0 from the neutral axis. The reader should now carefully reread the whole derivation with the fact in mind that the isolated piece of Fig. 30 is now replaced by the one shown in Fig. 33. Every sentence of the argument still holds, up to and including Eq. (14), which represents the shear stress uniformly distributed over the bottom face $b\,dx$ of the element. But now, when we make the step to transfer this stress to the front and back faces in the manner of Fig. 11, something new appears. If we do exactly as we did for the rectangular section, we get what is shown in the left half of Fig. 34, where at the periphery of the shaft there is a shear stress neither tangential to the boundary, nor normal to it, but having components in both directions. Now it can be seen in Fig. 35 that, at the periphery, a normal cross section of a beam cannot possibly have a shear-stress component normal to the

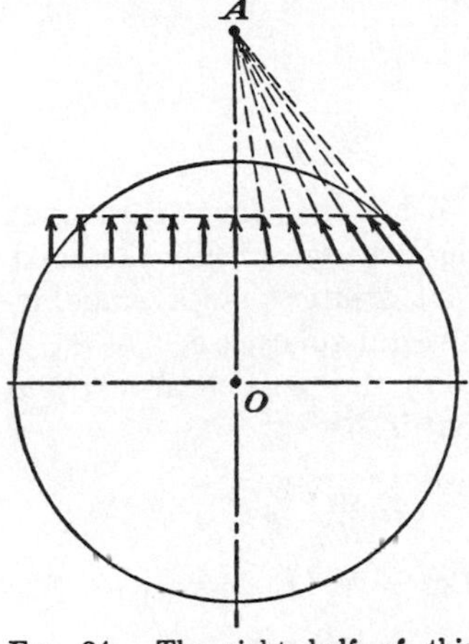

FIG. 34. The right half of this figure shows the distribution of shear stress across the section of a beam of circular cross section.

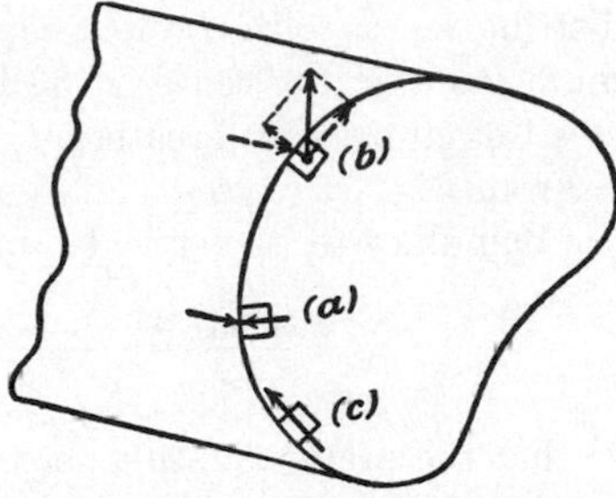

FIG. 35. Three cases of shear stress near the periphery of a cross section, of which (a) and (b) are impossible because they entail corresponding shear stresses on the free periphery of the beam.

boundary, because, by Fig. 11, such a component would be associated with an equal stress on the free periphery of the beam, which we observe as being without stress. Therefore the shear stress at the outside boundary of a normal cross section must be tangential. Equation (14) applied to Fig. 34 refers to the vertical component of s_s, and we can add a horizontal component to it without disturbing the equilibrium in the x, or axial, direction. We know how much horizontal component we have to add at the boundary, *i.e.*, sufficient to make the total s_s tangential, and we have no recipe for adding to the interior stress. Under the circumstances we are accustomed to make the assumption shown in the right half of Fig. 34, *i.e.*, that the total stress passes through point A, being the intersection of the tangent to the curve and its vertical center line. (Of course, point A varies with the location y_0 of the stress.) Only we take care to have the proper vertical component, as shown in the figure.

The only thing left for us to do in connection with the circular section is the evaluation of Eq. (14). The width of a horizontal strip at distance y above the center is $2\sqrt{R^2 - y^2}$, and further:

$$\int y\, dA = \int y \times 2\sqrt{R^2 - y^2}\, dy = \int \sqrt{R^2 - y^2}\, dy^2$$

$$= -\left.\frac{(R^2 - y^2)^{3/2}}{3/2}\right|_{y_0}^{R} = +\frac{2}{3}(R^2 - y_0^2)^{3/2};$$

$$s_s = \frac{S}{bI}\int y\, dA = \frac{S}{2\sqrt{R^2 - y_0^2}\,\pi R^4/4}\cdot\frac{2}{3}(R^2 - y_0^2)^{3/2}$$

$$= \frac{4}{3}\frac{S}{\pi R^2}\left(1 - \frac{y_0^2}{R^2}\right).$$

This is the vertical component only; it has the same "parabolic' distribution as with the rectangular cross section; only the maximum value in the center is 33 per cent greater than average, instead of 50 per cent greater as in the rectangular case.

From Fig. 34 we see that the ratio of the total shear stress at the boundary to its vertical component is:

$$\frac{s_{s\ \text{total}}}{s_{s\ \text{vert}}} = \frac{R}{\sqrt{R^2 - y_0^2}} = \left(1 - \frac{y_0^2}{R^2}\right)^{-1/2},$$

so that the previous result becomes:

$$(s_s)_{\text{total on boundary}} = \frac{4}{3}\frac{S}{\pi R^2}\sqrt{1 - \frac{y_0^2}{R^2}},$$

which gives a distribution curve somewhat different from Fig. 32. It is 133 per cent at its peak instead of 150 per cent and a little fuller at the fibers far from the center. The reader should sketch the curve for himself and verify in this case also that the integrated effect (or the area of the diagram) equals the shear force S.

As a last example we consider the idealized I beam of Fig. 36, of which the web height equals the flange width b and in which the thickness t is considered "small" with respect to b; that is, in the calculations only first powers of the ratio t/b will be retained, and higher powers will be neglected. In Eq. (14) the moment of inertia I appears, which first has to be calculated. Since t is small, the flanges appear as concentrated areas with contribution $2bt(b/2)^2$, and the web is a rectangle with $tb^3/12$, so that $I = 7b^3t/12$. To find the shear stress in the center of the web where $y_0 = 0$, we find:

$$\int_0^{\frac{b}{2}} y\, dA = bt\,\frac{b}{2}\ (\text{flange}) + \frac{bt}{2}\,\frac{b}{4}\ (\text{web}) = \frac{5}{8}\,b^2t.$$

The width b appearing in Eq. (14) here has the value t, so that:

$$(s_s)_{y_0=0} = \frac{S}{t \times 7b^3t/12} \cdot \frac{5}{8} b^2t = \frac{45}{14} \cdot \frac{S}{3bt} = 3.21 \frac{S}{A}.$$

At the top of the web, just under the flange, where the width is still t, we find:

$$\int_{\frac{b}{2}-t}^{\frac{b}{2}} y\, dA = bt \frac{b}{2} \text{ (flange alone)} = \frac{1}{2} b^2t,$$

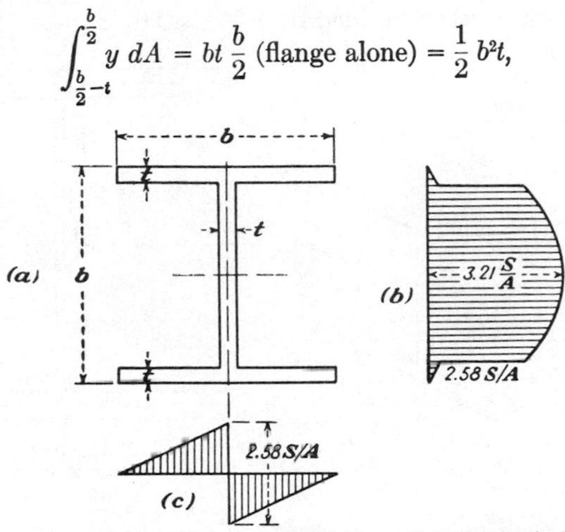

FIG. 36. I beam with its shear-stress distribution; (b) is the vertical component, mostly in the web; and (c) is the horizontal component in the flanges.

or 80 per cent of the previous result. All other quantities in Eq. (14) being the same, we have:

$$(s_s)_{\text{top of web}} = \frac{4}{5} \times 3.21 \frac{S}{A} = 2.58 \frac{S}{A}.$$

At the bottom of the flange, just above the top of the web, the width has suddenly jumped from t to b, everything else being the same. Thus the shear stress (or rather the vertical component of it) suddenly diminishes by a factor t/b, which we have dubbed "negligible" and which hence cannot be plotted. In Fig. 36*b*, however, this has been done just the same.

Now we should be ready to apply to this case the type of correction illustrated in the right half of Fig. 34 as compared with the left half. The only part of the periphery of Fig. 36*a* that is not vertical is the bottom of the flange. Here the slope with respect to the horizontal is "infinity," and the vertical component of shear stress is "small," almost zero. Hence the horizontal component of shear stress comes out as $\infty \times 0$, which, as we know from the calculus, may mean anything.

To find the horizontal component of shear stress in the flange we repeat the entire reasoning of Fig. 30, leading to Eq. (14), but this time we make our cut vertical, through the flange, as shown in Fig. 37, isolating a sliver of the right portion of the flange. We repeat the whole story that led from Fig. 30 to Eq. (14) and come

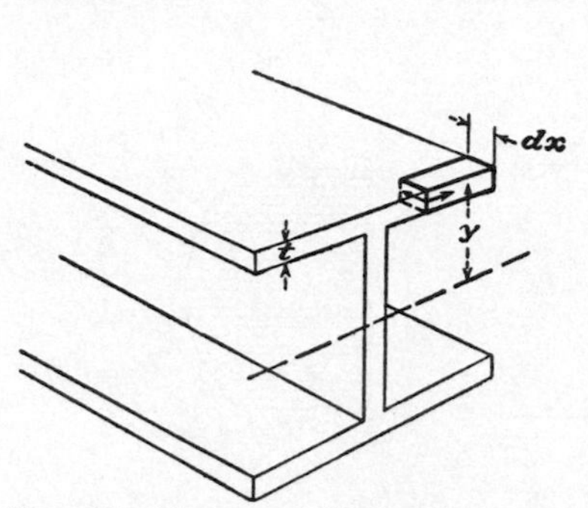

FIG. 37. Element of "length" dx and "width" t isolated from the flange of an I beam for the purpose of determining the *horizontal* component of shear stress in the cross section.

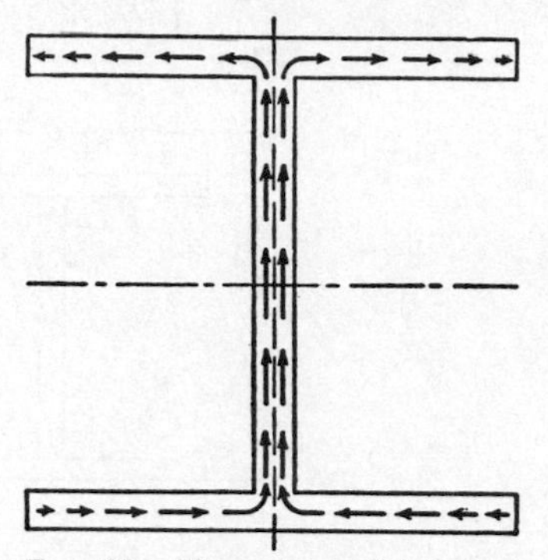

FIG. 38. Shear-stress distribution across the section of an I beam in bending. Quantities are shown in Fig. 36b and c.

to the identical result (except that the general symbol b for width in that equation here has to be interpreted as t, which is the width of the cross section, vertically, instead of horizontally as before). The shear stress along the x direction necessary to compensate for the difference in tensile stresses (Fig. 31) in the case of Fig. 37 acts in a vertical section, and its 90-deg counterpart thus becomes a horizontal shear stress in the flange. Again Eq. (14) gives the magnitude of this shear stress. Now if we isolate in Fig. 37 a complete half width of flange $t \times \frac{1}{2}b\ dx$, then Eq. (14) gives us numerically half the answer we found for the top of the web previously. In other words, in the middle of the flange the horizontal shear stress is half as large as the vertical shear stress is in the upper part of the web. The situation is illustrated in Fig. 36b, c and also in Fig. 38. The reader should reason out for himself why the horizontal shear stress diminishes *linearly* toward the extremities of the flanges.

Problems 101 *to* 110.

13. Applications. *a. Moments of Inertia.* In calculating the area moment of inertia of a cross section, a useful tool is the "parallel-axis theorem," which is known to some readers from previous courses in calculus or dynamics, but probably not to all. We shall now give a short derivation of it and then apply it to a specific case.

Consider in Fig. 39 an arbitrary cross section, and draw two parallel axes through it, one through the center of gravity G and

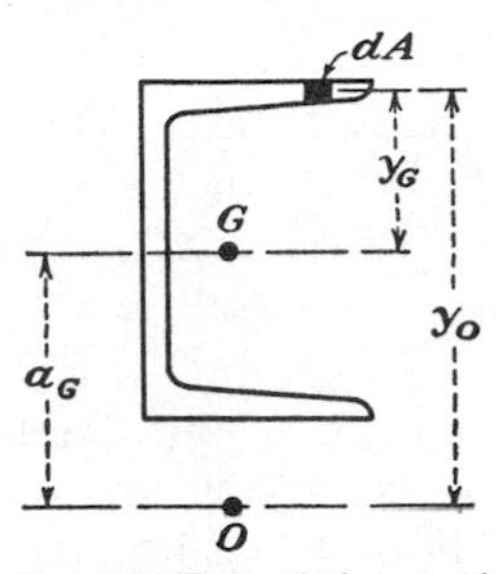

FIG. 39. Toward the proof of the parallel-axis theorem of moments of inertia.

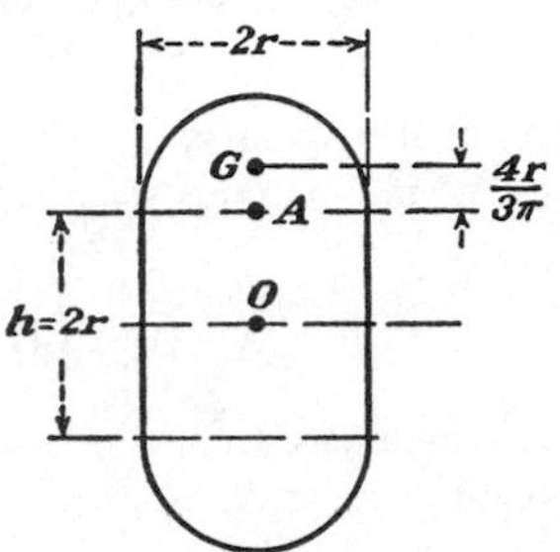

FIG. 40. In the calculation of the moment of inertia of this cross section, the parallel axis theorem is used to advantage.

the other through O, at distance a_G from the first one. Then we have, by definition of the symbol I:

$$I_O = \int y_O^2\,dA \qquad \text{and} \qquad I_G = \int y_G^2\,dA.$$

But $y_O = y_G + a_G$, so that:

$$I_O = \int (y_G + a_G)^2\,dA = \int y_G^2\,dA + 2a_G \int y_G\,dA + a_G^2 \int dA.$$

Of these three integrals the first one is I_G, the last one is $a_G^2 A$, and the middle one we recognize to be zero, because it is the definition of center of gravity. Thus:

$$I_O = I_G + a_G^2 A. \tag{15}$$

This equation is known as the parallel-axis theorem. One of its consequences is that, among all I's about all possible parallel axes, the one about the axis through the center of gravity is the smallest one.

For an application we turn to Fig. 40, a cross section of which the neutral line obviously passes through O. The moment of inertia of a circle or half circle about its diameter A is easy from Eq. (13*a*), but here we need it about a parallel axis through O. For the upper half circle we have:

$$I_O = I_G + \frac{\pi r^2}{2}\left(r + \frac{4r}{3\pi}\right)^2$$

$$= I_A - \frac{\pi r^2}{2}\left(\frac{4r}{3\pi}\right)^2 + \frac{\pi r^2}{2}\left(r + \frac{4r}{3\pi}\right)^2$$

$$= \frac{\pi r^4}{8} - \frac{8r^4}{9\pi^2} + \frac{\pi r^4}{2}\left(1 + \frac{8}{3\pi} + \frac{16}{9\pi^2}\right)$$

$$= \left(\frac{5\pi}{8} + \frac{4}{3}\right)r^4.$$

For the center square, by Eq. (12*a*) we find $(2r)^4/12$. The total moment of inertia consists of this square plus two half circles, or:

$$I = r^4\left(\frac{4}{3} + \frac{5\pi}{4} + \frac{8}{3}\right) = r^4\left(4 + \frac{5\pi}{4}\right) = 7.92r^4.$$

To check whether this is reasonable from a standpoint of order of magnitude, we replace Fig. 40 by a rectangle $2rH$ by drawing horizontal lines through the semicircles. Its moment of inertia is $2rH^3/12 = \frac{1}{6}r^4(H/r)^3$. If this is set equal to the previous answer, we find:

$$\frac{H}{r} = \sqrt[3]{6 \times 7.92} = \sqrt[3]{47.5} = 3.62,$$

which when sketched into Fig. 40 looks reasonable, and hence the answer is probably correct.

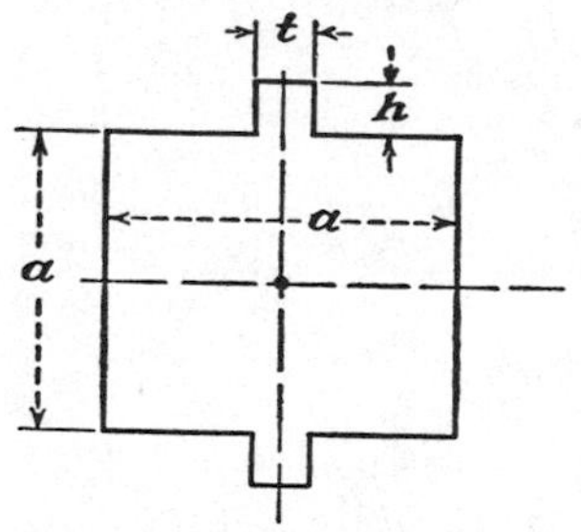

FIG. 41. The addition of ribs ht to a square cross section weakens that section if the ribs are too thin.

b. Weakening a Beam by Increasing Its Cross Section. The stress in a beam for a given bending moment is determined by the section modulus $Z = I/y_{max}$. When extra area is added to a given cross section, the quantity I always becomes greater, but so may y_{max}; and if y_{max} increases more, percentagewise, than I, the section modulus Z diminishes and the beam is weakened. As an example, consider a square beam of width and height a (Fig. 41) to which we add ribs in the center of width t and of height h. If these ribs are made thin, they hardly add to I but they do increase the distance of the extreme fiber y_{max} and hence weaken the section. We now ask for the critical thickness t of the rib below which Z diminishes. When the height h of the rib is "small," we can write:

$$I = \frac{a^4}{12} + 2th\left(\frac{a}{2}\right)^2,$$

$$y_{\max} = \frac{a}{2} + h,$$

$$Z = \frac{I}{y_{\max}} = \frac{a^3}{6}\,\frac{1 + 6th/a^2}{1 + 2h/a} = \frac{a^3}{6}\,\frac{1 + (2h/a)(3t/a)}{1 + (2h/a)}.$$

The section modulus is seen to be $a^3/6$ multiplied by a factor. For $h = 0$, that is, in the absence of ribs, the factor is unity, as it should be. Now for any small value of h the factor is smaller than unity if $3t/a$ is smaller than 1. Hence for

$$t < \frac{a}{3}$$

the section is weakened.

c. Relative Magnitude of Bending, Tensile, and Shear Stress. The stresses in a beam in bending are expressed by Eqs. (11) and (14), the first one giving the tensile or compressive stress in the extreme fibers and the second one giving the shear stress in the neutral fiber. We shall now show in an example that the shear stress is of practical importance only in short beams and that for the usual dimensions the tensile-compressive stress [Eq. (11)] is the determining factor.

Consider a beam on two supports of length l with a single load P in the center so that the bending-moment diagram is triangular with a central peak $M = Pl/4$ and with a shear force $S = P/2$. Let this beam have the I-shaped cross section illustrated in Fig. 36. We now state that in general a shear stress is twice as dangerous as a tensile or compressive one (see page 75) and ask how long the beam must be (what the ratio l/b must be) in order that the tensile stress in the lower fiber be twice as large as the shear stress in the neutral fiber.

We have, by Eqs. (11) and (14):

$$s = \frac{(Pl/4)(b/2)}{I} \qquad \text{and} \qquad s_s = \frac{P/2}{tI}\int_0^{\frac{b}{2}} y\,dA.$$

The ratio of these stresses is:

$$2 = \frac{s}{s_s} = \frac{lbt}{4\int y\,dA} = \frac{lbt}{4(5b^2t/8)},$$

by the last line of page 46.

Hence:

$$2 = \frac{2}{5}\,\frac{l}{b} \qquad \text{and} \qquad l = 5b.$$

We conclude that, for a beam of the section of Fig. 36, the shear stress is more important than the tensile one if the beam is shorter than five times its height; conversely, the tensile stress is the determining factor if, as usual, the beam is more than five times longer than it is high.

d. Rivet Design in Built-up Beams. Consider the cross section of Fig. 42, being an I beam, built up of three plates and four angles, riveted together. Let that beam be 20 ft long, on two supports with a central load P, as in the previous example. Let the permissible tensile stress be 15,000 lb/sq in., and let the permissible average shear stress across the rivets be 6,000 lb/sq in. What is the requisite spacing between the rivets in the top or bottom plates?

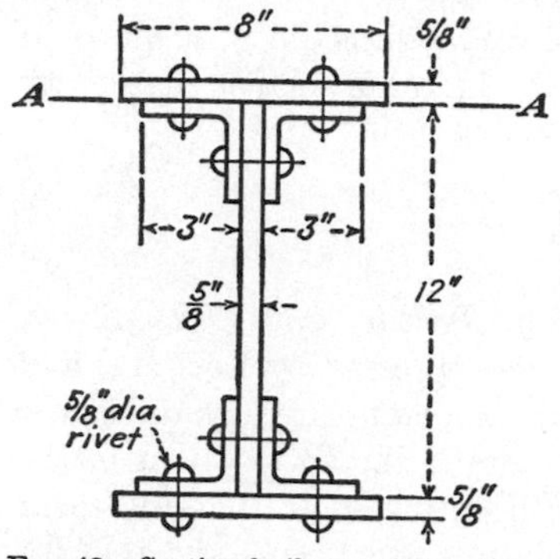

FIG. 42. Section built up of flat plates and angles riveted together.

To answer this question, we assume that the rivets hold the section together as a solid body, and we calculate the shear stress between the top plate and the angles, a width b of $6\frac{5}{8}$ in. Then:

$$s_s = \frac{P/2}{bI} \int y \, dA,$$

$$s = \frac{My_{\max}}{I} = \frac{Ply_{\max}}{4I}.$$

Hence:

$$\frac{s_s}{s} = \frac{2\int y \, dA}{bly_{\max}}.$$

The integral extends over the top plate only and is $6\frac{5}{16}$ in. $\times$ 5 sq in. = 31.56 cu in. The other values are $b = 6\frac{5}{8}$ in., $l = 240$ in., and $y_{\max} = 6\frac{5}{8}$ in., so that:

$$\frac{s_s}{s} = \frac{1}{167}.$$

If the beam were actually solid, as assumed so far, then the shear stress across section AA would be 167 times smaller than the tensile stress in the upper fiber. But the beam is not solid; and if the rivets in question are spaced x in. apart, the shear stress must be carried by two rivets instead of by a solid section of $6\frac{5}{8}x$ sq in. The shear stress across the rivets should be 6,000 lb/sq in., or

$$\frac{s_{s\ \text{rivet}}}{s} = \frac{6{,}000}{15{,}000} = \frac{1}{2.5}.$$

Hence we can afford to make the cross-sectional area of two rivets 167/2.5 = 67 times smaller than the corresponding solid area:

$$67\,\frac{\pi}{4}\left(\frac{5}{8}\right)^2 \times 2 = 6\frac{5}{8}x.$$

Thus $x = 6.2$ in. is the necessary spacing of the rivets.

Problems 111 *to* 120.

CHAPTER IV

COMPOUND STRESSES

14. Bending and Compression. In the first three chapters we have discussed the three most important cases of stress in beams or shafts, *i.e.*, compression, torsion, and bending with its attendant shear. Now we proceed to situations where two or more of these loads act on a beam simultaneously, and we shall have to find rules and procedures whereby the stresses can be added together. Without special consideration, it is not at all obvious how a tensile stress and a shear stress should be "added," and in fact this is a fairly complicated business that will be taken up in article 15. However, there is one case of compound stress that is easy and straightforward, and that is the addition of bending and axial compression in a beam, because in each of the two constituent cases the stress acts in the same cross-sectional face in the same direction and hence the two can be added algebraically.

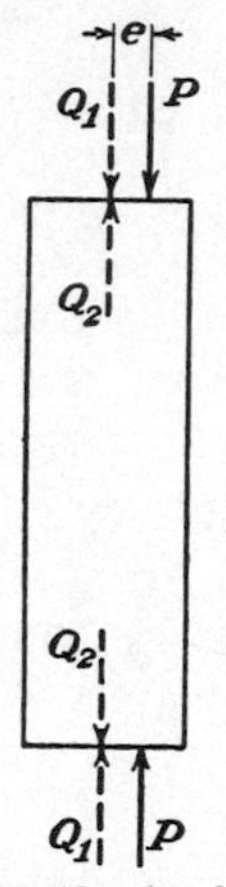

FIG. 43. A column loaded by eccentric compressive forces P is subjected to simultaneous pure compression P and pure bending $M = Pe$.

Consider the column of Fig. 43, loaded with a pair of forces P with the same line of action, not passing through the center of gravity of the cross section. We can apply four forces Q_1, Q_1, Q_2, Q_2 through the center of gravity, all equal in magnitude to P. The pair of forces Q_1 cause pure compression with a uniform stress distribution, because Q_1 passes through the center of gravity. (We remember that the definition of "center of gravity" is the location of the resultant of a set of parallel forces proportional to the elements of area.) On the other hand, the combination of the P's and Q_2's is a pair of bending moments, which put the beam into pure bending. Thus the column is subjected to a combination of compression and bending and is stressed accordingly. The compressive stress caused by P alone is $-P/A$ and is illustrated by the straight line AA' in Fig. 44. The bending stress is $My/I = Pey/I$, where e is the eccentricity measured in inches.

This stress is shown in Fig. 44 as BGB', having the center of gravity for a neutral point. The total stress is:

$$s = -\frac{P}{A} + \frac{Pey}{I}$$

and is illustrated by the line CC', passing through zero at a point N, the new location of the neutral line. Let us now calculate the location a of that neutral line, *i.e.*, the location $y = a$ of the point of zero stress:

$$s = 0 = -\frac{P}{A} + \frac{Pea}{I}$$

so that:

$$a = \frac{I}{Ae}$$

or

$$ae = \frac{I}{A} = k^2, \qquad (16)$$

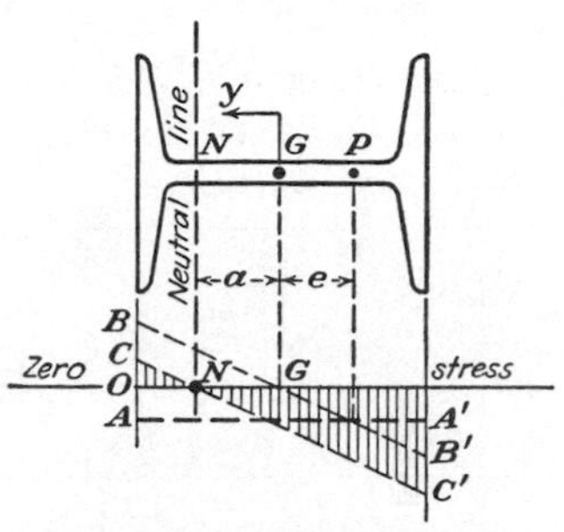

FIG. 44. Stress distribution in a cross section of a column loaded by a force P, off-center by distance e, but lying on one of the axes of symmetry.

where k is the "radius of gyration of the cross section." From the last form of the equation we see that the quantities a and e are interchangeable, which means that the points P and N in Fig. 44 are interchangeable, or, in words: If a load at point P causes a neutral point at N, then a load at N causes a neutral point at P. In Eq. (16) we can recognize two special cases:

1. For pure compression $e = 0$, and consequently $a = \infty$, which means that the neutral line is infinitely far away or that the stress diagram is a line parallel to the base.

2. For pure bending $a = 0$, and hence $e = \infty$, which means that the compressive force P in Fig. 43 is far away, and the (finite) moment $M = Pe$ takes the form $0 \times \infty$, so that the force P is zero.

So far we have put the eccentric load on an axis of symmetry, but now we shall investigate the more general case where P is located somewhere in the section not on any axis of symmetry. We choose a rectangular cross section $2a \times 2b$ (Fig. 45), and the eccentricity is now expressed by two numbers e_x and e_y, both measured in inches. The load P can be displaced parallel to itself, introducing a moment in the process, as was done in Fig. 43. Here, in Fig. 45, this is done in two steps, first from P to A, introducing a moment Pe_y about the x axis, and then from A to O, introducing the moment Pe_x about the y axis. Thus the eccentric loading P, and consequent stress, can be regarded as the sum or superposition of a central compression P and bending moments Pe_x and Pe_y about the two axes. The stress in an arbitrary point x, y of the cross section thus is:

$$s = -\frac{P}{A} - \frac{Pe_x x}{I_{yy}} - \frac{Pe_y y}{I_{xx}}.$$

To find the neutral line across the cross section, we simply set the stress $s = 0$ in the above equation and obtain a linear equation in the variables x and y, signifying a straight neutral line. In general that neutral line will not be perpendicular to the direction OP, as might be supposed, but it will be on the other side from O as the load P is, and the tensile stress, if any, will be a maximum in the corner C of the opposite quadrant. If the load P is sufficiently close to O, the corner C will have a compressive stress; but if P is far from O, there will be tension at the corner. Now we ask for which locations e_x, e_y of the load P the corner C will just be stressless, *i.e.*, on the border between tension and compression. The answer is written in the above equation; only now we have to make $x = -a$, $y = -b$ constant, and we must consider e_x and e_y as our variables, while, as before, s is zero. This is again an equation representing a straight line, *i.e.*, the locus of the load P for which the corner C is stressless, and the entire section has compressive stress. Substituting the special values $I_{xx} = \frac{1}{12}(2a)(2b)^3$, $A = 4ab$, etc., the equation becomes:

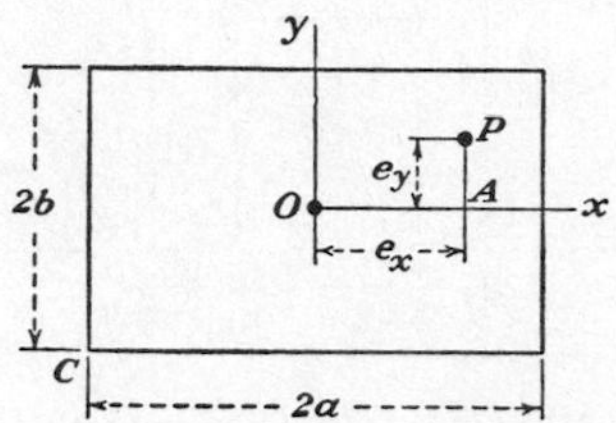

FIG. 45. A column of rectangular cross section with a load P having eccentricities e_x and e_y.

$$\frac{e_x}{a/3} + \frac{e_y}{b/3} = 1,$$

which is the straight line AA of Fig. 46, intersecting the x and y axes at $a/3$ and $b/3$, respectively. A load P placed above and to the right of line AA will cause tensile stress in corner C. Now, by symmetry we can repeat the argument for a load placed in the second quadrant $-x$, $+y$ of Figs. 45 or 46, having C_1 in Fig. 46 for its dangerous corner and BB for the locus of the loads P causing zero stress in that corner. Completing the analysis for the third and fourth quadrants, we arrive at the shaded diamond-shaped figure. If the load is within this figure, all four corners are in compression; if the load is outside it, one or more corners are in tension. The shaded figure is known as the *core** of the cross section, and any cross section possesses such a core. For a circular cross section, for example, it follows from rotational symmetry that the core must be a circle itself, and its radius is found from Eq. (16) where a, the distance of the neutral line, is the radius of

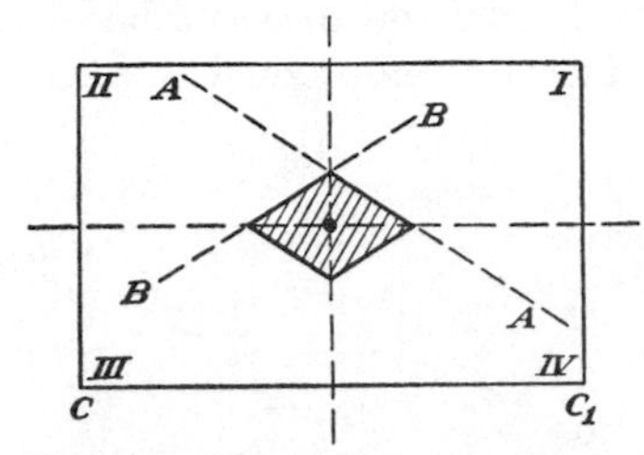

FIG. 46. Shows the core of a rectangular cross section. A load P placed within the core will cause compressive stress in the entire cross section.

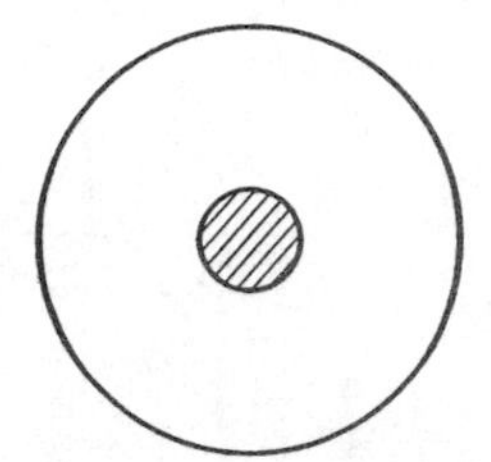

FIG. 47. For a column of circular cross section r, the core is a circle of radius $r/4$.

the cross section, and e, the eccentricity, is the radius of the core. Hence:

$$e = \frac{I}{Aa} = \frac{\pi r^4/4}{\pi r^2 \cdot r} = \frac{r}{4},$$

as illustrated in Fig. 47.

The theory of these cores is of importance in the design of structures made of materials that are very much stronger in compression than they are in tension, such as brickwork or non-reinforced concrete. For such materials, we usually assume that they have no tensile strength at all, to be on the safe side. If, then, such a structure is eccentrically loaded, the loading must be designed to be within the core or else cracks are likely to appear on the side opposite the eccentricity.

Problems 121 *to* 127.

15. Mohr's Circle. All cases of compound stress, except the one just discussed, are of some complication, and for their understanding it is necessary to study the relations between the stresses

* In some books, including Marks' Handbook, the word *kern* is used, which is a German expression meaning *core*.

in various directions at one point in the material. Let us look at an elemental piece $dx\ dy\ dz$ of the material, as shown in Fig. 48. On the sides of this little block, forces will act that represent the pushes and pulls from the neighboring material, and these forces can be resolved into components in the three directions. When these forces are divided by the area on which they act, they are called stresses. We distinguish between normal stresses (tensile-compressive stresses) and tangential or shear stresses. Now if a part of this block is cut off by an inclined plane, there will be stresses on that plane which are determined by the original stresses and

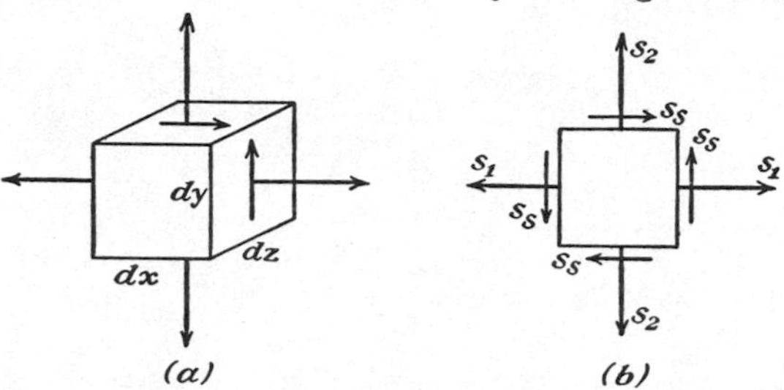

FIG. 48. State of plane stress at a point.

by the direction of the plane. Since the element was taken infinitely small ($dx\ dy\ dz$), we can consider it to be a point and thus we speak of the "stresses at a point." For the different points of our beam these stresses differ; they vary from point to point. The general study of the relations between the stresses at a point is very involved, and fortunately the most important practical results can be obtained by making a simplifying assumption at the outset, namely, that we deal with a "state of plane stress." This means that on our little element $dx\ dy\ dz$ stresses act only on four of the six faces, so that all stress vectors lie in one plane, as is indicated in Fig. 48. We shall see later that all combinations of tension, bending, shear, and torsion produce such plane stresses at each point. The most general case of plane stress is determined by three quantities: the two normal stresses s_1 and s_2 and the shear stress s_s, which has to satisfy the relation of Fig. 11*a* (page 15). The analysis which we are about to begin can be greatly simplified by assuming for the time being that the shear stress s_s in Fig. 48*b* is zero, so that only s_1 and s_2 appear. We shall soon see that no limitation to the generality of the argument is involved in this assumption, because it is always possible to pick two perpendicular directions at a point (at some angle with respect to the directions of Fig. 48*b*) so that the shear stress is zero for those directions. These directions later will be called principal directions and the stresses, principal stresses.

In Fig. 49, where the stresses s_x and s_y are such principal stresses, we cut the little cube by a plane perpendicular to the plane of zero stress and at angle α with the x axis, thus forming a small triangular prism. The dimension perpendicular to the paper is "small," the area of the inclined face is called dA, and hence the areas of the other two faces are $dA \sin \alpha$ and $dA \cos \alpha$. The principal stresses are s_x and s_y, and hence the forces on the two rectangular faces are as shown in the sketch, containing no shear components. The unknown stress components on the inclined face are called s_n and s_s, and the forces on that face then are $s_n\, dA$ and $s_s\, dA$. Any little piece cut out of a structure at rest is in equilibrium so that we can write in the horizontal and vertical directions:

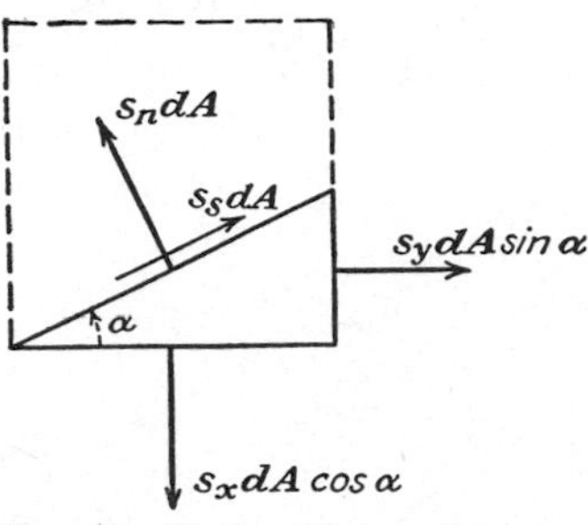

FIG. 49. The equilibrium of a prism cut out of the parallelepiped is expressed by Mohr's equation, Eq. (17).

$$s_y\, dA \sin \alpha + s_s\, dA \cos \alpha - s_n\, dA \sin \alpha = 0,$$
$$s_x\, dA \cos \alpha - s_s\, dA \sin \alpha - s_n\, dA \cos \alpha = 0.$$

First we can divide these by the element of area dA. Then, multiplying the top equation by $\sin \alpha$ and the bottom one by $\cos \alpha$ and adding, we get rid of s_s and find:

$$s_n = s_x \cos^2 \alpha + s_y \sin^2 \alpha. \qquad (a)$$

Similarly, multiplying the first equation by $\cos \alpha$, the second one by $\sin \alpha$, and subtracting, we find:

$$s_s = (s_x - s_y) \sin \alpha \cos \alpha. \qquad (b)$$

The last two equations express the stress s_n, s_s on the inclined face in terms of the principal stresses s_x, s_y and the angle α and thus constitute a solution to our problem. However, the equations can be written in another form:

$$\left.\begin{aligned} s_s &= \frac{s_x - s_y}{2} \sin 2\alpha, \\ s_n &= \frac{s_x + s_y}{2} + \frac{s_x - s_y}{2} \cos 2\alpha. \end{aligned}\right\} \qquad (17)$$

The first of these is obviously the same as the previous expression (*b*) for the shear stress, but for the second equation this is not immediately evident. The reader should work back from Eq. (17), remembering that $\cos 2\alpha = \cos^2 \alpha - \sin^2 \alpha$, and so check the correctness of it. Why anybody should write the more complicated equations (17) instead of the direct-forward simpler ones (*a*) and (*b*) is a question that might well be asked. The answer to that is that the form (17) lends itself to a beautiful and easily remembered graphical interpretation, invented by the German engineer Otto Mohr in 1880. In Fig. 50 we plot in a diagram the normal stress horizontally and the shear stress vertically. The two principal stresses are represented by points on the horizontal axis, Os_x and Os_y being those stresses. Then the distance OC to the mid-point C between s_x and s_y is represented by $(s_x + s_y)/2$, which is the first term in the expresssion for s_n in Eq. (17). Also, the other combination $(s_x - s_y)/2$ is represented by the distance $s_yC = Cs_x$. If we now draw a circle on C as center and draw a radius at angle 2α from Cs_x, we see that $PQ = r \sin 2\alpha = s_s$, by Eq. (17). Also $CQ = r \cos 2\alpha$, and $OQ = OC + CQ = s_n$, by Eq. (17). Thus the ordinate and abscissa of a point P on the circle are equal to the shear and normal stress on a plane at angle α with respect to the plane s_x, that is, point P in Fig. 50 represents the state of stress on the plane α in Fig. 49. If, in Fig. 49, we let the angle α slowly increase from 0 to 90 deg, the point P in Fig. 50 runs through $2\alpha = 180°$ along the circle, in the same counterclockwise direction, starting at s_x, representing the state of stress on the horizontal plane of Fig. 49, and ending up as s_y, representing the state of stress on the vertical plane of Fig. 49.

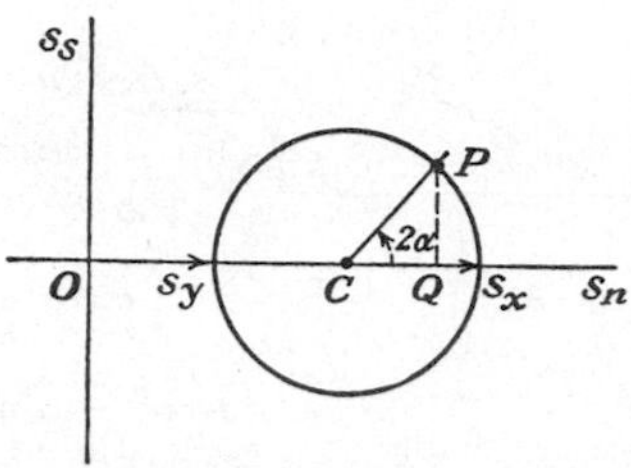

FIG. 50. Mohr's circle for stresses, showing shear stresses as ordinates and normal stresses as abscissae.

Before applying the Mohr-circle diagram to other examples, some remarks about signs are in order. In the first place we note that the directions of rotation of the angle α in Fig. 49 and of 2α in Fig. 50 are the same, although, of course, in Fig. 50 the angle turns twice as fast. Then the sign of a normal stress is as usual: positive when tensile, plotted to the right of O in Fig. 50; negative when compressive, plotted to the left of O. But the sign of the shear stress s_s is not obvious or familiar. We now define a shear stress acting on a face of an element as positive when it tends to turn that element in a clockwise direction. Figure 51 shows an element and four adjacent pieces with eight shear stresses, which are all alike, partly because of Fig. 11*a* and partly because of the theorem of action equals reaction. Four of these shear stresses are positive, and four are negative by our definition. The reader should look at this figure carefully and then note that in Fig. 49 the shear stress as drawn is positive, which was assumed in the analysis. Of course, in Fig. 50 a positive shear stress is plotted above the horizontal axis, and a negative shear stress is plotted below that axis.

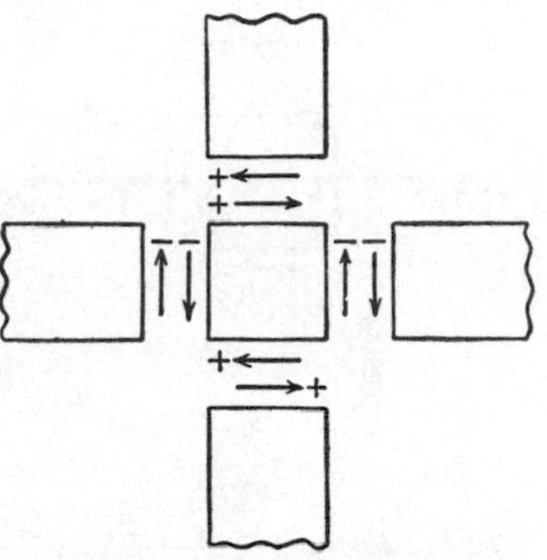

FIG. 51. A shear *stress* is defined as positive when it tends to turn the element clockwise, as negative when it exerts a counterclockwise couple. This is a new definition, independent of the sign definition of the shear *force* in a beam, Fig. 19.

Now we are ready to apply the Mohr-circle diagram to a specific case: the element shown in Fig. 52*a*, which is stressed with a tensile stress of 15,000 lb/sq in., a compressive stress of 5,000 lb/sq in., and shear stresses of 10,000 lb/sq in. In Fig. 52*b* we start constructing the Mohr diagram by first drawing the horizontal and vertical axes. In Fig. 52*a* the face marked 1 has a 15,000 normal stress and a −10,000 shear stress, and thus its state of stress is depicted by point 1 in Fig. 52*b*. The other piece of information given to use in Fig. 52*a* is face No. 2 with −5,000 normal stress and 10,000 shear stress and hence depicted by point 2 in Fig. 52*b*. The two faces in Fig. 52*a* are 90 deg apart; hence the two points in Fig. 52*b* must be 180 deg apart. By Mohr's theorem they must lie on a circle, which now can be constructed. In the completed circle we note points *A* and *B* having no shear stress and which we therefore call principal stresses. The points *A* and *B* are 180 deg apart in Fig. 52*b* so that the corresponding faces in Fig. 52*a* must be 90 deg apart. From this example we recognize that:

For any arbitrary plane state of stress at a point there exist two mutually perpendicular directions where there is no shear stress and of which the normal stresses are the largest and the smallest normal stress anywhere at that point. These extreme stresses are called the principal stresses.

This is our justification for starting Mohr's derivation on the simplified case of Fig. 49 lined up with the principal directions. In the more general case we could first find where the principal stresses are, then lay our coordinate axes in those directions and proceed with the proof as given. This is acting on the premise that it is better to make the coordinate system our slave than the other way round. If Mohr's proof had been started from Fig. 48*b*, instead

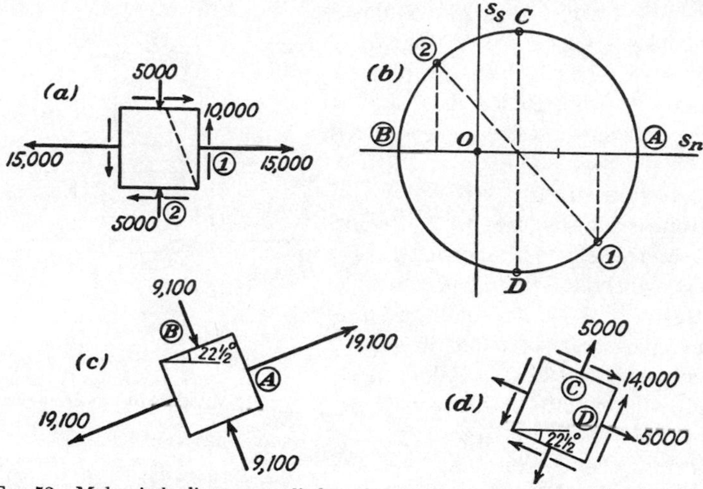

FIG. 52. Mohr-circle diagram applied to the case (*a*) and leading to the principal stresses (*c*). Figure 52*d* is just another possible representation of the same state of stress (*a*) or (*c*).

of from Fig. 49, it would have been considerably more complicated and would yield the same result. Readers interested in algebra are welcome to verify this statement.

Now we return to Fig. 52*b* and notice that point *A* is 45 deg counterclockwise from 1, or 135 deg clockwise from 2. Hence in Fig. 52*a* the plane of maximum principal stress must be 22½ deg counterclockwise from face 1, or 67½ deg clockwise from 2, which is the same thing. A similar conclusion can be drawn from face *B*, and all this information is assembled in Fig. 52*c*. The magnitude of the principal stresses can be deduced graphically, or, more accurately, they can be calculated quickly from the diagram using Pythagoras' theorem. The reader is advised to check the magnitudes shown in Fig. 52*c* from Fig. 52*b*. Figures 52*c* and *a* represent the same state of stress at the same point; it just depends how we cut our little square out of that point what sort of stresses we get.

For example, we could also have picked points C and D in Fig. 52b, which are 90 deg from A and B and hence 45 deg from the corresponding faces in Fig. 52c. The reader should deduce Fig. 52d from c and b by turning the faces in the proper directions and by remembering the definition of positive sign of Fig. 51.

Now we turn to an entirely different subject: the application of Mohr's circle to moments of inertia. The reader may study this now as another application of the Mohr-circle process, or he may postpone it to page 116 and at this time proceed to article 16, page 66.

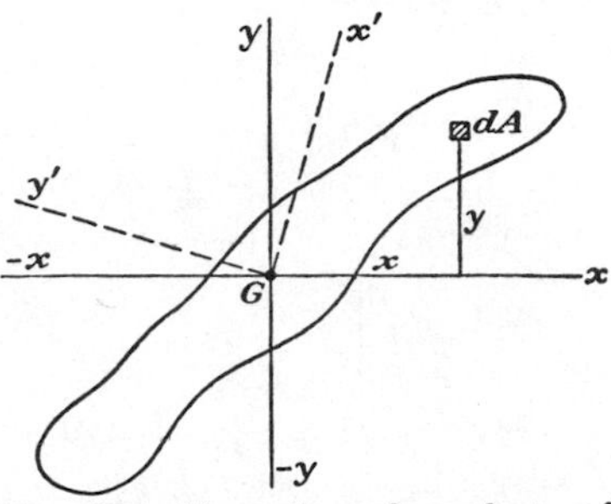

FIG. 53. Moments and products of inertia of an arbitrary cross section.

Consider in Fig. 53 an arbitrary outline, representing the cross section of a beam in bending, and draw two perpendicular axes x, y through the center of gravity in an arbitrary direction. The (area) moment of inertia about the x axis as neutral line was defined on page 39 as:

$$I_x = \int y^2 \, dA.$$

Similarly, about the y axis as neutral line we define:

$$I_y = \int x^2 \, dA.$$

Without having any particular application in mind, just as a mathematical fancy, we define the "product of inertia":

$$I_{xy} = \int xy \, dA.$$

Now, each of the two moments of inertia must be a positive quantity, becasue x^2 or y^2, being squares, must be positive for each element dA. However, this is not the case for I_{xy}; here the product xy for an element dA may be positive or negative, depending on where it is located. The product xy is positive in the first and third quadrants; it is negative in the second and fourth. In Fig. 53, I_{xy} is obviously positive for the fully drawn x and y axes because most of the area lies in the first and third quadrants. But suppose we had drawn another set of axes through the section, such as x' and y', dotted in the figure. Now most of the area lies in quadrants where either x' or y' is negative so that $I_{x'y'}$ is surely negative. By slowly turning the axes from xy to $x'y'$, we pass from a positive product of inertia to a negative one, continuously: hence we must pass through a position where $I_{x'y'} = 0$. This, we observe, is roughly at 45 deg, or where the coordinate axes are roughly axes of symmetry. Through any arbitrary point O, not necessarily the center of gravity G, in any arbitrary cross section, it is possible to find a set of two perpendicular directions xy for which the product of inertia vanishes, and these are called the principal axes of that point. With this in mind, we turn to Fig. 54, where the axes x, y have been taken in these "principal" directions, and we shall now derive formulae for the moments and product of inertia about another set of skew axes x', y', in terms of the two moments of inertia about the original axes. An element of area dA at point P is shown with its xy coordinates in wavy line and its $x'y'$ coordinates in heavy line. The projection of the wavy broken line OP on the x' axis has the value x', and its projection on the y' axis has value y'. This is expressed as follows:

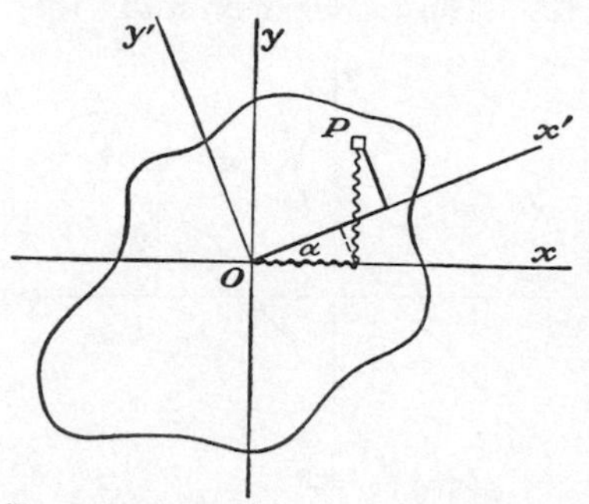

FIG. 54. Moments of inertia about various axes through a point O, not necessarily the center of gravity.

$$x' = x \cos \alpha + y \sin \alpha,$$
$$y' = y \cos \alpha - x \sin \alpha.$$

With these we calculate $I_{x'}$:

$$I_{x'} = \int y'^2 \, dA = \int (y \cos \alpha - x \sin \alpha)^2 \, dA$$
$$= \cos^2 \alpha \int y^2 \, dA + \sin^2 \alpha \int x^2 \, dA - 2 \sin \alpha \cos \alpha \int xy \, dA.$$

The last integral is zero because we chose our x-y axis system accordingly. Hence:

$$I_{x'} = I_x \cos^2 \alpha + I_y \sin^2 \alpha.$$

Similarly:

$$I_{x'y'} = \int x'y' \, dA = \sin \alpha \cos \alpha \left(- \int x^2 \, dA + \int y^2 \, dA\right),$$
$$I_{x'y'} = (I_x - I_y) \sin \alpha \cos \alpha.$$

If we compare these expressions with those for the stresses at a point [(a) and (b), page 59], we see that they are identical in form. Hence we can follow up in the same manner and write Mohr's equations:

$$
\left.\begin{aligned}
I_{x'} &= \frac{I_x + I_y}{2} + \frac{I_x - I_y}{2}\cos 2\alpha, \\
I_{x'y'} &= \frac{I_x - I_y}{2}\cos 2\alpha.
\end{aligned}\right\} \qquad (17a)
$$

The Mohr-circle graphical representation again applies, with moments of inertia plotted horizontally and products of inertia vertically, as shown in Fig. 55. The only difference is that in Fig. 55

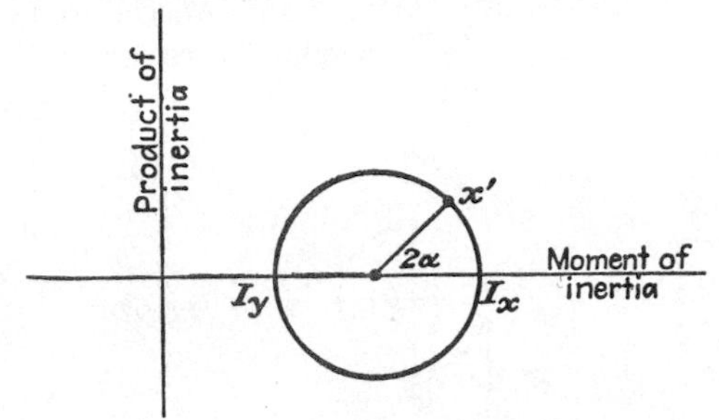

FIG. 55. Mohr's circle for moments of inertia.

the circle never, never can lie partly to the left of the ordinate axis, whereas in Fig. 50 it can. (Why?) As before with stresses, the moments I_x and I_y for the special case where $I_{xy} = 0$ are called the principal moments of inertia, and we observe that they are the largest and the smallest possible moments for any axis passing through that point. An important conclusion we can draw from Fig. 55 is that, if for three different axes through a point the mo-

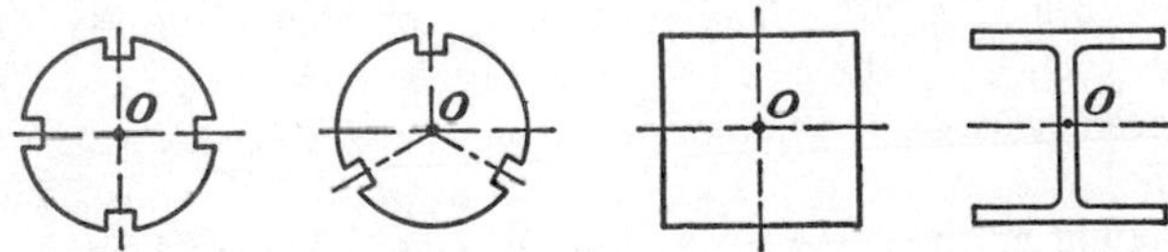

FIG. 56. Four cross sections of beams for which the moments of inertia about every axis through O is the same, so that every such axis is a principal one and no product of inertia exists about any axis through O.

ments of inertia are the same, then the circle shrinks to a point and the moments of inertia about all possible axes through that point are the same. Also, if we recognize the two principal moments of inertia to be the same, the circle becomes a point and all moments of inertia are the same. All these can be called principal moments of inertia, because the products of inertia all disappear. Examples of this are shown in Fig. 56, which the reader should examine carefully.

Problems 128 *to* 145.

16. Bending, Shear, and Torsion. In this article we shall investigate with the aid of Mohr's circle the states of stress at a point caused by the individual loadings of the preceding chapters, first separately, then in combination, ending up with the most general case of a shaft subjected simultaneously to tension, bending, shear, and torsion.

We begin with pure tension, illustrated by Fig. 57. We start in Fig. 57*a* to mark the faces 1 and 2 and proceed to find the corresponding points in Fig. 57*b*. As soon as these are found, Mohr's

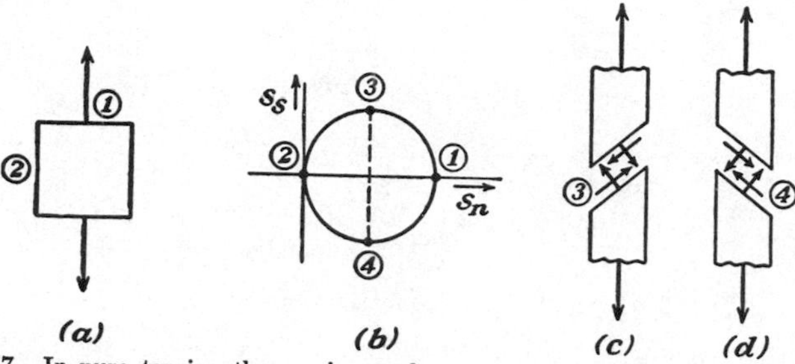

FIG. 57. In pure tension the maximum shear stress is half the maximum tensile stress and occurs at 45 deg.

circle can be drawn, which we see is tangent to the vertical axis. We recognize immediately that the maximum shear stress is half as large as the tensile stress. For example, point 3 is 2 × 45 deg clockwise from 2 (or counterclockwise from 1). This gives us the cut shown in Fig. 57*c*. The shear stress of point 3 in Fig. 57*b* is positive; hence, by the convention of Fig. 51, it is supposed to turn our piece clockwise so that we can draw the 45-deg shear arrows in Fig. 57*c*. Also we note in Fig. 57*b* that at point 3 there is a tensile stress equal to the shear stress (and of course half as large as the maximum or principal stress of point 1). This has been shown by the proper arrows in Fig. 57*c*. The reader should now for himself check the equilibrium of the pieces of Fig. 57*c*, and then he should check everything in Fig. 57*d*, which is the state of stress depicted by point 4 of Fig. 57*b*.

The case of pure uniaxial compression is virtually the same as that of tension. It is illustrated in Fig. 58, which the reader should carefully construct for himself, step by step. The numbering of the faces and the direction of the stress purposely have been made different from Fig. 57 to show that the answer is, of course, independent of how we number our faces. It is necessary only to see that the direction of rotation in the element diagram is the same as in the Mohr-circle diagram and also to keep in mind the sign rule of Fig. 51.

The next case we investigate is shown in Fig. 59, usually referred to as "hydrostatic compression" or as "two-dimensional hydro-

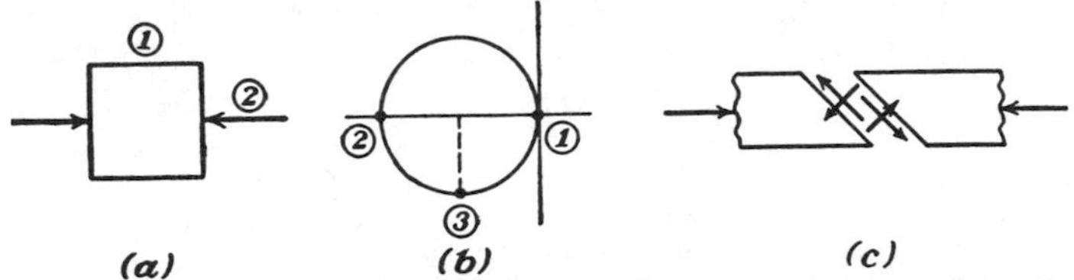

Fig. 58. For uniaxial tension or compression Mohr's circle is tangent to the vertical axis.

static compression." A true hydrostatic case would, of course, require pressure on the third set of faces as well, and this we have ruled out from the outset (page 58), as being too complicated for the time being.

In Fig. 59*a* the two faces 1 and 2 have identical stresses, which therefore are represented by the same point in Mohr's circle. But this same point is 180 deg along the circle removed from itself so

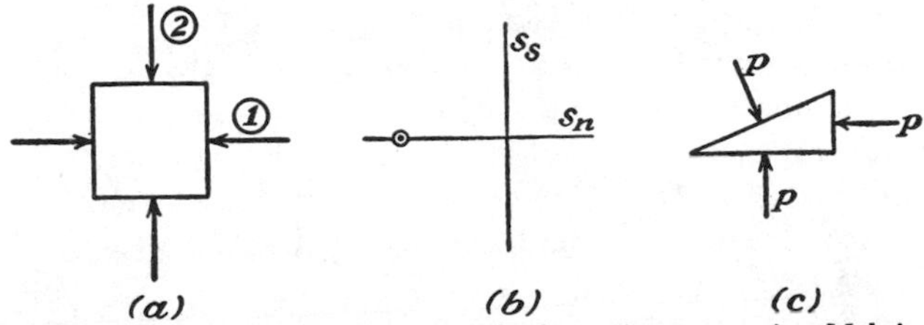

Fig. 59. In the case of two-dimensional hydrostatic compression Mohr's circle shrinks to a point and the compression stress is the same on all planes passing through the point in question, perpendicular to the plane of zero stress.

that the circle becomes a point, and all compressive stresses are the same, while no shear stresses appear anywhere. This is illustrated in Fig. 59*c*, which is identical with one of the first figures appearing in any text on hydrostatics.

Our next example is "pure shear," as it occurs in torsion. The sequence of operations is shown in Fig. 60. In Fig. 60*a* the faces have been marked 1 and 2 at random, and we note that face 1, trying to turn the cube counterclockwise, experiences a negative shear stress, which is depicted in Fig. 60*b*. Face 2 has an equal positive shear stress. We now construct the circle of Fig. 60*b* and note that the principal stresses 3 and 4 are tension and compression of equal magnitude. Face 3 is twice 45 deg clockwise from face 2, while the compressive face 4 is twice 45 deg counterclock-

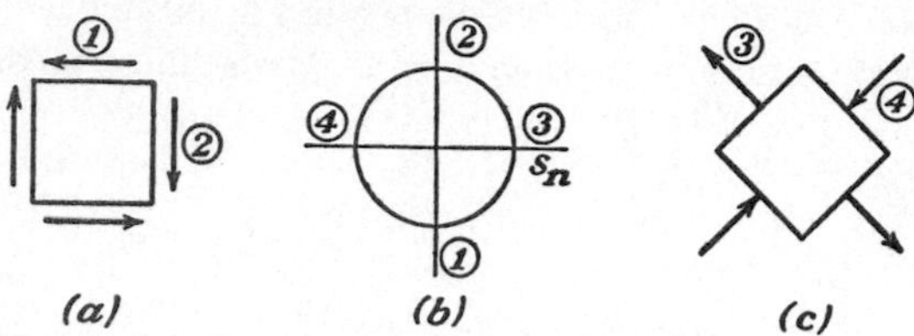

FIG. 60. Mohr's circle for pure shear. The principal stresses are push-pull of intensity equal to the "pure" shear stress.

wise from 2. This is shown in Fig. 60*c*, and the reader should try to visualize clearly the equivalence of the push-pull situation of Fig. 60*c* with the shear diagram, Fig. 60*a*, by taking the block of Fig. 60*a* in his hands and by applying the shear stresses by friction of his fingers.

From Fig. 60 we conclude that on a circular shaft in torsion the lines of principal stress are 45-deg spirals, as illustrated in Fig. 61. Such lines, following the directions of the principal stress, are called "stress trajectories." What are the trajectories for the cases of pure tension and compression in a bar?

FIG. 61. The principal stress trajectories on the outer surface of a twisted shaft are 45-deg spirals.

Pure bending of a beam results in a state of stress of tension or compression at each point. Hence Figs. 57 and 58 represent pure bending, and the trajectories are simple straight lines parallel to the neutral plane.

The stress at a point of a beam subjected to nonpure bending, *i.e.*, to bending and shear combined, is determined by Eqs. (11) and (14). The tensile or compressive stress [Eq. (11)] at any point is parallel to the center line of the beam, while the shear stress [Eq. (14)] at any point has a certain definite direction in the plane of the cross section, such as is illustrated in the right half of Fig. 34 (page 45), for example. Figure 62*a* is a little block cut out of the beam with its stressless face (*i.e.*, the plane of the paper of Fig. 62*a*) parallel to the beam's center line and also parallel to the shear stress at that point. This stressless plane in the case of a beam of rectangular section (Fig. 32) is a vertical plane; in other cases it may not be vertical. On the periphery of a circular section, Fig. 34, for example, the stressless plane is tangent to the circle. But in any case it is possible to find a stressless plane at any point of the beam; hence we have a state of plane stress, and Mohr's circle can be applied.

From Fig. 62*a* we proceed in the usual way, first finding in Mohr's diagram (Fig. 62*b*) the images of the two faces 1 and 2 and then constructing the circle. If s_b and s_s are the original bending

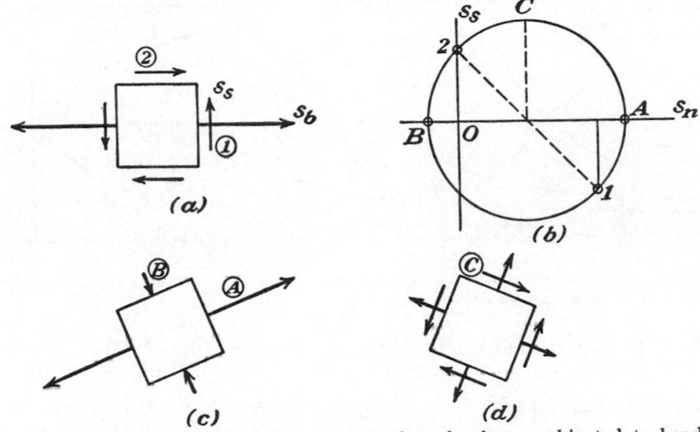

FIG. 62. Stresses and Mohr's circle for a point of a beam subjected to bending and shear combined.

and shear stress, as indicated in Fig. 62*a*, then we see in Fig. 62*b* that the maximum tensile stress OA is:

$$s_{\max} = \tfrac{1}{2}s_b + \sqrt{(\tfrac{1}{2}s_b)^2 + s_s^2},$$

and that this stress occurs on a plane which is turned counterclockwise from face 1 through an angle half as large as $\tan^{-1}(2s_s/s_b)$. This principal stress is shown in Fig. 62*c*. The maximum shear stress occurs at point C in Fig. 62*b* on a face turned 45 deg counterclockwise from face A. This gives the state of stress shown in Fig. 62*d*, consisting of a combination of a shear stress of magnitude $\sqrt{(s_b/2)^2 + s_s^2}$ and a hydrostatic tension of $s_b/2$.

The angle of diagram 62*c*, and hence the direction of the trajectory in the beam, depends on the ratio s_b/s_s. It is illustrated in Fig. 63 for the example of a cantilever beam. In the neutral plane of bending $s_b = 0$, so that there is pure shear, and the trajectory angle is 45 deg. At the upper and lower fibers $s_s = 0$, and the trajectories are horizontal. At an intermediate fiber, say midway between neutral and top side, the stress s_b is proportional to the distance from the free end, while s_s is independent of the location along the beam.

The last and most complicated case of compound stress that we shall consider is the general combination of tension, bending,

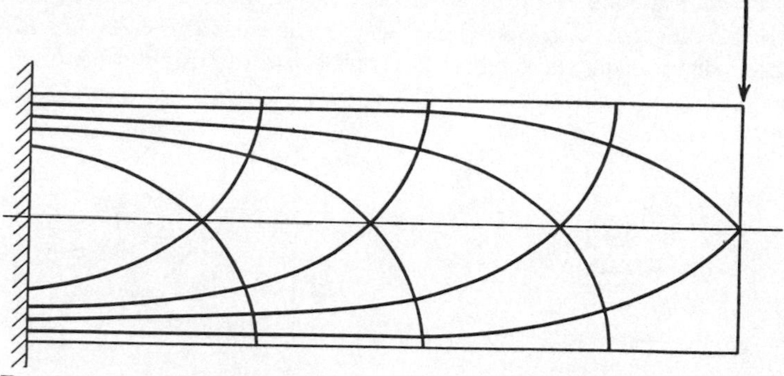

FIG. 63. Stress trajectories for a cantilever beam, as derived from repeated Mohr-circle diagrams, Fig. 62b.

shear, and torsion in the same bar. Since our knowledge of torsion at this stage is limited to circular sections only (solid or hollow), we shall have to choose such a section for our example, but the main conclusion, *i.e.*, that the compound stress still is a "plane" one, holds true for any cross section.

The stress, then, is made up of components determined by Eqs. (1), (5), (11), and (14). The stresses (1) and (11), *i.e.*, pure tension and pure bending, act on the same face and are directly additive, as was done on pages 54 to 57. The stresses (5) and (14) again act on the same face, but they are shear stresses not in the same direction generally, as indicated in Fig. 64. These stresses can be added together, but the addition must be vectorial, i.e., by the parallelogram construction. The maximum shear stress due to (5) and (14) undoubtedly occurs at point A in Fig. 64, because there they are both maximum and add together arithmetically. To the picture of Fig. 64 we have to

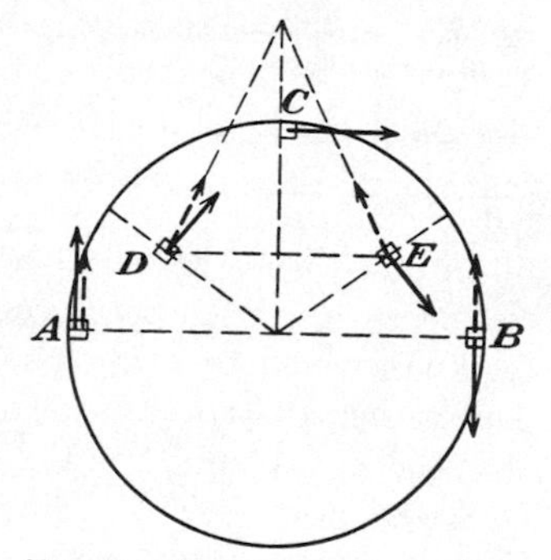

FIG. 64. Shear stresses caused by torsion (full-lined arrows) and by shear (dotted arrows) in the cross section of a beam (see Fig. 34).

add normal stresses perpendicular to the paper with a distribution such as shown in Fig. 44. Now at any point in the cross section we can cut out a little element with sides parallel to the center line of the beam and to the resultant of the two local vectors of Fig. 64. Such an element is in the state of stress shown in Fig. 62 and hence is in a plane state of stress. The analysis differs from the case of Fig. 62 only in that s_b and s_s are each made up of two components that have to be calculated in advance. Once that is done for any point we perform the same steps as with Fig. 62. Where in the cross section of Fig. 64 does the maximum stress appear? That question is even more complicated than it sounds. In the first place we have not yet made up our minds what we mean by "maximum stress." We may mean maximum tensile stress, or maximum shear stress, or some combination of the two. This will be the subject of the next article. Even when we know what we mean and ask, for instance, for the maximum principal stress, then it might be at A when the torsion is large and the bending small, or it might be at C when the bending is more important than the torsion. The only way to settle the question is to construct Mohr's diagram for a number of suspected points and take the largest answer.

Before we pass on to the next subject, some remarks on three-dimensional stress are in order. All cases discussed so far were plane or two-dimensional, the third pair of opposite faces being without stress. Unfortunately, the theory of three-dimensional stress at a point is too involved for decent explanation at this stage so that a statement without proof will now be given with the poor consolation that it is fully explained in other books one will read (maybe) when one gets older and wiser.

The statement is that in three dimensions we can always find three mutually perpendicular "principal directions" in which no shear stresses appear, as in Fig. 65*a*. We can plot the three principal stresses in Fig. 65*b* and draw three circles as shown. The stress on any skew plane at the point in question can then be represented by a point in the shaded region of Fig. 65*b*, points on the circles being included.

A two-dimensional or plane state of stress is a three-dimensional stress, of which one of the principal stresses is zero. Hence, if the fully drawn circle of Fig. 66 represents a plane state of stress, then in order to complete the three-dimensional diagram we have to draw in the two dotted circles. As a particular example, let us reconsider the case of two-dimensional hydrostatic compression

(Fig. 59), where Mohr's circle shrinks to a point. In that case the compressive stress is the same, and the shear stress is zero for all

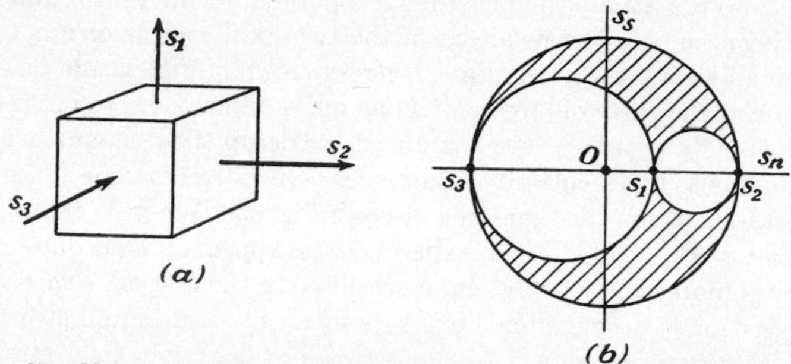

FIG. 65. Mohr's diagram for three-dimensional stress consists of three circles having their centers on the horizontal axis.

planes passing through that point *perpendicular to the plane of zero stress.* We never found out what happens in planes that are *not* perpendicular to that plane. This is shown in Fig. 67, where first the Mohr diagram is completed by two dotted circles. The fully drawn point circle has been given some small size for purposes of illustration only. Figure 67*b* shows the element, and now we make a cut $A'ACC'$, which is not perpendicular to the plane of zero stress. The equilibrium of the prism $A'B'C'ABC$ is shown in Fig. 67*c*, and the reader should check that it requires compressive and shear stresses on the face AC which are each half as large as the original compressive stress. Hence the face $A'ACC'$ is represented by point P in Fig. 67*a*.

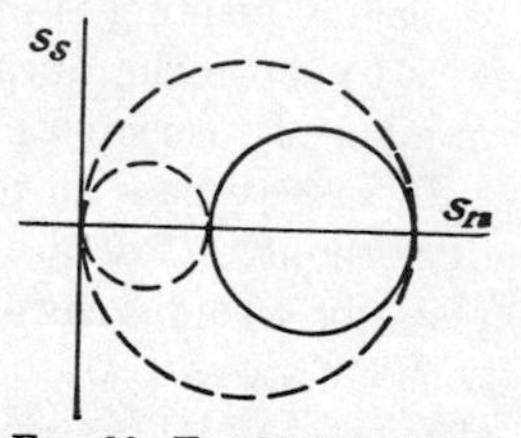

FIG. 66. To a Mohr's circle for two-dimensional stress, drawn in full, two other circles, passing through the origin, must be added in order to complete the diagram for three dimensions.

Problems 146 *to* 152.

17. Theories of Strength. After the loads on an engineering structure have been determined or properly estimated and when from these loads and the chosen dimensions of the structure the stresses have been calculated, the sixty-four-dollar question presents itself of whether the construction is safe or not. Since a structure will always start to fail at one spot (the weakest) first, the question reduces to that of the safety of an element of material $dx\ dy\ dz$ subjected to a given stress pattern. During the last two centuries a number of assumptions have been made concerning the criterion of failure of such an element, and these assumptions have been honored with the designation "theory of strength." The most important of them, in historical order, are:

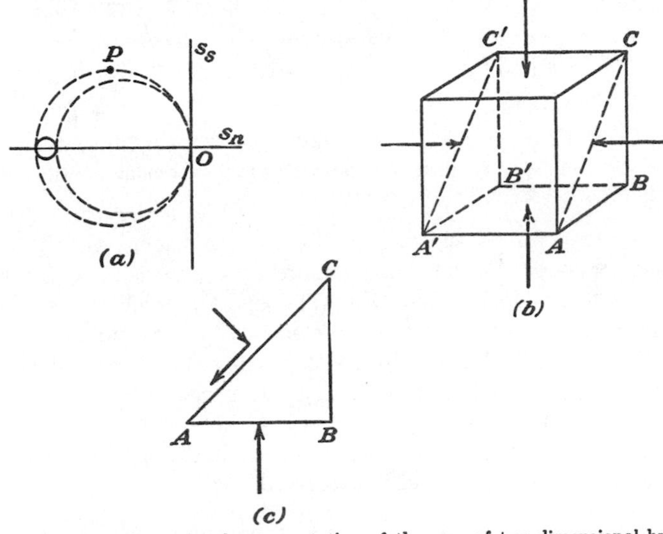

FIG. 67. Three-dimensional representation of the case of two-dimensional hydrostatic compression.

a. The maximum-stress theory
b. The maximum-strain theory
c. The maximum-shear-stress theory
d. The theory of Mohr
e. The theory of maximum distortion energy

We shall now discuss these assumptions one by one and then compare them with each other.

The *maximum-stress theory* is easiest of all; it simply states that the element will fail when the maximum principal stress reaches a certain critical value. It sounds obvious, and on first thought it seems as if nothing more reasonable could possibly be proposed. However, it does not seem quite so obvious when we look at Fig. 68, showing a case of pure shear (*a*), pure tension (*b*), and hydrostatic compression (*c*), all having the same maximum principal stress. All three cases are equally strong according to the principal-stress theory, but case a has twice the maximum shear stress of cases b or c.

Now we turn to the *maximum-strain theory*, which states that the element fails when the maximum strain ϵ in any direction

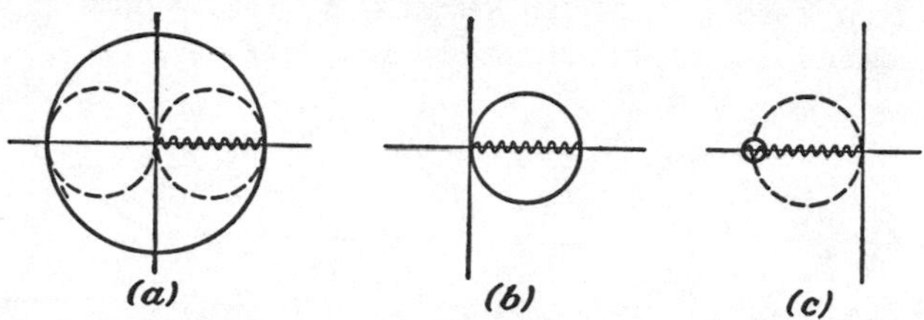

Fig. 68. Three cases of equal strength, according to the maximum stress theory; (a) is torsion, (b) tension, and (c) compression.

reaches a certain critical value. Before this theory can be explained in detail, we must know more about strain. So far we have seen only Hooke's law [Eq. (1), page 4], which is a correct statement for a single stress on an element. At the time we were not interested in the change in length of the element in a direction perpendicular to the direction of pull. Experience shows us that an element which is being pulled by a stress s_1 will extend in the direction of pull and at the same time shrink in the across direction. This is shown in Fig. 69 and is expressed by:

$$\epsilon_1 = s_1/E, \qquad \epsilon_2 = -\mu s_1/E,$$

in which ϵ_2 is the lateral strain. It is negative, because the piece contracts sidewise and is about one-quarter as large as the main

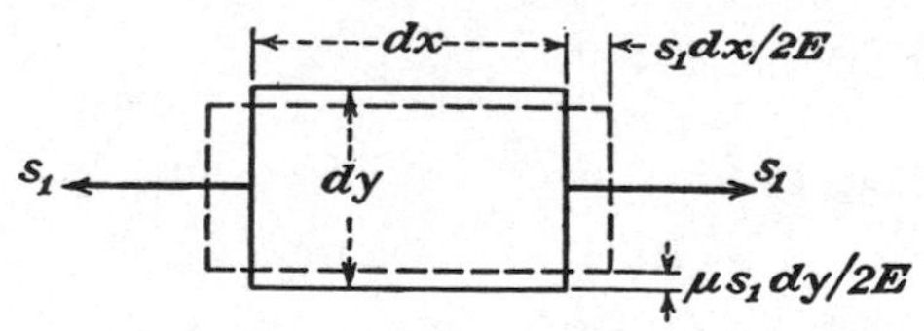

Fig. 69. Illustrates the action of Poisson's ratio. The element in solid outline is strained to the dotted figure by the single stress s_1.

strain ϵ_1. The factor μ, a pure number, is called *Poisson's ratio;* it is the ratio of the lateral contraction to the longitudinal extension. The value of μ is different for various materials; for steel it varies from 0.25 to 0.33; for rubber it is slightly less than 0.50.

When an element is subjected simultaneously to two normal stresses s_1 and s_2 on perpendicular faces, the strain in any direction is the sum of the strains caused by each stress individually. In a formula this is:

$$\left.\begin{aligned}\epsilon_1 &= \frac{1}{E}(s_1 - \mu s_2),\\ \epsilon_2 &= \frac{1}{E}(s_2 - \mu s_1).\end{aligned}\right\} \qquad (18)$$

In case an element is acted upon by three principal stresses s_1, s_2, s_3, as in Fig. 65*a*, we have:

$$\left.\begin{aligned}E\epsilon_1 &= s_1 - \mu(s_2 + s_3),\\ E\epsilon_2 &= s_2 - \mu(s_1 + s_3),\\ E\epsilon_3 &= s_3 - \mu(s_1 + s_2).\end{aligned}\right\} \qquad (18a)$$

A large strain occurs in direction 1 if the stress s_1 is tensile while s_2 and s_3 are compressive, thus making all three contributions additive. If $s_1 = -s_2 = -s_3$, then the main strain is $(1 + 2\mu)s/E$, while the lateral contraction is s/E. On the other hand, when $s_1 = +s_2 = +s_3$ (three-dimensional hydrostatic compression or tension), the strain in all three directions is $(1 - 2\mu)s/E$. These two cases have the same maximum stress, while their maximum strains differ by the ratio $(1 + 2\mu)/(1 - 2\mu)$. Therefore different strains can occur with the same maximum stress, and the maximum-strain theory of strength is different from the maximum-stress theory.

The *maximum-shear theory* of strength states that the criterion of failure of an element consists in reaching a certain critical shear stress. Thus, returning to Fig. 68, we find that the case of Fig. 68*a* is stressed twice as highly as cases *b* or *c*, while by the maximum-stress theory they are all equally stressed. The two theories are nicely illustrated in Fig. 70, showing the diagram in which Mohr's circles are usually drawn. The maximum-stress theory declares an element to be unsafe if some points of Mohr's circle fall outside a vertical strip of a width equal to twice the safe stress; and, similarly, an element is unsafe by the maximum-shear theory if its Mohr's circle penetrates outside a horizontal strip of width twice the safe shear stress and of infinite length. This infinite length of the strips has no practical meaning in the maximum-stress theory, because the center of Mohr's circle lies on the horizontal axis so that the height of the circle can never be greater than its radius. This automatically limits the height of the vertical safe strip to no more than its width. But for the maximum-shear theory the infinite length of the horizontal strip has a very important meaning. It implies that an element in which the three principal stresses are very large, but almost equal to each other, is safe. According to the maximum-shear theory, the state of hydrostatic compression

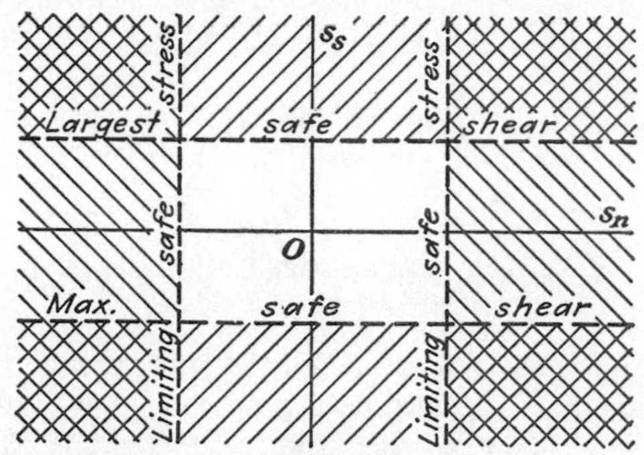

FIG. 70. The s_n, s_s-diagram in which Mohr's circles are drawn. The maximum-stress theory guarantees safety in a vertical strip of infinite height; the maximum-shear theory gives safety in a horizontal strip of infinite length. The central, unshaded region is safe by both standards.

or hydrostatic tension is safe for any value of the stress, because Mohr's circle for those states shrinks to a point very far away from the origin horizontally.

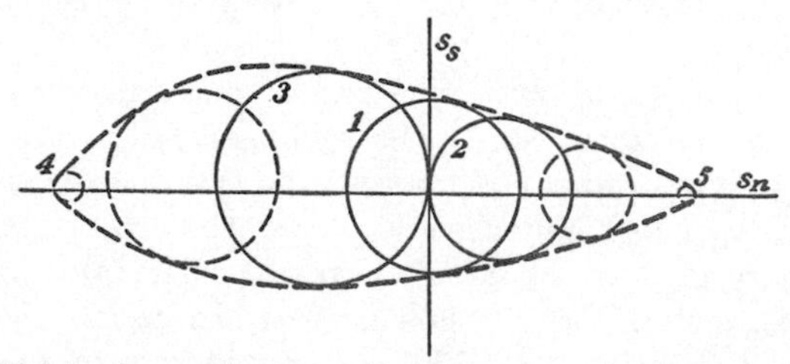

FIG. 71. Mohr's theory directs us to make many experiments with various combinations of stress until failure occurs and to plot the circles for such failure. Then the element is safe if its Mohr's circle lies entirely within the envelope of all previous circles.

Now we come to the next theory of failure, the *theory of Mohr*. It differs in character from all previous theories in that these previous theories were definite and Mohr's is quite indefinite, ordering us to perform numerous experiments before deciding what is safe or unsafe. We must run a pure-shear test (circle 1 in Fig. 71), a pure-tension test 2, a compression test 3, a hydrostatic-compression test 4, a hydrostatic-tension test 5, and some more tests of a mixed nature, plotting the circles when the stress has reached the value where the undesirable thing occurs. That undesirable thing may be complete failure or beginning of yield, depending on what we want. Then Mohr draws the enveloping curve of all the circles and declares the region inside safe and outside unsafe.

The last theory mentioned is that of *maximum distortion energy*. It is not yet used in engineering practice but is considered one of the best theories by researchers in the field of properties of materials. Its explanation will be postponed to the last chapter of this book (page 223).

Now after all the theories have been explained, which one are we going to use, and when? The answer is that Mohr's theory is beautiful but too complicated, that the maximum-strain theory, historically the oldest, has been discredited by experiments, that *the maximum-stress theory applies well to the ultimate failure of brittle materials, such as cast iron, and that the maxmium-shear theory applies well to the beginning of yielding in ductile materials, such as steel.*

In a brittle material the deformations remain small (of the order of 1 or 2 parts per 1,000) until final failure, which takes place suddenly, without warning, by separation in a plane across the pull. Thus it is reasonable to expect that the maximum principal-stress theory applies. In steel we first reach a yield point, after which plastic elongations occur locally up to about 30 per cent, or 300 parts per 1,000, before the final break. The deformations can be seen to consist in a slipping along the planes of maximum shear; the edges of a broken tensile-test piece show angles of 45 deg. For such ductile materials the ultimate failure is of secondary interest to us; in many cases the structure becomes useless when yielding starts. And this yield starts when the shear stress attains a certain value so that the material starts to slip along these shear planes. Hence the maximum-shear theory applies with a good approximation. One consequence of it is that, when a tensile-test piece yields at a certain tensile stress, a shear stress of only half that amount will also cause yielding. This is illustrated in Fig. 57 or 68*b*. From Fig. 65 we recognize that the maximum shear stress equals half the algebraic sum of the two extreme principal stresses and that the maximum shear stress is entirely independent of the intermediate principal stress. Thus, according to this theory, yielding will start in an element if the largest of the three Mohr's circles of Fig. 65 reaches a certain size, and this yielding is completely independent of how the two smaller circles inside are arranged.

Further remarks on this subject, including hints on how to make a proper choice of *working stress*, will come in the chapter on Experimental Elasticity (page 221).

Problems 153 *to* 157.

CHAPTER V

DEFLECTIONS OF BEAMS

18. The Differential Equation of Flexure. The sags or deflections of beams bent by various concentrated or distributed loads are calculated by integrating a differential form that expresses the local deformation of a small element of the beam. When we derived Eq. (11) for the bending stresses in the beam, it was necessary to consider the local deformations in the process, for which the reader is referred to Fig. 25 (page 37) and to the story leading up

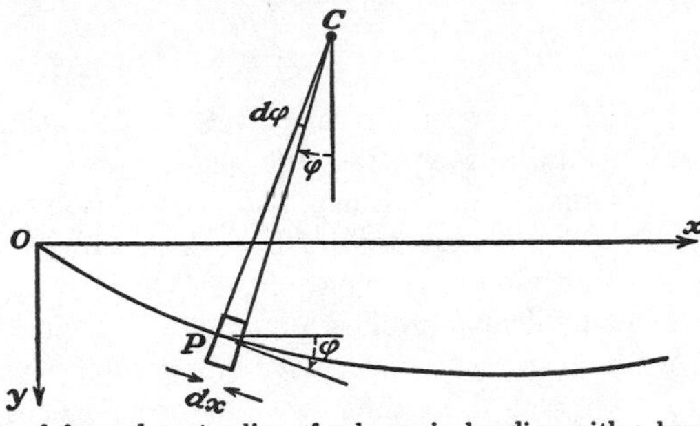

Fig. 72. The deformed center line of a beam in bending with a locally deformed element of length dx.

to Eq. (b) (page 39). The latter equation is practically the equation we need; we only have to reduce the factor to $d\varphi/dx$ to a form more convenient for integration.

In Fig. 72 we have drawn a coordinate system with the sign convention of Fig. 19, the deflections y being considered positive downward. The neutral plane of the beam, *i.e.*, the horizontal plane passing through the centers of gravity of all cross sections, is supposed to coincide with the x-axis when the beam is unloaded and hence unstressed; as a result of the loading this neutral plane assumes a trace $y = f(x)$ as sketched. The angle of slope of this deflection curve we denote by φ. If an element dx of the beam has zero bending moment in it, its two cross sections, dx apart, both turn through the same angle φ, remaining parallel to each other. If the element dx is subjected to a bending moment M, then by Eq. (b) (page 39) its two faces turn with respect to each other through a small angle $d\varphi$. Comparing Fig. 72 with Fig. 25 and Eq. (*b*) of page 39, we see that $d\alpha$ of Fig. 25 is identical with $d\varphi$ of Fig. 72, except for sign. In Fig. 72 the angle φ at the right face of the element is smaller than at the left face, so that with increasing x we have a decreasing φ, or $d\varphi/dx$ is negative. Hence we must replace $d\alpha/dx$ in Eq. (*b*) by $-\,d\varphi/dx$ to make that equation applicable to Fig. 72:

$$\frac{d\varphi}{dx} = \frac{M}{EI}.$$

Now, in Fig. 72 the angle φ satisfies the relation $\tan \varphi = dy/dx$, and if the beam bends through only slightly, and hence if the slope φ is very small, we may write approximately $\tan \varphi = \varphi = dy/dx$. Substituting this into the above expression we obtain:

$$M = EI\,\frac{d^2y}{dx^2}. \tag{19}$$

This most important result is called the "differential equation of flexure of a beam." As written, the bending moment M, variable along the beam, the deflection y, and the location coordinate x all must obey the sign convention of Fig. 19. It must not be forgotten that the equation is true only for small slopes, a condition which is practically always fulfilled for structural beams but which often does not hold for steel springs with large deflections. Then Eq. (19) is *not* a good approximation, but problems in this class fortunately are seldom of practical importance. (Fortunately, indeed, because the complete theory where $\tan \varphi$ can no longer decently be replaced by φ is very much more complicated and beyond the scope of this book.)

For small slopes the quantity y'' is the curvature of the beam; and the quantity EI is often called the "bending stiffness" or "flexural stiffness" of the beam. Thus Eq. (19) states that *the local curvature equals the local bending moment divided by the local flexural stiffness of the beam.* Since the equation was derived for one element only, it is true even for a beam in which the stiffness EI varies along the length. In that case M, EI, and y'' all three are functions of x. However, in most practical cases the stiffness EI is constant along the beam or at least constant along considerable portions of it.

Most of the remainder of this chapter consists in finding the deflection y of an arbitrary beam with arbitrary loading, by integrating Eq. (19). This can be done by the ordinary methods of the calculus; but since an engineer must solve a great many beam problems during his life, short-cut methods are imperative. In this article we shall integrate Eq. (19) for a few cases by ordinary calculus, without short cuts or special tricks.

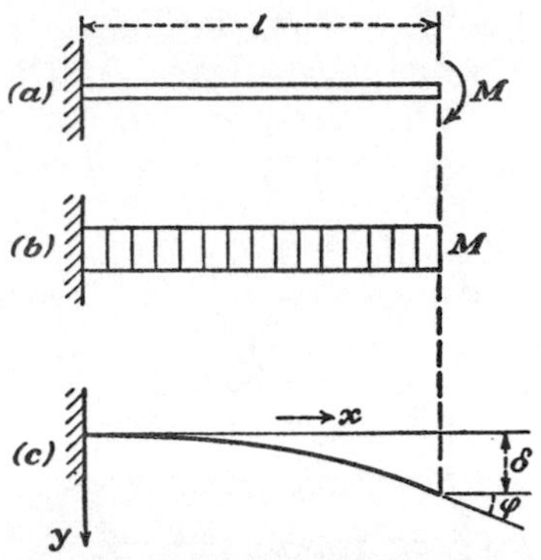

FIG. 73. A cantilever beam subjected to an end-bending moment with its bending-moment diagram (b) and its deflection curve (c).

The first example we choose is a cantilever beam of constant section EI, and length l, loaded by a bending moment M at its free end (Fig. 73a). The bending moment is constant along the beam, as shown in Fig. 73b. The first integration of Eq. (19) then becomes:

$$y' = \frac{dy}{dx} = \int \frac{d^2y}{dx^2}\,dx = \int \frac{M}{EI}\,dx = \frac{M}{EI}\int dx = \frac{Mx}{EI} + C_1.$$

The integration constant C_1 is found from the condition that, at $x = 0$, $y' = 0$, because the beam there is built in horizontally. Hence $C_1 = 0$. A second integration leads to:

$$y = \int y'\,dx = \int \frac{Mx}{EI}\,dx = \frac{M}{EI}\int x\,dx = \frac{Mx^2}{2EI} + C_2.$$

The second integration constant C_2 is also zero because at $x = 0$ the deflection $y = 0$. Hence the deflection curve (Fig. 73c) is a parabola, and at the free end we have:

$$\left.\begin{aligned} \varphi &= y'_{x=l} = \frac{Ml}{EI}, \\ \delta &= y_{x=l} = \frac{Ml^2}{2EI}. \end{aligned}\right\} \qquad \text{(Fig. 73)}$$

In drawing the deflection curve (Fig. 73c) and in all subsequent deflection curves, we make the y scale about 100 times larger than it actually is, so that we can plainly see the shape of the curve. In the actual case the curve can barely be distinguished from a straight line by the naked eye; on the other hand, if the deflection curve were really as shown in Fig. 73c with φ about 30 deg, Eq. (19) would be but a poor approximation of the truth.

The next example is a cantilever beam loaded by a single end load P as shown in Fig. 74. The bending moment is $P(l - x)$. The first integration of Eq. (19) is:

$$y' = \int \frac{M\,dx}{EI} = \frac{P}{EI}\int (l - x)\,dx = \frac{P}{EI}\left(lx - \frac{x^2}{2}\right) + C_1.$$

Again, at $x = 0$ the slope $y' = 0$ at the fixed end of the cantilever, so that C_1 must be zero. The second integration gives:

$$y = \frac{P}{EI}\int\left(lx - \frac{x^2}{2}\right)dx = \frac{P}{EI}\left(l\frac{x^2}{2} - \frac{x^3}{6}\right) + C_2.$$

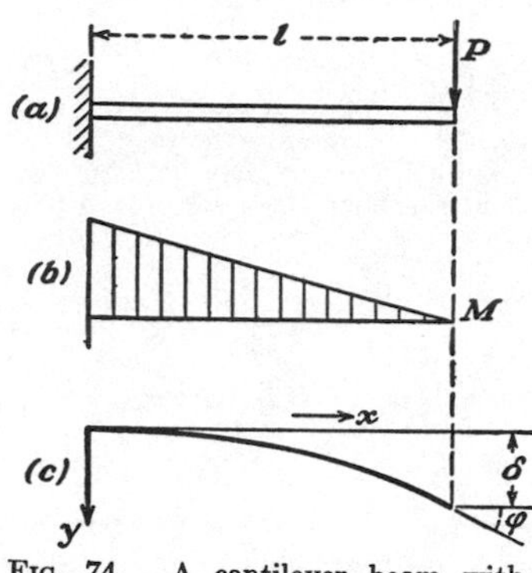

Fig. 74. A cantilever beam with an end load P; its bending moment diagram (b); and deflection curve (c).

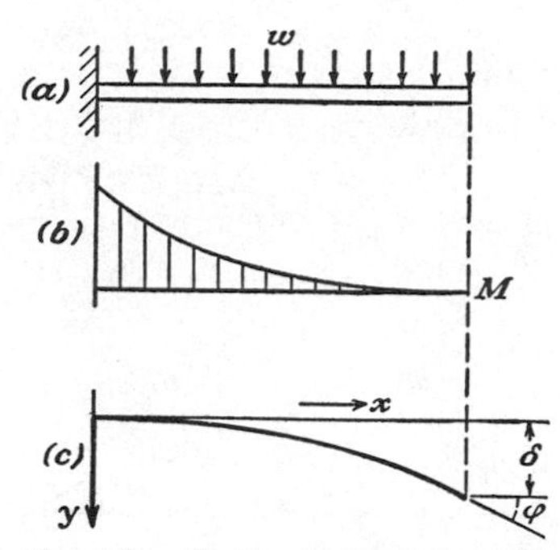

Fig. 75. Uniformly loaded cantilever beam.

For $x = 0$, $y = 0$ so that $C_2 = 0$. The slope and deflection at the end $x = l$ then are:

$$\left.\begin{aligned} \varphi = y'_{x=l} &= \frac{Pl^2}{2EI}, \\ \delta = y_{x=l} &= \frac{Pl^3}{3EI}. \end{aligned}\right\} \qquad \text{(Fig. 74)}$$

The third example is a cantilever beam loaded with a uniform load w all along its length l (Fig. 75). At a point x, which is distance $l - x$ from the free end, the bending moment is $w(l - x)^2/2$. Hence by integration:

$$y' = \int \frac{M\,dx}{EI} = \frac{w}{2EI}\int (l - x)^2\,dx = \frac{w}{2EI}\left(l^2x - lx^2 + \frac{x^3}{3}\right) + C_1.$$

As in the previous examples the constant C_1 is zero, and

$$y = \int y'\,dx = \frac{w}{2EI}\left(\frac{l^2x^2}{2} - \frac{lx^3}{3} + \frac{x^4}{12}\right) + \text{zero}$$

The deflection curve is no longer a simple parabola but an algebraic curve of the fourth order. Substituting the end point $x = l$ gives:

$$\left.\begin{aligned} \varphi = y'_{x=l} &= \frac{wl^3}{6EI}, \\ \delta = y_{x=l} &= \frac{wl^4}{8EI}. \end{aligned}\right\} \quad \text{(Fig. 75)}$$

These three examples are quite simple. Now we proceed to a more complicated case: a beam on two supports loaded by a single

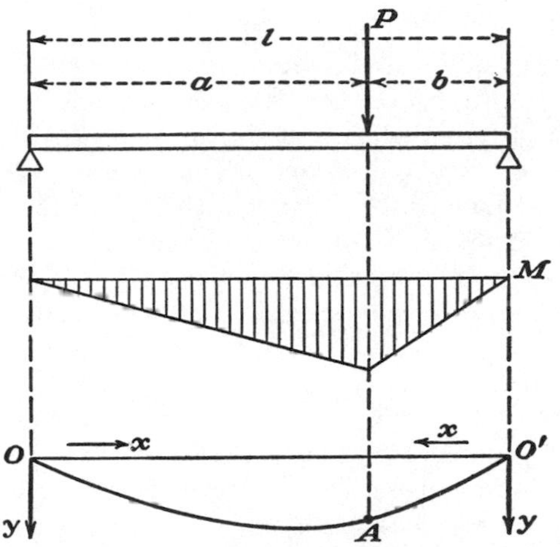

FIG. 76. Beam on two simple supports with a load P off center.

load P, not in the center (Fig. 76). The bending moment is negative according to the convention of Fig. 19, and its diagram is a simple triangle. The analytical complication now is that this triangle cannot be represented by a single formula. The formula for M is simple enough in each of the two sections of beam, but the two formulae are different. Hence we can integrate for the two sections separately only. We start with the left section of length a:

$$M = -P\frac{b}{l}x,$$

$$y' = -\frac{Pb}{EIl}\int x\,dx = -\frac{Pbx^2}{2EIl} + C_1,$$

$$y = \int y'\,dx = -\frac{Pbx^3}{6EIl} + C_1x + C_2.$$

The only boundary condition we have is that at $x = 0$, that is, the left end, the deflection $y = 0$, but the slope y' is not zero. Hence we conclude that $C_2 = 0$ but C_1 is not zero. We thus have found the equation of the deflection curve of the left section, but only subject to an unknown integration constant C_1. In particular the slope and deflection at the point A under the load ($x = a$) are:

$$\left.\begin{aligned} \varphi_A &= -\frac{Pba^2}{2EIl} + C_1, \\ \delta_A &= -\frac{Pba^3}{6EIl} + C_1 a. \end{aligned}\right\} \qquad \text{(from the left)}$$

This is all we can deduce from the left section of the beam; the integration constant C_1 can only be found by considering the right section of length b, to which we now turn.

If we had no imagination and insisted on retaining the same xy-coordinate system with the origin O at the left-hand end, we should have to integrate an expression for the bending moment having two terms, obtain answers twice as large as the above, and have a complicated boundary condition on our hands. It is better to choose a new coordinate system with O' at the right end as shown. Then the only difference between the right and left ends is in the end reaction, which is Pa/l at right and Pb/l at left, and in the length, which is b at right and a at left. This difference amounts to an interchange of the letters b and a so that we can almost copy our previous result:

$$\left.\begin{aligned} \varphi_A &= -\frac{Pab^2}{2EIl} + C_3, \\ \delta_A &= -\frac{Pab^3}{6EIl} + C_3 b. \end{aligned}\right\} \qquad \text{(from the right)}$$

Since this is a new integration, we have a different integration constant C_3, which cannot be expected to be equal to C_1. Now the deflections at A are obviously the same from the left or right:

$$\delta_A = -\frac{Pba^3}{E6Il} + C_1 a = -\frac{Pab^3}{6EIl} + C_3 b.$$

The slopes are likewise the same, because the beam is continuous, but what is called a positive slope in the O system is negative in the O' system, so that:

$$\varphi_A = -\frac{Pba^2}{2EIl} + C_1 = +\frac{Pab^2}{2EIl} - C_3.$$

Now we have two equations with the two unknowns C_1 and C_3, which can be solved easily with the result:

$$C_1 = \frac{Pab}{EIl^2}\left(\frac{a^2}{6} + \frac{b^2}{3} + \frac{ab}{2}\right),$$

and with C_3 equal to the expression for C_1 with a and b reversed. Substituting and solving for the slope and deflection under the load, remembering that $a + b = l$, we finally obtain:

$$\left.\begin{aligned} \varphi_A &= \frac{Pab}{3EIl^2}(b^2 - a^2), \\ \delta_A &= \frac{Pa^2b^2}{3EIl} \end{aligned}\right\} \quad \text{(Fig. 76)}$$

If the load happens to be central, $a = b = l/2$, the deflection is:

$$\delta = \frac{Pl^3}{48EI}$$

(mid-span deflection of centrally loaded beams on 2 supports).

Although none of the steps in this last example are difficult, there are many of them and the probability of making an error in so many algebraic operations is great. This becomes even more serious for beams of greater complication. Therefore the method of direct integration of Eq. (19) and of calculating the integration constants from the boundary conditions is so cumbersome that it is unfit for practical work. Several short-cut methods to arrive at the result exist, of which the simplest and best one is that of repeated application of cantilever formulae, to which we now proceed.

Problems 158 *to* 166.

19. The Myosotis Method. From the previous article (Figs. 73 to 75) we now quote six little formulae for the angular and linear deflections of a cantilever beam under three different loadings:

	Angle	Deflection	
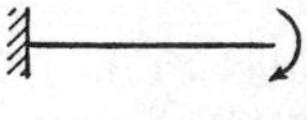 (cantilever, end moment)	$\frac{Ml}{EI}$	$\frac{Ml^2}{2EI}$	
 (cantilever, end load)	$\frac{Pl^2}{2EI}$	$\frac{Pl^3}{3EI}$	(20)
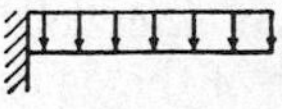 (cantilever, uniform load)	$\frac{wl^3}{6EI}$	$\frac{wl^4}{8EI}$	

These most important expressions are called the "Myosotis formulae," after the renowned Greek scholar *Myosotis Palustris*, whose pedigree the reader is advised to look up in Webster's dictionary. The expressions should be memorized from the start, which is not difficult if we only remember the sequence 122368. If the exponent of the length l is forgotten, it can be reestablished in each case by dimensional reasoning. The reader should check all six expressions for dimensional correctness before proceeding.

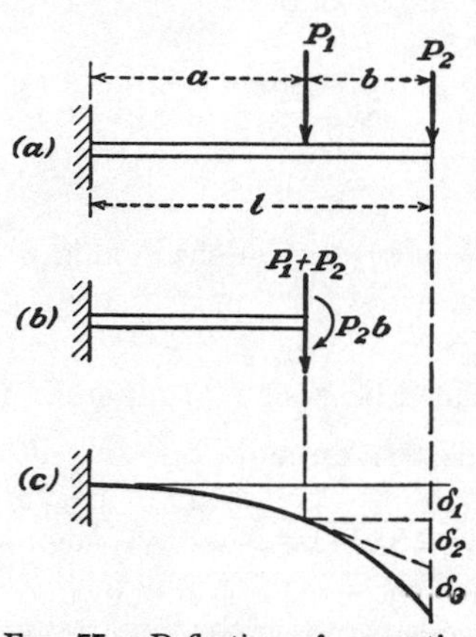

FIG. 77. Deflection of a cantilever beam subjected to two concentrated loads.

The method consists in repeated application of these six formulae, and with it the most complicated beam-deflection problems can be solved, provided that the loading can be represented by bending moments, concentrated, and uniformly distributed loads. We shall now illustrate the method by application to several examples.

First we consider Fig. 77, a uniform cantilever loaded with P_1 and P_2. We imagine a section just to the left of P_1, where the shear force is $P_1 + P_2$ and the bending moment is P_2b. For this case we can write down the deflection δ_1 and angle φ at P_1. Now we sketch the deflection curve (Fig. 77c) and remark that the end deflection consists of three parts: the deflection δ_1 at P_1, the deflection $\delta_2 = \varphi_1 b$, which is the linear extrapolation of the slope at P_1, and the deflection δ_3 of a cantilever of length b built in at P_1 at the slightly downward angle φ_1. With Myosotis the answer can be written immediately:

$$\delta = \left[\frac{(P_1 + P_2)a^3}{3EI} + \frac{(P_2b)a^2}{2EI}\right] + \left[\frac{(P_1 + P_2)a^2}{2EI} + \frac{(P_2b)a}{EI}\right]b + \left(\frac{P_2b^3}{3EI}\right).$$

The expression for δ_3 has been simply added to the other two, because the diagram 77c has a negligibly small slope throughout, so that the question of whether the deflection is perpendicular to the deformed bar or to the original horizontal line has no significance.

The answer can be obtained still more simply by using the "principle of superposition," which states that the deflection curve caused by two or more loadings is the sum of the deflection curves caused by the individual loadings separately. Thus in Fig. 77 the deflection at the end is the sum of the deflections caused there by the two loads separately. Load P_1 alone will deform the left portion of the bar, while the right portion remains straight but is inclined downward slightly. Thus the total end deflection is:

$$\delta = P_1\left(\frac{a^3}{3EI} + \frac{a^2}{2EI}b\right) + \frac{P_2l^3}{3EI},$$

which is the same as the above, when we remember that $l = a + b$.

Before going on to the next example the reader might obtain the

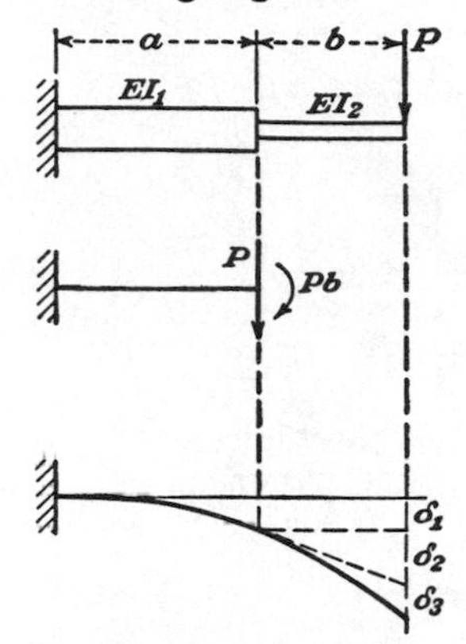

FIG. 78. A cantilever of two different stiffnesses, loaded by a single end load.

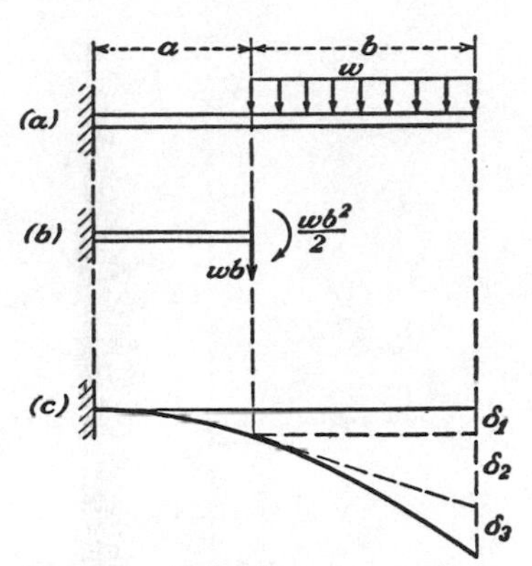

FIG. 79. Deflection of a cantilever loaded uniformly along part of its length.

above result by direct integration of Eq. (19). The analytical expressions for the two sections of the deflection curve are different, and they must be fitted together at the joint P_1 by properly adjusting the integration constants of the two sections. This certainly is no small job, and the actual carrying out of it will impress anyone with the enormous simplification involved in the Myosotis method.

The second example we take up is shown in Fig. 78. There is but a single load on the cantilever, but it consists of two parts of unequal stiffness. Again we immediately write the solution for the deflection at the end:

$$\delta = \delta_1 + \delta_2 + \delta_3 = \left[\frac{Pa^3}{3EI_1} + \frac{(Pb)a^2}{2EI_1}\right] + b\left[\frac{Pa^2}{2EI_1} + \frac{(Pb)a}{EI_1}\right] + \frac{Pb^3}{2EI_2}.$$

For a third example we choose a cantilever loaded uniformly over part of its length (Fig. 79). Again we cut it into two parts and calculate the shear force and bending moment at the joint by statics, as shown in Fig. 79*b*. Then the end deflection is, as before:

$$\delta = \left[\frac{wba^3}{3EI} + \frac{(wb^2/2)a^2}{2EI}\right] + b\left[\frac{wba^2}{2EI} + \frac{(wb^2/2)a}{EI}\right] + \frac{wb^4}{8EI}.$$

The fourth example, shown in Fig. 80, is no longer a cantilever beam, but the beam on two supports we solved so laboriously in the previous article. Now we sketch the deflection curve and note

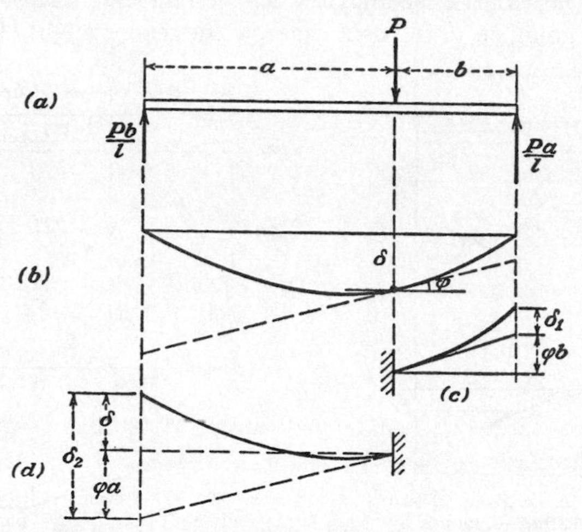

FIG. 80. Solution of the problem of Fig. 76 by the Myosotis method.

that at point P there will be some slope φ of which we do not know the value. We then break up the beam into two cantilevers, built in at P at the unknown angle φ. Each of these cantilevers is loaded by a single end load only: Pa/l at right and Pb/l at the left. The two cantilevers are sketched separately in Fig. 80*c* and *d*, although these sketches are superfluous, all information being shown in Fig. 80*b*. We read off the figure:

$$\delta = \delta_1 + \varphi b = \delta_2 - \varphi a.$$

From this we solve for the auxiliary variable φ:

$$\varphi(a + b) = \delta_2 - \delta_1 \qquad \text{or} \qquad \varphi = \frac{\delta_2 - \delta_1}{l}.$$

Substituting:

$$\delta = \delta_1 + \frac{b}{l}(\delta_2 - \delta_1) = \frac{b}{l}\delta_2 + \frac{a}{l}\delta_1$$

The deflections δ_1 and δ_2 can be written down directly with Myosotis:

$$\delta = \frac{b}{l}\frac{(Pb/l)a^3}{3EI} + \frac{a}{l}\frac{(Pa/l)b^3}{3EI} = \frac{Pb^2a^2}{3EIl^2}(a+b) = \frac{Pb^2a^2}{3EIl}.$$

This is the same result found previously (Fig. 76).

As a fifth example consider Fig. 81, where we ask for the deflection in the middle of the beam. The end reactions are calculated by statics to be $3wl/8$ and $wl/8$, as shown. We sketch in a

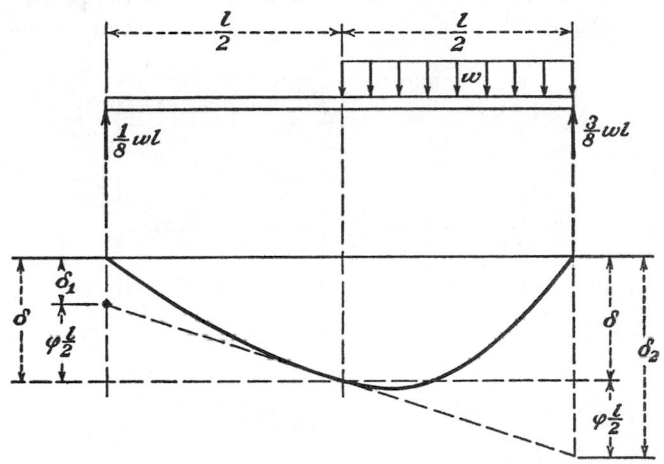

Fig. 81. Beam on two supports, uniformly loaded along half its length.

deflection curve and draw the tangent to it in the center, having the unknown slope φ. Now we consider the beam clamped at the center at angle φ, and we have two cantilever beams, one to the right, one to the left. Again:

$$\delta = \delta_1 + \varphi\frac{l}{2} = \delta_2 - \varphi\frac{l}{2},$$

so that:

$$\varphi = \frac{\delta_2 - \delta_1}{l},$$

and the desired deflection in the center is:

$$\delta = \delta_1 + \frac{\delta_2 - \delta_1}{l}\frac{l}{2} = \frac{1}{2}(\delta_1 + \delta_2),$$

a result we could have seen immediately from the figure. The deflections δ_1 and δ_2 are:

$$\delta_1 = \frac{(wl/8)(l/2)^3}{3EI} = \frac{wl^4}{192EI},$$

$$\delta_2 = \frac{(3wl/8)(l/2)^3}{3EI} - \frac{w(l/2)^4}{8EI} = \frac{wl^4}{128EI},$$

so that:

$$\delta = \frac{\delta_1 + \delta_2}{2} = \frac{5wl^4}{768EI}.$$

A quick check on this result can be obtained by the principle of superposition. If we had a beam like Fig. 81, but loaded on the

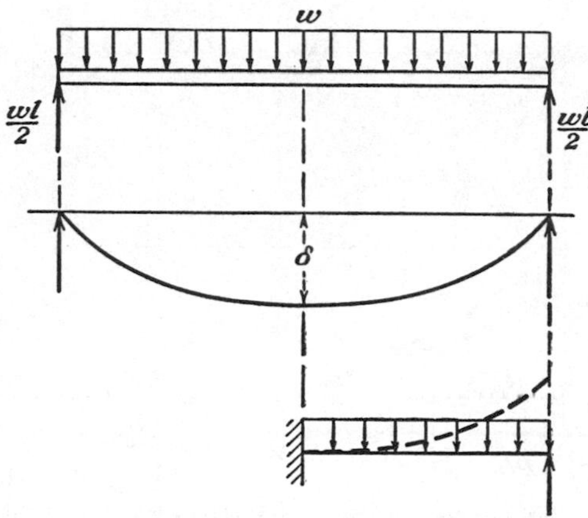

FIG. 82. The uniformly loaded beam on two supports must have a deflection curve symmetrical about the center.

left half only instead of on the right half, the deflection curve would be the mirrored image and the central deflection the same. Adding these two loads together produces the case of a beam uniformly loaded along its entire length, and its deflection should be twice the answer found above. But if in Fig. 82 we clamp that beam at its center, and consider half of it as a cantilever, then:

$$\delta = \frac{(wl/2)(l/2)^3}{3EI} - \frac{w(l/2)^4}{8EI} = \frac{5wl^4}{384EI},$$

which is twice the above figure, as it should be.

For our last example we return to Fig. 80 and ask for the location of maximum deflection along the beam and for the value of that maximum deflection. This is a complicated question, and in the straight calculus method it would involve first finding the analytical expression for the deflection curve, then differentiating it to get the horizontal tangent and then solving for the location. The problem can be handled by the Myosotis process in a much simpler way. In Fig. 83 we do not know the location A of maximum deflection, but we can always give it a name: x. Breaking the beam at A into two cantilevers, these have horizontal tangents at A and hence have the same deflection δ:

$$\delta_{\text{left}} = \frac{(Pb/l)x^3}{3EI} = \delta_{\text{right}} = \frac{(Pa/l)(l-x)^3}{3EI} - \frac{P(a-x)^3}{3EI} - \frac{P(a-x)^2 b}{2EI}.$$

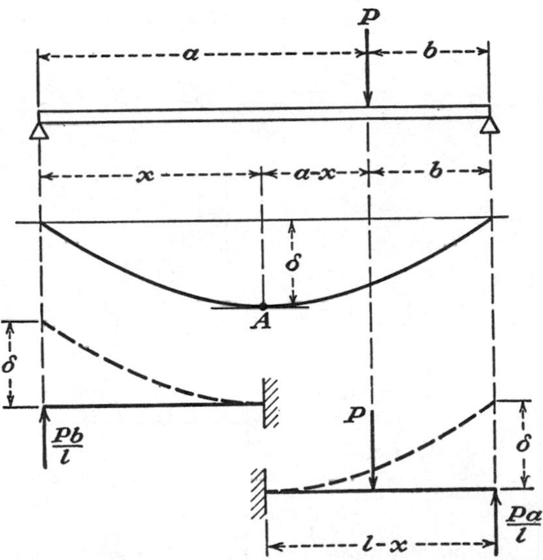

Fig. 83. To find the location x of maximum deflection by the Myosotis method.

This is an equation in which the only unknown is x, and from it x must be solved. That this solution involves a lot of algebra is not the fault of Myosotis; his method enables us to write the equation down at once. The further algebraic process is left to the reader. In working it out, the terms proportional to x^3 and to x^1 cancel out, and we find simply:

$$x = \sqrt{\frac{a(a+2b)}{3}} \qquad \text{and} \qquad \delta = \frac{Pb}{l}\frac{x^3}{3EI}.$$

Problems 167 *to* 187.

20. Statically Indeterminate Beams. In the preceding sections we have learned how to calculate the deflections of beams, but it was admitted from the start that these deflections are quite small, in fact insignificantly small in most cases. Moreover, they do not affect the stresses in the beam, which can be calculated from Eqs. (11) and (14), as soon as we know the bending moment and shear force and without knowledge of the deflection curve. Then why so much ado about nothing? Why Myosotis and all his works, when he gives us only some small deflections we hardly

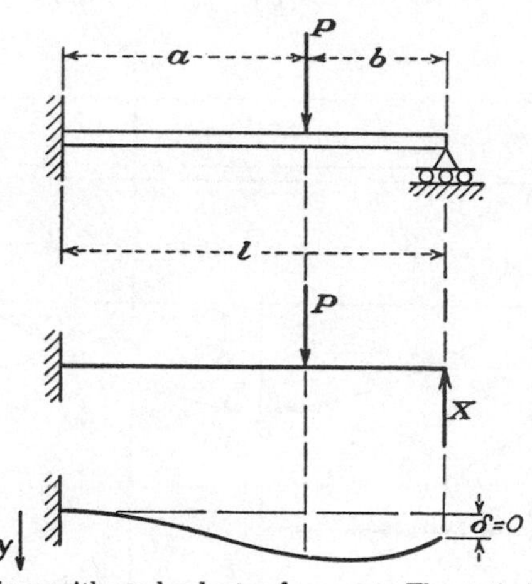

Fig. 84. A cantilever with a redundant end support. The statically indeterminate reaction X is calculated by equating to zero the end deflection of the beam caused by P and X combined.

care about, while the important thing, the stress, can be found without him? The answer is that so far we have considered only statically determined beams where the support reactions can be calculated by statics and where the answers are independent of the deformations. As soon, however, as the beam has one or more unnecessary, or "redundant," supports, then the reactions can no longer be calculated by statics alone and they depend on the deformations. Then the bending moments and with them the stresses in the beam depend on the deformations, and it becomes of vital importance to be able to calculate them. For the calculation of these redundant reactions the Myosotis process is very suitable, as will now be shown in some examples.

First we consider a cantilever beam that has an additional support at its free end (Fig. 84). The reaction force from this support may have any value as far as the static equilibrium of the beam is concerned: the clamped end at the left by itself can take care of any loading up to the breaking point of the beam. We call the right end reaction X and consider it for the time being as another external load on the beam. Then we can calculate the end deflection of the beam under the loads P and X, and from the conditions of the problem this deflection must be zero:

$$0 = \delta = \frac{Xl^3}{3EI} - \frac{Pa^3}{3EI} - \frac{Pa^2b}{2EI}.$$

Solving for the unknown X we find:

$$X = P\left(\frac{a}{l}\right)^2 \frac{a + \frac{3}{2}b}{l} = P\left[\frac{3}{2}\left(\frac{a}{l}\right)^2 - \frac{1}{2}\left(\frac{a}{l}\right)^3\right].$$

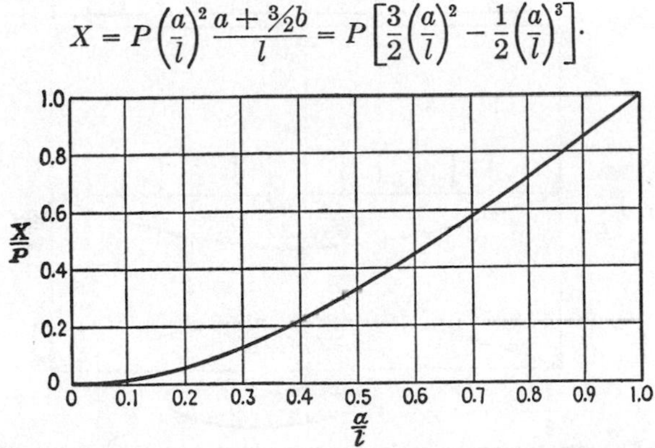

FIG. 85. The influence curve of the end reaction of the cantilever of Fig. 84, being a plot of that reaction as a function of the location of the load P along the beam.

In particular, if P is in the middle of the beam, where $a = l/2$, we find $X = \frac{5}{16}P$. If the value X/P is plotted against the location a/l of the load P, the diagram so obtained (Fig. 85) is called an *influence diagram.* We read from it that, when the load P gradually moves from the built-in end ($a = 0$) to the end support ($a = l$), the end reaction gradually grows from zero to the full load P.

As a second example of a statically indeterminate beam we take Fig. 86, showing a beam of length l on three equidistant supports loaded uniformly along its length. Any one of the three supports can be removed without having the beam collapse, but it is natural to think of the middle support as the redundant one. Then that support is replaced by an unknown upward load X. The downward deflection due to w must be equal to the upward deflection due to X in the middle of the beam. Both deflections were calculated in this chapter. The w deflection was found in connection with Fig. 82, and the X deflection was found on page 85 (Fig. 76). Thus:

$$\frac{Xl^3}{48EI} = \frac{5}{384}\frac{wl^4}{EI},$$

or, solved:

$$X = \tfrac{5}{8}wl.$$

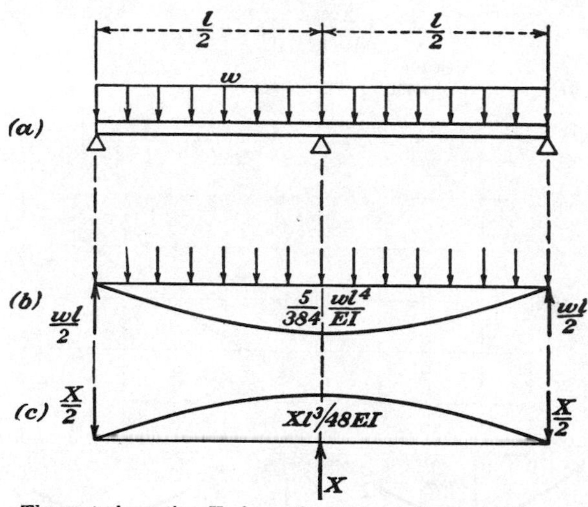

Fig. 86. The central reaction X of a uniformly loaded beam on three equidistant supports is found by equating the downward w deflection to the upward X deflection.

Thus 62.5 per cent of the total load is carried by the central support and 18.75 per cent by each of the end supports. With the reactions known, the bending moments can be calculated. At distance x from the left support the moment is:

$$-\tfrac{3}{16}wlx + \tfrac{1}{2}wx^2,$$

which expression holds in the left half of the beam, while in the right half the mirrored picture occurs. The moment has been plotted in Fig. 87*a*, and the reader should verify the location and magnitude of the maximum and zero moments as they are shown in the figure. Right below it is plotted the approximate shape of the deflection curve, in which the points I are "inflection points," *i.e.*, points where the curvature reverses from concave to convex by virtue of the differential equation (19). The calculation of the location and magnitude of the maximum deflection (by considering the right half of the beam as a horizontally clamped cantilever loaded by w and by $\frac{3}{16}wl$ at its end) is left as an exercise to the reader (Problem 189).

Figure 87a shows that the maximum stress in this beam occurs at the center support with tension in the upper fibers. If the middle support is thought of as a means of shoring up the beam, which was originally on the two end supports only, we have overdone the job.

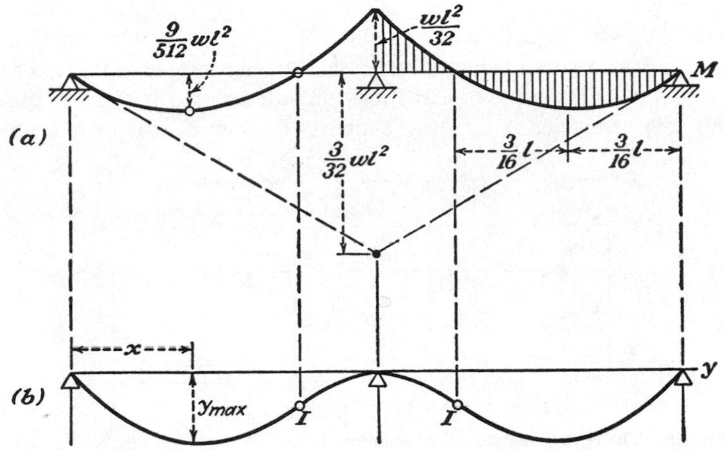

FIG. 87. Bending-moment and deflection diagrams for the beam on three supports of Fig. 86.

The center support never can be made entirely stiff, giving zero deflection there, as was assumed. We therefore now reexamine the problem of Fig. 86, replacing the center support by a spring of stiffness k (page 29). In that case there will be a central deflection δ_c and the support force X, being furnished by the spring, must be $X = k\delta_c$. In Fig. 86 the downward deflection due to w is no longer equal to the upward deflection due to X but must exceed it by the amount δ_c:

$$\frac{5wl^4}{384EI} - \frac{k\delta_c l^3}{48EI} = \delta_c.$$

This equation is linear in δ_c and can easily be solved for it and hence for $X = k\delta_c$:

$$X = k\delta_c = \frac{\frac{5}{8}wl}{1 + (48EI/kl^3)}.$$

The answer is seen to be smaller than $\frac{5}{8}wl$, owing to the softness of the spring. The expression $48EI/l^3$ in the denominator can be interpreted by Fig. 86c as the stiffness of the beam itself considered as a spring, and we might write for it k_b. Our original k for the support we might now denote by k_s. Then the above result can be written very neatly as:

$$X = \frac{\frac{5}{8}wl}{1 + (k_b/k_s)},$$

and this relation is shown plotted in Fig. 88. The central reaction becomes smaller and smaller as the central support becomes softer.

As a third example of a statically indeterminate beam we take Fig. 89a, considering all three supports to be stiff, and we ask for

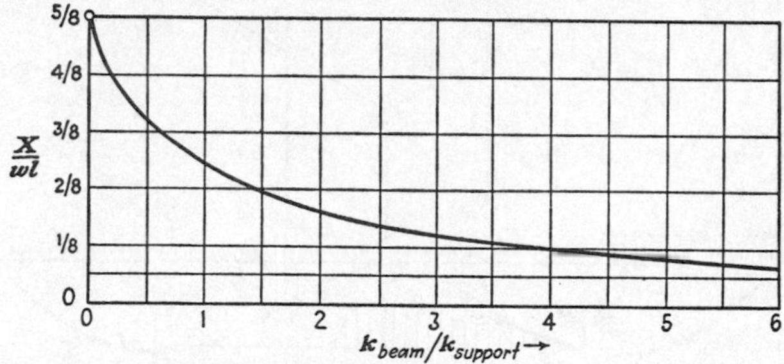

Fig. 88. The value of the central reaction in Fig. 86 as a function of the stiffness of the central support.

the magnitude and location of the maximum bending moment. In solving this problem one of the three support reactions has to be considered the indeterminate quantity, and it has to be replaced by an unknown loading X. It does not matter which one of the three we choose, and since in the previous example we took the middle support we shall now take the left end support for this purpose. Figure 89b shows the deflection curve under load P if the left support is omitted, and Fig. 89c shows the deflection when only X and no P act on the beam. The actual deflection curve is the sum of these two cases and has zero deflection at the left support. Therefore, $\delta_1 = \delta_2$, and we must calculate these individual deflections by the Myosotis method. The deflection $\delta_1 = \varphi l/2$, in which φ is not known. We can find φ most easily by remarking that the right half span of Fig. 89b is symmetrical, so that the tangent under P is horizontal. Then a cantilever of length $l/4$ to the right of P, loaded by an end load $P/2$, must give a terminal angle φ:

$$\varphi = \frac{(P/2)(l/4)^2}{2EI}.$$

Hence:

$$\delta_1 = \varphi \frac{l}{2} = \frac{Pl^3}{128EI}.$$

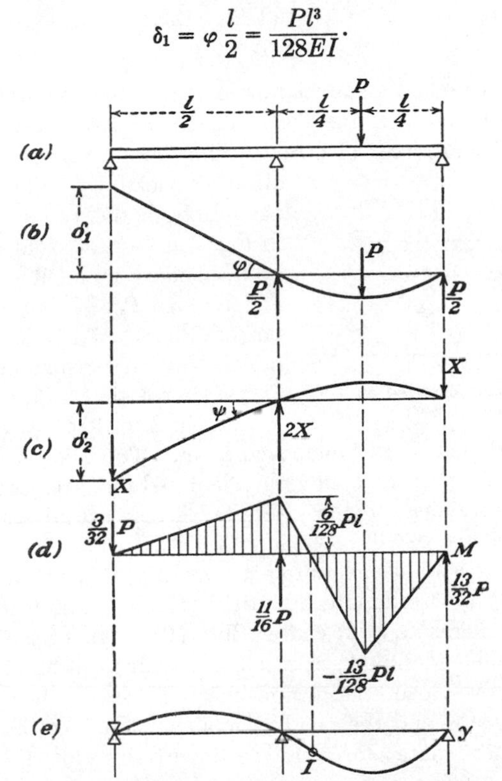

FIG. 89. A continuous beam on three supports loaded by a single force in the middle of one span.

To find δ_2 we consider Fig. 89c as two cantilevers built in at angle ψ in the center, one protruding to the left, the other to the right. The cantilever to the right has an end deflection zero, or

$$\delta = 0 = \psi \frac{l}{2} - \frac{X(l/2)^3}{3EI},$$

so that:

$$\psi = \frac{Xl^2}{12EI}.$$

The cantilever to the left has an end deflection δ_2, or

$$\delta_2 = \psi\frac{l}{2} + \frac{X(l/2)^3}{3EI} = \frac{Xl^3}{24EI} + \frac{Xl^3}{24EI} = \frac{Xl^3}{12EI}.$$

Now:

$$\delta_1 = \delta_2 = \frac{Pl^3}{128EI} = \frac{Xl^3}{12EI}.$$

$$\therefore \quad X = \tfrac{3}{32}P.$$

The other two reactions must now be calculated by statics, with the result indicated in Fig. 89*d*. The bending-moment diagram consists of three pieces of straight line, because there is no distributed loading. The reader should check the values indicated in Fig. 89*d* and note that the maximum bending moment occurs under the load P with tension in the bottom fibers. There is a reverse moment (with tension in the top fibers) above the middle support. Finally in Fig. 89*e* the deflection curve is sketched in without calculation. There is an inflection point at I. At this stage, having calculated the stresses and reactions, we are no longer much interested in the exact values of the deflections of Fig. 89*e*.

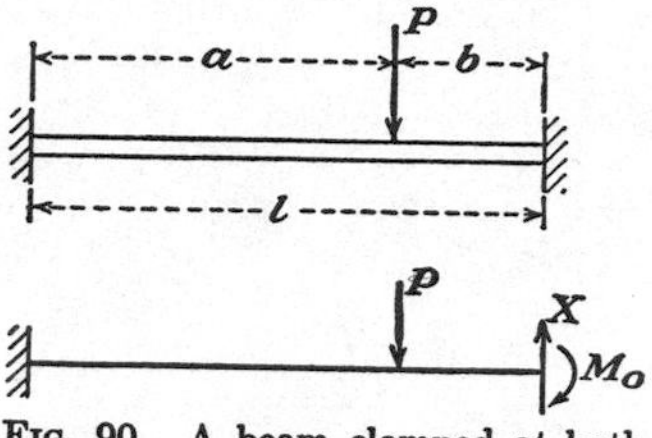

FIG. 90. A beam clamped at both ends with a single eccentric load.

For our fourth and last example we take a case in which there are two statically indeterminate quantities: the beam built in at both ends with a single load, off center (Fig. 90). One of the clamped end supports is redundant; as usual we eliminate it and replace it by the unknown load. Only now there are two such unknowns, the shear force X and the bending moment M_0. We then have a statically determinate simple cantilever with three loads, P, X, and M_0, and we know that the end deflection and end slope of that beam must be zero:

$$\delta = \frac{Pa^3}{3EI} + \frac{Pa^2b}{2EI} + \frac{M_0 l}{2EI} - \frac{Xl^3}{3EI} = 0,$$

$$\varphi = \frac{Pa^2}{2EI} + \frac{M_0 l}{EI} - \frac{Xl^2}{2EI} = 0.$$

These are two equations, linear with respect to the two unknowns X and M_0. The problem is therefore reduced to simple algebra and arithmetic, and the reader should have no difficulty in finishing it.

Problems 188 *to* 198.

21. The Area-moment Method. The differential equation of flexure (19) states that the bending moment M is proportional to the second derivative of the deflection. Equations (10a) and (10b) together state that the distributed loading w is the second derivative of the bending moment. It is instructive to look at these relations written down all together in a schedule:

$$y = f(x) = \text{deflection}$$

$$y' = f'(x) = \frac{dy}{dx} = \text{slope}$$

$$y'' = f''(x) = \frac{M}{EI} = \text{bending moment/stiffness}$$

$$y''' = f'''(x) = \frac{S}{EI} = \text{shear force/stiffness}$$

$$y^{(4)} = f^{(4)}(x) = \frac{w}{EI} = \text{loading/stiffness}$$

For the usual beams of constant cross section, the stiffness EI is a constant, not depending on x, and then we have a series of five quantities, y to d^4y/dx^4, that are all derivable from each other by differentiation or integration, depending on whether we step up or down the line. Examples of some of this we have seen in the previous pages, but it is again systematically illustrated in Figs. 91 and 92. The first of these is a uniformly loaded cantilever beam, and the reader should start from the top and successively deduce the sketches below by integration, noting that the slope of each diagram is the ordinate of the one above it. The sign conventions observed are those used throughout this book, defined in Figs. 19 and 21 (page 33). Each new sketch in the series is derived from the one above it by integration, *i.e.*, subject to an integration constant, that has to be chosen to fit boundary conditions. This is indicated by the

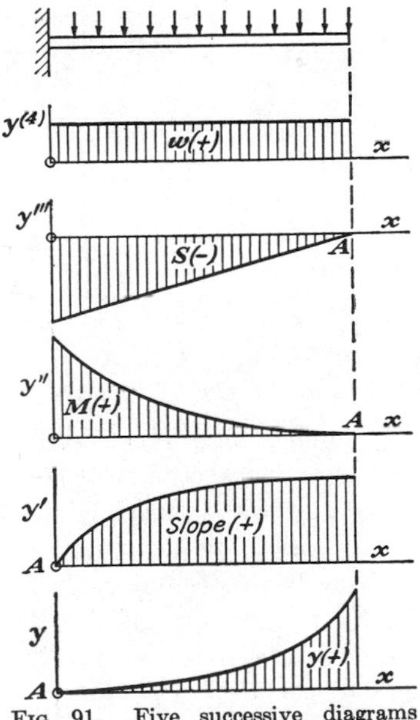

FIG. 91. Five successive diagrams of the deflection y and its derivatives for a uniformly loaded cantilever of constant cross section.

letter A, appearing four times in Fig. 91 in locations where we know that the quantity to be plotted must be zero, so that we start plotting from A. In all diagrams, including the last one, positive quantities are plotted up and negative ones down; the actual meaning of these quantities is then determined by Figs. 19 and 21. In the

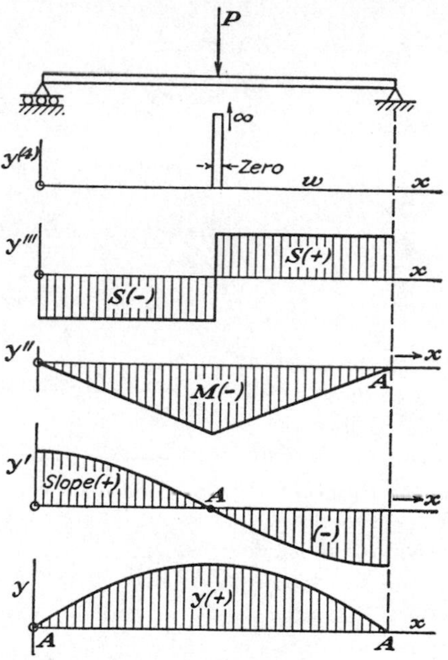

FIG. 92. The diagrams of y and its derivatives for a simply supported beam of constant cross section EI. See Figs. 19 and 21, page 33 for sign definitions.

bottom diagram of Fig. 91, for example, y is shown positive, which means a downward displacement by Fig. 19.

As a second example the centrally loaded, simply supported beam of Fig. 92 is shown, and again the reader should spend considerable time checking every detail of this figure before proceeding. The $y^{(4)}$ diagram is peculiar; it is zero everywhere except in the center, where it is $0 \times \infty = P$. The slope of the y''' diagram at that spot is infinitely large.

From the general properties of the calculus we know that the ordinate of each diagram equals the slope of the diagram below it (differentiation) and also that the area of each diagram between two parallel vertical lines equals the difference in the ordinates at those parallel lines in the diagram below it (integration).

The five equations of the schedule on page 99 can be rolled into a single expression:

$$w = \frac{d^2}{dx^2}\left(EI\,\frac{d^2y}{dx^2}\right). \tag{21}$$

In the way this equation is written it applies to a most general beam in which the stiffness EI itself is a function of x: a beam of variable cross section. In the simpler, more common case that EI is constant, the stiffness can be brought before the differentiation sign, and Eq. (21) simplifies to:

$$w = EI\,\frac{d^4y}{dx^4} \qquad \text{(only for } EI = \text{constant)}$$

Equation (21) and its implications, Figs. 91 and 92, have been made the basis for a method of finding deflections in beams, known as the *area-moment method*. It consists of repeated applications to the beam of one or more of the following three statements:

1. The deflection diagram $y = f(x)$ of a beam is the same as the bending moment diagram of that beam loaded by a (fictitious) distributed loading of intensity M/EI, where M is the local bending moment caused by the true actual loading and EI is the local stiffness, which may be variable with x.

2. The difference in slopes between two points of a beam equals the area of the M/EI diagram between those two points, where M is the actual local bending moment and EI the local stiffness, both being functions of x in the general case.

3. The deflection of a point, measured with respect to the extrapolated tangent of another point, equals the moment of the area of the M/EI diagram between the two points taken about the first point, *i.e.*, about the point where the deflection is measured. This statement as well as the other two applies to beams of variable cross section.

The first of these three statements follows immediately from Eqs. (21), (10*a*), and (10*b*). It reduces the problem of finding deflections to that of constructing a bending-moment diagram. All known methods of finding bending-moment diagrams thus become applicable to the deflection problem. For example, there exists a graphical method, involving a "funicular curve," which is often applied to this case but which will not be treated in detail because the Myosotis process is so much simpler.

The second statement of the area-moment method is a first

integration of the flexure equation (21) between two points A and B of the beam:

$$\frac{d^2y}{dx^2} = \frac{M}{EI}, \qquad d\left(\frac{dy}{dx}\right) = \frac{M}{EI}\,dx,$$

$$\int_A^B d\left(\frac{dy}{dx}\right) = \frac{dy}{dx}\Big|_A^B = \int_A^B \left(\frac{M}{EI}\right) dx \qquad q.e.d.$$

The third statement of the method is a double integration, and its proof is slightly more complicated. In Fig. 93 we see a piece of the beam between points A and B, and, with the same under-

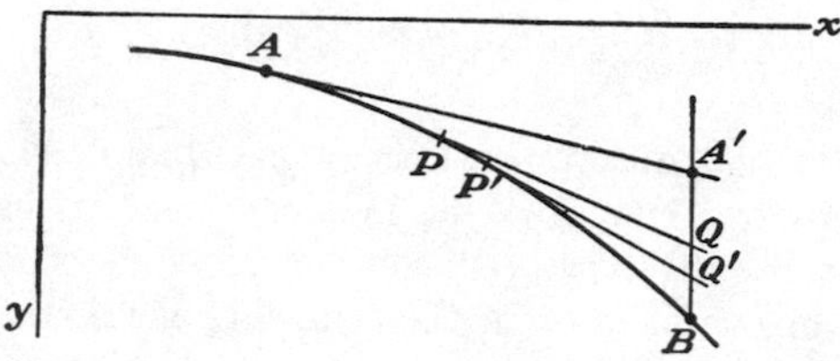

FIG. 93. Detail of a bent beam for the proof of the third statement of the area-moment method.

standing we had in all previous cases, the curve is "horizontal," all angles being "small." Then the distance $A'B$ is the "deflection of B measured with respect to the extrapolated tangent of point A," and we think of $A'B$ as built up of small contributions QQ', being the intercepts of the vertical $A'B$ by two tangents PQ and $P'Q'$ at two neighboring points on the curve. The angle $d\theta$ between PQ and $P'Q'$ is, by the previous statement:

$$d\theta = d\left(\frac{dy}{dx}\right) = \frac{M}{EI}\,dx.$$

The distance $QQ' = d\theta \cdot PQ$. Now PQ is the horizontal distance from the moving point P to the fixed point B, and we call it x_B. Then:

$$QQ' = x_B\,d\theta = x_B\left(\frac{M}{EI}\,dx\right) = x_B\,d\text{ area}.$$

Integrated:

$$\delta = A'B = \int x_B\,d\theta = \int_A^B x_B\,d\text{ area} \qquad q.e.d.$$

We shall now apply the method to a number of cases, of which the first one is the simple cantilever with end load (Fig. 74, page 82). The bending moment at the fixed end is Pl, so that the triangle is $\frac{1}{2} \cdot Pl \cdot l$. The area of the M/EI diagram thus is $Pl^2/2EI$. By the second statement of the area-moment method this is equal to the difference between the slopes at the two ends, of which the one at the fixed end is zero. Hence $Pl^2/2EI$ is the slope at the free end, which is one of our well-known Myosotis results. To find the deflection at the end we use the third statement with respect to the free end. The moment of the area is the

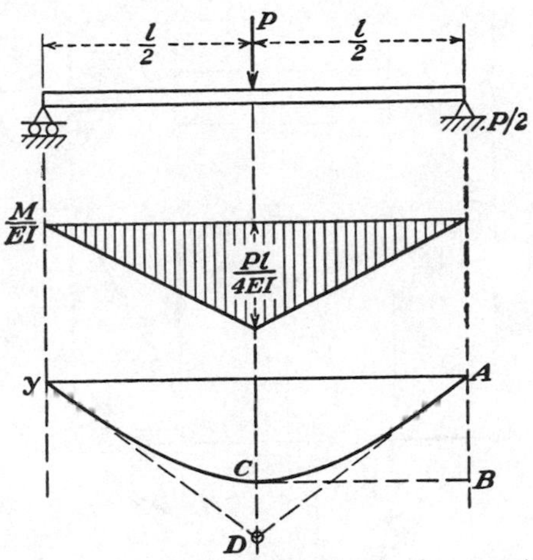

FIG. 94. Calculating the center deflection of a simply supported beam by the area-moment method.

area multiplied by the distance of its center of gravity which is at $\frac{2}{3}l$ from the free end. Hence:

$$\frac{Pl^2}{2EI}\frac{2l}{3} = \frac{Pl^3}{3EI}$$

is the "deflection at the free end measured with respect to the extrapolated tangent from the fixed end." Since the latter tangent is horizontal, the above result is the correct end deflection.

The next example is the beam on two supports (Fig. 94), of which we want to know the center deflection. The area of the M/EI diagram extending over half the beam is:

$$\text{Area} = \frac{1}{2}\frac{l}{2}\frac{Pl}{4EI} = \frac{Pl^2}{16EI}.$$

This, then, is the difference in slope between the center and end points, and since the center slope is zero, the above value gives the end slope. Taking the moment of the same triangular area about the end gives:

$$\frac{Pl^2}{16EI} \cdot \frac{2}{3} \frac{l}{2} = \frac{Pl^3}{48EI}.$$

This is the "deflection of the supported end measured with respect to a tangent at the center," is the distance AB in the figure, and

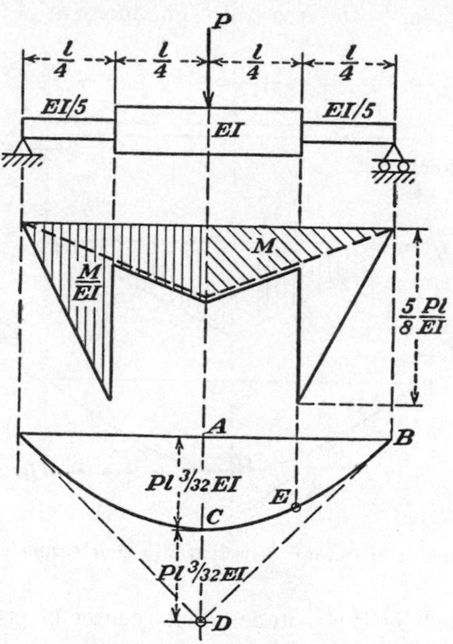

FIG. 95. Slopes and deflections of a centrally loaded shaft of variable cross section.

hence is the value we set out to calculate. Suppose we had taken the moment of the triangular area about the center point, instead of that about the end point. Then the moment arm would have been half as large, and the result would be half as large, or $Pl^3/96EI$. This is the "deflection of the center measured with respect to the extrapolated tangent from a fixed end"; it is shown as CD in the sketch and has no particular meaning.

The method can be applied successfully to more complicated cases, but a reader familiar with the Myosotis process will use that in preference to the area-moment method, so that we shall not pursue the matter further.

Rework problems 167 *to* 187 *by the area-moment method.*

22. Variable Cross Sections; Shear Deflection. We consider in Fig. 95 a simply supported beam with a central section five times as stiff as the end pieces and ask for the deflection in the center, to be calculated by the area-moment as well as by the Myosotis method. The bending moment in the center is $Pl/4$, and the bending-moment diagram is triangular, independent of the stiffness distribution, as indicated by the dotted line in the diagram. The area-moment method concerns itself with the combination M/EI, rather than with M itself. We plot the quantity M/EI in the same diagram as M, to a different scale, so that the center sections of the two diagrams are identical. Then the outer sections of M/EI are five times larger than M itself, as shown in the diagram.

To find the central deflection we calculate the moment of half this figure about one of the supports as moments center. We take the right half of the figure and consider separately the 45-deg shaded triangle and the unshaded triangle. The first triangle has an area

$$\frac{1}{2} \cdot \frac{Pl}{4EI} \cdot \frac{l}{2},$$

and the unshaded triangle has a height of $(5Pl/8EI) - (Pl/8EI) = Pl/2EI$, so that its area is

$$\frac{1}{2} \cdot \frac{Pl}{2EI} \cdot \frac{l}{4} = \frac{Pl^2}{16EI}.$$

The total area of the right half thus is $Pl^2/8EI$, which by the second statement of the area-moment method equals the difference in slope φ between the center and the right-hand end. The tangent BD thus has an intercept $AD = \varphi l/2 = Pl^3/16EI$, as marked in the figure. The moment of the area about B is the area times its center-of-gravity distance, or

$$\frac{Pl^2}{16EI} \cdot \frac{2}{3}\frac{l}{2} = \frac{Pl^3}{48EI},$$

for the 45-deg shaded triangle plus

$$\frac{Pl^2}{16EI} \cdot \frac{2}{3}\frac{l}{4} = \frac{Pl^3}{96EI}$$

for the white triangle, or $Pl^3/32EI$ in total. This is the "deflection of the end B, measured with respect to the extrapolated tangent at C," or the distance we asked for. In order to be able to sketch in the deflection curve nicely, the reader should work Problem 199 and find the deflection at E.

By the Myosotis method we consider the right half of the beam as a cantilever clamped horizontally in the center and write for the deflection of the end B:

$$\delta = \frac{(P/2)(l/4)^3}{3EI} + \frac{(Pl/8)(l/4)^2}{2EI} + \frac{l}{4}\left[\frac{(P/2)(l/4)^2}{2EI} + \frac{(Pl/8)(l/4)}{EI}\right]$$

$$+ \frac{(P/2)(l/4)^3}{3EI/5} = \frac{Pl^3}{EI}\left(\frac{1}{384} + \frac{1}{256} + \frac{1}{256} + \frac{1}{128} + \frac{5}{384}\right)$$

$$= \frac{24}{768}\frac{Pl^3}{EI} = \frac{Pl^3}{32EI}.$$

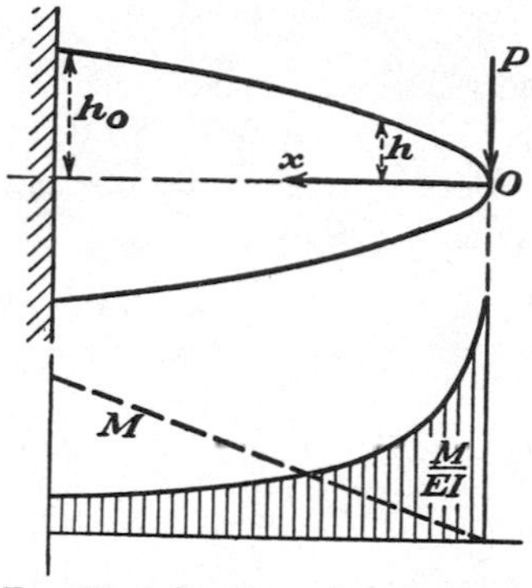

FIG. 96. Cantilever beam of constant strength with rectangular cross section bh, in which b is constant and $h = f(x)$.

The next example we study is the design of a cantilever beam (Fig. 96) of rectangular cross section bh, in which the width b is constant but h is made variable with x so as to keep the maximum top-fiber stress constant along the length. Such a beam is called one of constant strength.

Taking the origin of coordinates at the free end and counting x to the left, the bending moment is Px and the (constant) stress at the top fiber is $s_0 = Px/(bh^2/6)$, so that:

$$h^2 = \frac{6P}{bs_0}x$$

is the equation of the (parabolic) contour of the beam as shown in Fig. 96. At the clamped end $x = l$ we call $h = h_0$ so that $h_0^2 = 6Pl/bs_0$, and

$$\left(\frac{h}{h_0}\right)^2 = \frac{x}{l}$$

is another expression of the contour curve. To find the deflections we must first compute the value of M/EI:

$$\frac{M}{EI} = \frac{Px}{Ebh^3/12} = \frac{12Px}{Ebh_0^3}\left(\frac{h_0}{h}\right)^3 = \frac{12Px}{Ebh_0^3}\left(\frac{x}{l}\right)^{-3/2}$$

$$= \frac{Px}{EI_0}\left(\frac{x}{l}\right)^{-3/2} = \frac{Pl^{3/2}}{EI_0}x^{-1/2},$$

where $I_0 = bh_0^3/12$ is the moment of inertia at the clamped end.

Now, by the moment-area method the slope at the free end (or rather the difference in slope between the free and clamped ends) is:

$$\varphi = \int_0^l \frac{M}{EI}\,dx = \frac{Pl^{3/2}}{EI_0}\int^l x^{-1/2}\,dx = \frac{Pl^{3/2}}{EI_0}\frac{l^{1/2}}{1/2}$$

$$= 2\,\frac{Pl^2}{EI_0} = 4\cdot\frac{Pl^2}{2EI_0},$$

or four times the deflection the beam would have had if it had been made of constant height h_0 throughout.

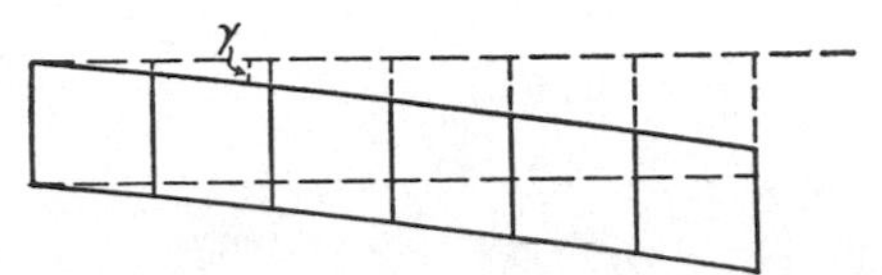

FIG. 97. Fictitious deformation of a region near the neutral line of a beam with a constant shear force S along its length.

The end deflection δ (or rather the deflection of the free end with respect to the extrapolated tangent of the clamped end) is:

$$\delta = \int^l \frac{M}{EI}\,x\,dx = \frac{Pl^{3/2}}{EI_0}\int x^{1/2}\,dx = \frac{Pl^{3/2}}{EI_0}\frac{l^{3/2}}{3/2} = 2\,\frac{Pl^3}{3EI_0},$$

or twice the deflection of a cantilever of constant height h_0.

This last example is one in which only the area-moment method or the method of direct double integration will lead to a result: the Myosotis process is powerless in this case, but fortunately the steel companies do not make many beams of parabolic contour.

So far all the effects considered in this chapter were caused by the bending stresses in the beam, while the deformations due to the shear stress [Eq. (14)] have never even been mentioned. There is a perfectly valid reason for this omission: the deflections and angles in the beam caused by the shear are negligibly small in comparison with those caused by bending, and by Eq. (10) a shear stress is always associated with bending stresses in the beam. Although it is not a very exact or convincing argument, we might visualize how a beam would deform in the (impossible) case that it had a shear force in it constant along the length, that is, S independent of x, associated with *zero* bending moment everywhere. The deformation in that fictitious case would be zero according to the formulae of this chapter so far, because, with $M = 0$, the curvature and its integrals are also zero. But there is a shear stress in the neutral line, so that little squares there are being distorted into little diamonds, as indicated in Fig. 97. The angle γ by Hooke's law equals s_s/G, and therefore the deflection at the end of a cantilever beam of length l would be $\delta = \gamma l = s_s l/G$. This is known as the *shear deflection*, and in the real case of a bending moment [Eq. (10)] we assume that the total end deflection equals the sum of the bending deflection [Eq. (20)] and the shear deflection:

$$\delta_{\text{shear}} = \frac{1}{G}\int_0^l s_s \, dx. \tag{22}$$

Now, for a cantilever of rectangular cross section bh, loaded with an end load P we have:

$$\delta_{\text{bending}} = \frac{Pl^3}{3EI}; \qquad \delta_{\text{shear}} = \frac{3Pl}{2AG}.$$

$$\frac{\delta_{\text{shear}}}{\delta_{\text{bending}}} = \frac{3}{2}\frac{Pl}{AG}\frac{3EI}{Pl^3} = \frac{9}{2}\frac{E}{G}\frac{I}{Al^2}$$

$$= \frac{9}{2}\frac{E}{G}\frac{bh^3}{12bhl^2} = \frac{3}{8}\frac{E}{G}\frac{h^2}{l^2} \approx \frac{15}{16}\frac{h^2}{l^2}.$$

Thus for beams of any decent length the shear deflection is completely negligible. Only for beams that are but little longer than they are high is the shear deflection a sensible fraction of the bending deflection. But for such beams all our theories are unsatisfactory, because at the beginning of our developments a beam was defined as a member "long with respect to its sidewise dimensions." Short beams are not true beams, and when we apply our beam formulae to them, we must expect bad agreement with tests. But for $l > 5h$ we can speak of a true beam, and then the shear deflection is less than 4 per cent of the bending deflection. We were therefore completely justified in neglecting it from the beginning.

A similar reasoning holds for the deflections in coil springs as discussed on page 28. The shear stress in a circular cross section by page 46 is 4/3 times the average stress. Thus for a coil spring with force P in it the shear stress is $s_s = 4P/3\pi r^2$, and the deflection is $s_s l/G$. But the total length of the spring is $N\pi D$ with the notations of page 28. Thus we have [Eq. (9)]:

$$\frac{\delta_{\text{shear}}}{\delta_{\text{torsion}}} = \frac{4PN\pi D}{3\pi r^2 G}\frac{Gd^4}{P \times 8D^3N} = \frac{2}{3}\frac{d^2}{D^2}.$$

For the usual springs the coil diameter is some 4 or more times the wire diameter, so that the shear deflection is less than 4 per cent of the deflection due to torsion.

Problems 199 *to* 201.

CHAPTER VI

SPECIAL BEAM PROBLEMS

23. Beams of Two Materials. In some constructions beams are used that are made of two or more different materials, such as wood and steel or bronze and steel. Then there is more than one modulus of elasticity E, and the formulae in the preceding pages have to be generalized correspondingly. The principal equation at stake is (11) (page 39), with its preceding ones (a) and (b). When we go over the derivation once more, this time having in mind that E is not constant across the section but may vary from fiber to fiber, the only change we have to make in the entire argument is to keep the symbol E under the integral sign of any integration involved, instead of putting it outside the integral sign, as we did. Equation (a) (page 38) holds for a single fiber only, hence is as good now as it was before; however, in Eq. (b), we must write:

$$\int Ey^2\,dA \qquad \text{instead of} \qquad E\int y^2\,dA.$$

When this integral is extended over the entire cross section, the answer can be written as:

$$\sum_n E_n I_n,$$

i.e., as the sum of the products EI for each part of the beam made of another material. In practice the number n is hardly ever larger than 2. Hence in Eq. (11) we must write:

$$\frac{1}{E}\sum_n E_n I_n \qquad \text{instead of} \qquad I,$$

and that equation becomes:

$$s_f = E_f \frac{My}{\sum_n E_n I_n} \tag{11b}$$

where the subscript f simply indicates that when the stress in a certain fiber is to be found we begin by writing the E of the material *of that fiber.* Of course, Eq. (11*a*) (page 39) generalizes to:

$$Z = \frac{\sum_n E_n I_n}{E_f y_{\max}}, \tag{11c}$$

where E_f is the modulus of the extreme fiber to which $y_{\max}$ refers.

The differential equation of flexure [Eq. (19), page 80] contains the combination EI directly so that it simply becomes:

$$M = \frac{d^2y}{dx^2} \sum_n E_n I_n, \tag{19a}$$

which the reader should verify by going over the derivation once more.

Another fundamental equation is that of the shear-stress distribution [Eq. (14), page 43]. Its derivation was based on Figs. 30 and 31, and the principal steps will now briefly be retraced for the case that E is variable across the section. The bending stress in any fiber f is:

$$s_f = \frac{MyE_f}{\Sigma E_n I_n},$$

and the difference in this stress across the face dx is:

$$\Delta s_f = \frac{SyE_f}{\Sigma E_n I_n}\, dx.$$

The difference in force is:

$$d\,\Delta F = \frac{SyE_f\, dA}{\Sigma E_n I_n}\, dx,$$

and in integrating this force over the element of Fig. 30 or 31 we remember that E_f is not constant, so that:

$$\Delta F = \frac{S\, dx}{\Sigma E_n I_n} \int_{y=y_0}^{y=\frac{h}{2}} Ey\, dA.$$

This force equals the shear force on the bottom section $s_s b\, dx$, so that finally

$$s_s = \frac{S}{b \Sigma E_n I_n} \int_{y_0}^{\text{top}} Ey\, dA \tag{14a}$$

is the generalization of Eq. (14) to more than one material in the cross section.

These results will now be illustrated by application to a numerical example. Consider a factory floor of 30 ft width supported by beams of 30 ft span, simply supported at the ends (the walls) and spaced 3 ft apart center to center. The cross section of the beams is as shown in Fig. 98, with E_w for wood $= 1 \times 10^6$ lb/sq in. and E_s for steel $= 30 \times 10^6$ lb/sq in. (*a*) If the permissible "working stress" in the steel of the beam is 10,000 lb/sq in., what is the permissible floor loading and what is the sag in the center of the floor? (*b*) If the lag bolts are to be designed for an average shear stress across their section of 5,000 lb/sq in., what is the necessary cross-sectional area of these bolts? In question *a* the weakening effect of the boltholes in the steel is supposed to be absorbed in the conservative figure for the working stress.

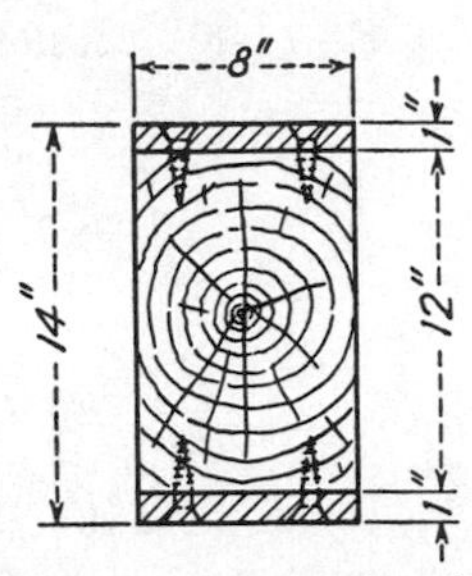

FIG. 98. Cross section of a beam with a wooden interior covered by steel plates, attached to it by lag bolts.

The computation proceeds as follows:

$$I_{\text{wood}} = \frac{bh^3}{12} = \frac{8 \times 12^3}{12} = 1{,}150 \text{ in.}^4,$$

$$I_{\text{steel}} = 2\left(\frac{bh^3}{12} + bhy_G^2\right) = 2\,[8/12 + 8(6\tfrac{1}{2})^2] = 680 \text{ in.}^4,$$

$$\Sigma EI = 1{,}150 \times 10^6 + 30 \times 680 \times 10^6 = 21{,}550 \times 10^6 \text{ lb in.}^2$$

For a load w lb per running inch of beam the maximum bending moment in the center of the span is:

$$M = \frac{wl^2}{8} = \frac{w}{8}(30 \times 12)^2 = 16{,}200w \text{ in.-lb.}$$

The stress

$$s = \frac{E_f M y}{\Sigma EI} = \frac{30 \times 10^6 \times 16{,}200w \times 7}{21{,}550 \times 10^6} = 158w.$$

This stress is to be 10,000 lb/sq. in so that:

$$w = \frac{10{,}000}{158} = 63.5 \text{ lb/in.} = 760 \text{ lb/ft.}$$

The beams are spaced 3 ft apart so that this load must be carried by 1 × 3 sq ft of floor. The permissible floor load thus is 253 lb/sq ft.

The sag in the center (Fig. 86, page 94) is:

$$\delta = \frac{5wl^4}{384\Sigma EI} = \frac{5 \times 63.5(360)^4}{384 \times 21{,}550 \times 10^6} = 0.65 \text{ in.}$$

This answers question *a*. For question *b* we calculate the maximum shear force near one of the supports from the loading just found:

$$S = 760 \text{ lb/ft} \times 15 \text{ ft} = 11{,}400 \text{ lb}.$$

By Eq. (14*a*), just derived, we have for the junction between wood and steel (Figs. 30 and 98):

$$\int_{y=6 \text{ in.}}^{y=7 \text{ in.}} Ey\, dA = 30 \times 10^6 (1 \times 8) \times 6\tfrac{1}{2} = 1{,}560 \times 10^6 \text{ lb-in.}$$

and

$$s_s = \frac{S}{b\Sigma EI} \int Ey\, dA = \frac{11{,}400 \times 1{,}560 \times 10^6}{8 \times 21{,}550 \times 10^6} = 103 \text{ lb/sq in.}$$

This would be the shear stress if it were uniformly distributed across the entire bottom of the steel plate. Instead of that it is taken by a number of bolts in which the permissible average shear stress is 5,000 lb/sq in. Hence the required bolt cross section is:

$$\frac{103}{5{,}000} = 0.0206 \text{ sq in./sq in.},$$

or slightly more than 2 per cent of the surface of the plate. This is necessary near the end supports of the beam, while in the center of the span, where the shear force is zero, no bolts at all would be required.

A problem very similar to the one just discussed is that of the torsion of a shaft made up of two materials. The "tail shaft," or propeller shaft, of almost every ocean-going ship is made of steel with a bronze sleeve shrunk over it, so that the problem is of considerable practical importance.

We return to the derivation of Eqs. (4) and (5) on page 19 and modify the analysis for the case that the elastic constant G varies from fiber to fiber. Equation (*a*) holds for one fiber only and hence is not modified. In integrating the stress to the torque we must keep G under the integral sign, or

$$M_t = \frac{d\varphi}{dx} \int Gr^2\, dA = \frac{d\varphi}{dx} \sum_n G_n I_{pn},$$

so that Eqs. (4) and (5) change to:

$$\varphi = \frac{M_t l}{\Sigma G I_p}, \tag{4a}$$

$$s_{sf} = G_f \frac{M_t r}{\Sigma G I_p}, \tag{5a}$$

where the subscript f denotes the fiber in question.

As an example we shall now calculate the torsional stiffness of the tail shaft and the horsepower of a ship's engine, in which that shaft (Fig. 99) has a solid core of 14 in. diameter on which a 1-in. bronze bushing is shrunk over the entire length of 30 ft. The shaft runs at 200 rpm; the constants are $G_{\text{steel}} = 12 \times 10^6$,

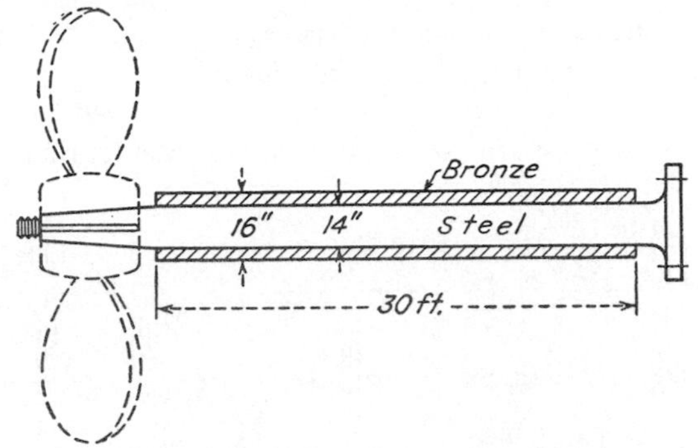

Fig. 99. Tail shaft of a very large ship.

$G_{\text{bronze}} = 6 \times 10^6$ lb/sq in.; and the shaft has been designed for a maximum shear stress in the steel of 1,500 lb/sq in.

The calculation is as follows:

By Eq. (6):

$$I_{p\ \text{steel}} = \frac{\pi}{32} \times 14^4 = 3{,}770 \text{ in.}^4,$$

$$I_{p\ \text{bronze}} = \frac{\pi}{32}(16^4 - 14^4) = 2{,}670 \text{ in.}^4,$$

$$\Sigma GI_p = 10^6(12 \times 3{,}770 + 6 \times 2{,}670) = 61{,}300 \times 10^6 \text{ lb in.}^2$$

The torsional stiffness of the shaft, or the torque required to twist it through 1 radian, is:

$$k = \frac{M}{\varphi} = \frac{\Sigma GI_p}{l} = \frac{61{,}300 \times 10^6}{12 \times 30} = 170 \times 10^6 \text{ in.-lb/radian.}$$

The stress in the steel is 1,500 lb/sq in.

$$s = G\frac{M_t r}{\Sigma GI_p} = 1{,}500 = \frac{12 \times 10^6 \times M_t \times 7}{61{,}300 \times 10^6},$$

or

$$M_t = 1.1 \times 10^6 \text{ in.-lb.}$$

$$\text{hp} \times 33{,}000 = M_t \times 2\pi \times \text{rpm} = 1.1 \times 10^6 \times 2\pi \times 200;$$

$$\therefore \quad \text{hp} = 42{,}000.$$

The shear stress in the bronze sleeve is $\frac{8}{7} \times \frac{6 \times 10^6}{12 \times 10^6}$ times the stress in the steel, or $4/7 \times 1{,}500 = 850$ lb/sq in.

Problems 202 *to* 207.

24. Skew Loads. When the theory of flexure of beams was developed on page 35 to 39, the argument was kept restricted to symmetrical cross sections, symmetrically loaded, as in Figs. 24*a* to *c*. Now we are ready to drop this assumption, and first

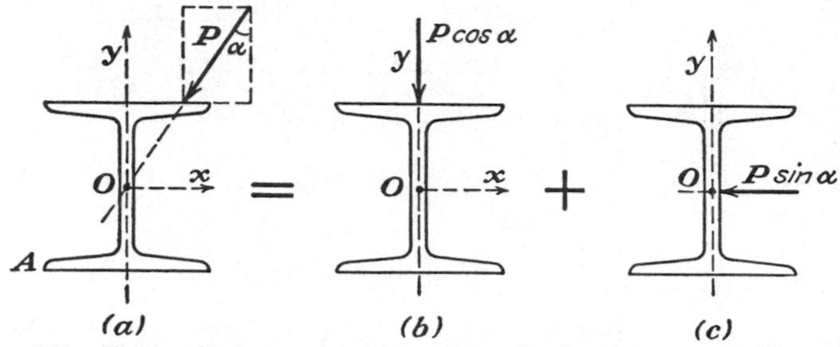

FIG. 100. The bending stress caused by a skew load on a beam of symmetrical cross section can be considered as the superposition of the stresses caused individually by the two symmetrical components of the load.

we consider a section with two perpendicular axes of symmetry like Fig. 24*a* or *b*, with a load still passing through the center of gravity, but with its direction no longer coinciding with an axis of symmetry, as shown in Fig. 100*a*. If we resolve this load into components along the axes of symmetry, we have two cases, each of which we understand by the previous theory. The case of Fig. 100*b* has a horizontal neutral line, while Fig. 100*c* has a vertical neutral line. The bending stress at any point in the section has the same direction in the two cases (*i.e.*, parallel to the axis of the beam) so that the stresses can be added algebraically without any complication of Mohr's circle. That stress then is, by Eq. (11):

$$s = \frac{M \cos \alpha}{I_x} y + \frac{M \sin \alpha}{I_y} x.$$

Both terms are linear functions, graphically representable in Fig. 100 by planes, the plane of the first term intersecting the paper along the x axis and the second one intersecting it along the y axis. The sum of two linear functions is again a linear function, representable by a plane, which intersects the paper in Fig. 100 along the line where $s = 0$, the neutral line. The equation of that neutral line thus is:

$$s = 0 = \frac{M \cos \alpha}{I_x} y + \frac{M \sin \alpha}{I_y} x = 0,$$

which runs at an angle with respect to both axes. In general this neutral line is *not* perpendicular to the load P; it would be so only if $I_x = I_y$. The maximum stress may appear at any of the four extreme corners of the I beam; in the case illustrated the maximum stress will be at corner A because there the stresses due to both components b and c (Fig. 100) reach their maxima with the same sign.

For calculating the deflections of the beam the same process of resolution of the loads into their two symmetrical components is applied. Here it is obvious that the deflection will not, in general, be in the direction of the load. If the two principal moments of inertia are different, we can distinguish a "stiff plane" and a "limber plane" in the beam, and for a given load the deflection in the limber plane will be relatively larger than that in the stiff plane. Therefore, a skew load will produce a deflection in a direction that does not coincide with the direction of the load, and is turned away from it toward the limber plane. For example, if, in Fig. 100, $\alpha = 45°$ and if $I_x = 5I_y$, then the deflection in the horizontal direction will be five times the vertical deflection and the total deflection will be off the horizontal direction by an angle $\tan^{-1} 1/5 = 11.3°$, while the force operates at 45 deg.

Now we proceed to the more difficult question of a non-symmetrical section, and for our first case we choose the Z shape of Fig. 101, loaded vertically, *i.e.*, along the web of the section. Suppose we try to do what our previous knowledge suggests and assume bending with a horizontal neutral line, *i.e.*, a neutral line perpendicular to the load. Then the top half of the Z section is in tension, and the bottom half in compression, both expressed by $s = Cy$, where C is some constant. The bending moment of these stresses above the horizontal neutral axis is:

$$M_x = \int sy \, dA = C \int y^2 \, dA = CI_x,$$

a familiar result. But, what is new here is that these stresses have a moment about the vertical axis as well:

$$M_y = \int sx \, dA = C \int xy \, dA = CI_{xy},$$

proportional to the product of inertia of the cross section. Such a moment is not imposed by the loading P: that force P has a bending moment about the horizontal x axis only. Hence the M_y moment must be zero. For a symmetrical cross section, $I_{xy} = 0$, so that no difficulty exists. For the case of Fig. 101, however, we conclude that our first assumption of a horizontal neutral line is incorrect, because it entails a bending moment about the vertical axis. We further conclude that the only way the old theory can

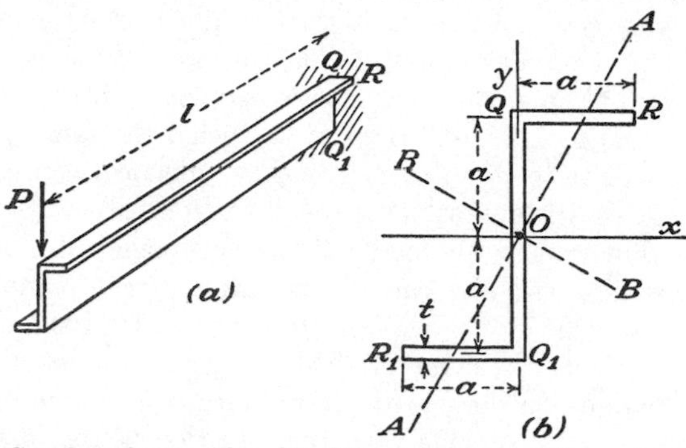

Fig. 101. Cantilever beam with end load P, passing through the center of gravity O of the Z-shaped cross section. Under the influence of a purely vertical load this beam will deflect sidewise to the right $1\frac{1}{2}$ times as much as it sags downward.

be made applicable to non-symmetrical sections is when $I_{xy} = 0$, or when the load applies along one of the two principal directions of inertia of the cross section. Now, in Fig. 101, the vertical load P is obviously not along a principal direction of inertia, so that the first thing we must do is to find the principal axes of the cross section, somewhere near AOA sketched in the figure. Then we must resolve the load P in the two principal directions. Our first job then is a Mohr-circle construction:

$$I_x = \int y^2 \, dA = \frac{1}{12} t(2a)^3 + 2 \cdot ta \cdot a^2 = \frac{8}{3} ta^3,$$

$$I_y = \int x^2 \, dA = \frac{1}{12} t(2a)^3 = \frac{2}{3} ta^3,$$

$$I_{xy} = \int xy \, dA = 2 \cdot ta \cdot a \cdot \frac{a}{2} = ta^3.$$

In Fig. 102 we now plot point x, depicting the inertia properties of the x axis with an abscissa $\frac{8}{3}ta^3$ and an ordinate ta^3. Then we plot point y, depicting the inertia properties of the y axis with an abscissa $\frac{2}{3}ta^3$ and an ordinate $-ta^3$. (The latter value is negative because we consider the y axis as our "first" axis, the second one being 90 deg counterclockwise from it: the negative x axis. For this set of axes the product of inertia is negative). With the points x and y known in Fig. 102 and with the additional information that these axes are 90 deg apart in Fig. 101b and hence 180 deg apart in Fig. 102, we can construct the circle. We read from it that its radius is $ta^3\sqrt{2} = 1.41ta^3$, and hence the principal moments

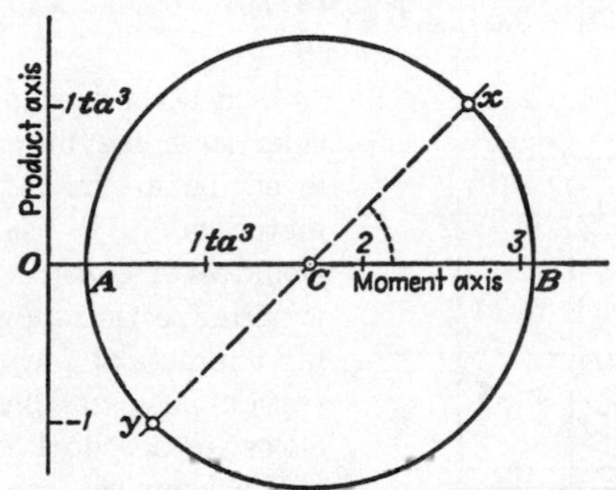

FIG. 102. Mohr's circle of inertia for the cross section of Fig. 101b.

of inertia are $3.08ta^3$ and $0.26ta^3$, and the angle xCB is 45 deg. Hence in Fig. 101b the axis cross AA, BB is at 22.5 deg with respect to the xy system, and the force P has to be resolved in these two directions. The deflection in the A direction downward is:

$$\delta_A = \frac{P \cos 22\frac{1}{2}\ l^3}{3E \times 3.08ta^3} = 0.10\,\frac{Pl^3}{Eta^3},$$

and in the B direction, to the right, it is:

$$\delta_B = \frac{P \sin 22\frac{1}{2}\ l^3}{3E \times 0.26ta^3} = 0.49\,\frac{Pl^3}{Eta^3}.$$

Resolving these deflections back to the x and y directions of Fig. 101b, we find:

$$\delta_y = \delta_A \cos 22\frac{1}{2} + \delta_B \sin 22\frac{1}{2} = 0.28\,\frac{Pl^3}{Eta^3},$$

$$\delta_x = \delta_B \cos 22\frac{1}{2} - \delta_A \sin 22\frac{1}{2} = 0.42\,\frac{Pl^3}{Eta^3}.$$

Thus not only does the beam of Fig. 101a sag down at its end, but it deflects horizontally to the right as well, by an amount 50 per cent greater than the downward sag.

To find the maximum stress we observe that in Fig. 101b the fibers above the line BB are in tension due to one component while the fibers to the left of AA are in tension due to the other component. Thus in the corners Q and Q_1 the stresses from the two components are additive, while at points R and R_1 they subtract from each other. The stress in Q is:

$$s = \frac{(Pl \cos 22\tfrac{1}{2})(a \cos 22\tfrac{1}{2})}{3.08ta^3} + \frac{(Pl \sin 22\tfrac{1}{2})(a \sin 22\tfrac{1}{2})}{0.26ta^3}$$

$$= \frac{Pl}{ta^2}(0.28 + 0.56) = 0.84 \frac{Pl}{ta^2}.$$

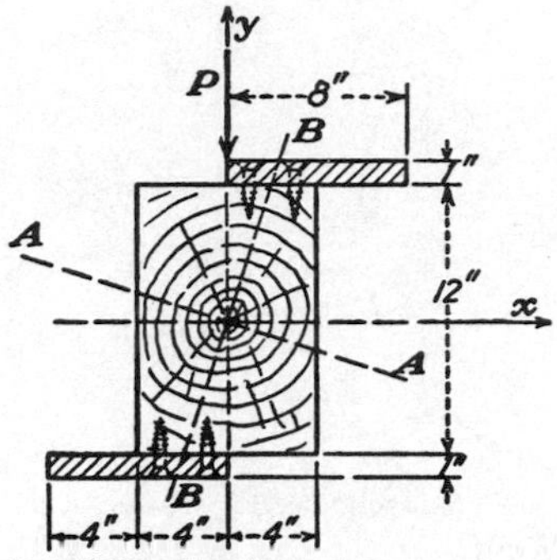

FIG. 103. Cantilever beam made up of a wooden 12- by 8-in. section, $E_w = 10^6$ lb/ sq. in., and two 1- by 8-in. steel plates, $E_s = 30 \times 10^6$ lb/ sq. in., not centrally attached.

As a second example we consider a non-symmetrical cantilever beam made up of two materials. No sane engineer would ever construct a beam of the cross section shown in Fig. 103 for a practical purpose, but the section possesses beautiful properties for a schoolteacher, as the reader soon will see. We ask in which direction this beam will deflect under a vertical load. The problem is about the same as the previous one, except that in the bending of beams of two materials E is not a constant and therefore cannot be brought outside of the integral sign. We thus calculate:

$$\Sigma EI_x = \int Ey^2\,dA = 21{,}550 \times 10^6 \text{ lb in.}^2, \qquad \text{(page 111, Fig. 98)}$$

$$\Sigma EI_y = \int Ex^2\,dA = 10^6\,\frac{12 \times 8^3}{12} + 30 \times 10^6\,\frac{1 \times 16^3}{12}$$
$$= 10{,}800 \times 10^6 \text{ lb in.}^2,$$

$$\Sigma EI_{xy} = \int Exy\,dA = 30 \times 10^6 \times 2\,(8 \times 4 \times 6\tfrac{1}{2}) + 10^6 \times \text{zero}$$
$$= 12{,}500 \times 10^6 \text{ lb in.}^2$$

With these quantities we construct a Mohr diagram (Fig. 104), first plotting the points x and y, then constructing the circle and finding the principal points A and B. From the circle we read:

$$EI_{A,B} = 16{,}175 \pm 13{,}600 = 29{,}775 \text{ and } 2{,}575 \times 10^6,$$

$$\tan \angle xCA = \frac{12{,}500}{5{,}375} = 2.33 \qquad \text{and} \qquad \angle xCA = 66.8°.$$

Hence the principal axes AA and BB in the cross section of Fig. 103 are 33.4 deg from the horizontal and vertical. The (downward) deflection in the BB direction will be proportional to:

$$\frac{P \cos 33.4}{29{,}775} = 0.028P,$$

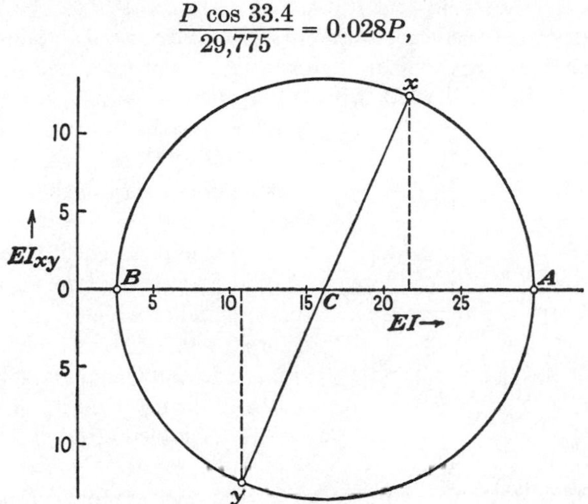

Fig. 104. Mohr's circle for the inertia properties of the cross section of Fig. 103.

and the deflection (to the right) in the A direction is proportional to:

$$\frac{P \sin 33.4}{2{,}575} = 0.214P.$$

These are in ratio 1 to 7.63, which is the tangent of the angle between AA and the direction of the deflection. That angle is found to be 7.5 deg, so that the beam deflects downward and to the right in a direction $7.5 + 33.4 = 41°$ below the horizontal x axis.

Problems 208 *to* 213.

25. The Center of Shear. If cantilevers of symmetrical cross section or of the cross sections of Figs. 101 and 103 are actually built and carefully tested in a laboratory, the experimental deflections and stresses are found to agree quite well with the calculated ones. However, if we take an angle section, as shown in Fig. 105, no such agreement is found. The principal directions of inertia in a symmetrical angle section are at 45 deg with respect to the legs, and the two moments of inertia differ considerably. Hence our theory predicts that, under a vertical load through G, the beam will deflect downward and sidewise, from the unloaded position in thin lines to the loaded dotted position of Fig. 105. The experiment, on the other hand, shows the beam deflected and twisted as shown in the heavy-line picture of Fig. 105. This discrepancy between theory and practice was noticed and described in the engineering literature before 1900, but it remained an unexplained riddle until 1922, when C. Weber in Germany published the theory of the "center of shear." Historically, this is very curious because the theory of the center of shear is fairly simple, and certainly no more complicated than many of the theories we have met so far in this text, all of which date back to before 1850. (Mohr's circle diagram was published in 1880 but represented only a more convenient manner of interpreting the knowledge that existed before in the shape of rather formidable formulae.)

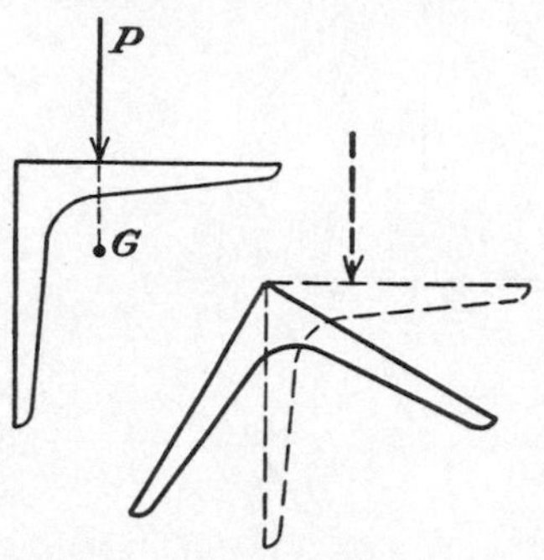

FIG. 105. An angle iron as a cantilever beam, loaded with an end load passing through the center of gravity. Without load the end is in the position shown in thin solid lines; by the pre-1922 theory it deflects to the dotted position, while the experiment shows it to go to the position in heavy outline.

Weber's explanation consisted in an examination of the equilibrium of the cantilever, cut off at its built-in end. It experiences at that built-in end a shear force P and a moment Pl and hence must develop stresses in that section to counteract P and Pl. The bending stresses, calculated by Eq. (11), as explained in the previous article, do form a moment $-Pl$, and the shear stresses, as calculated by Eq. (14) do indeed again form a force $-P$. The thing that nobody had thought about prior to 1922 was to see whether the resultant of all these shear stresses did or did not pass through the center of gravity of the section. For symmetrical sections that resultant certainly does pass through G, as is illustrated in Figs. 32, 34, or 38. But for many non-symmetrical sections the resultant of the shear stresses, calculated by the ancient, pre-1850 equation (14), does not pass through G, although it does give the correct magnitude and direction $-P$. Then if we look at the cutoff cantilever again, we see that the actual load P at the free end, through G, and the resultant $-P$ of the shear stresses in the built-in end, not through G but offset sidewise, add up primarily to a bending moment Pl but secondarily to a twisting moment P times the sidewise offset. This twisting moment causes the behavior of Fig. 105.

The effect exists only for non-symmetrical sections, such as are shown in Fig. 106, and we shall now calculate it in detail, starting

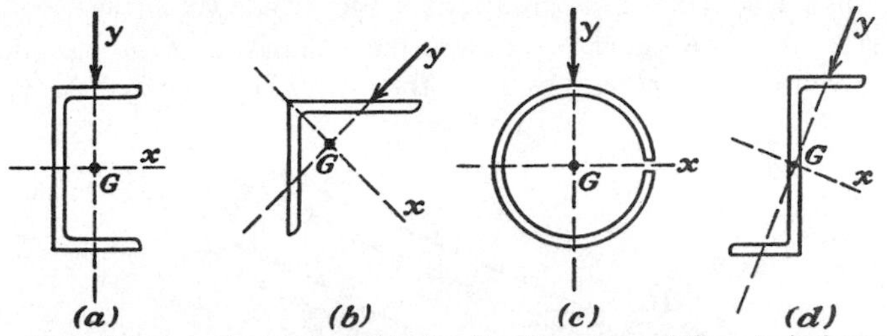

Fig. 106. Cantilever beams loaded in the direction of one of the principal axes of inertia of a cross section, which is *not* an axis of symmetry geometrically.

with the channel of Fig. 106*a*, illustrated again in Fig. 107. For simplicity we shall assume the thickness t to be "small" with respect to a, which means that the ratio t/a will be retained in the calculations but that $(t/a)^2$ and higher powers will be neglected.

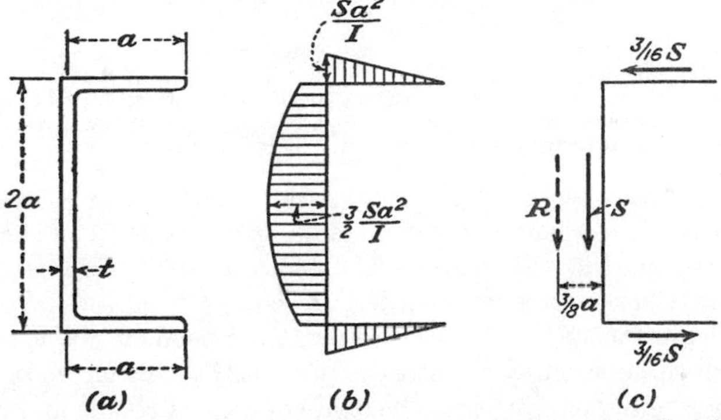

Fig. 107. Shear-stress distribution and center of shear in a channel section.

Application of Eq. (14) to this case leads to the result shown in Fig. 107*b*, which the reader should verify for himself. When the stresses of Fig. 107*b* are integrated separately for the three legs of the section and it is remembered that $I = 8a^3t/3$, the forces shown in Fig. 107*c* are found. We see that the vertical force is S, as it should be, and the resultant horizontal force is zero, again as it should be. But the two horizontal forces form a counterclockwise couple, and hence the three forces of Fig. 107*c* are equivalent to a single force R, of magnitude S, displaced distance x to the left of the vertical leg, where:

$$Sx = \frac{3}{16} S \times 2a,$$

so that $x = \frac{3}{8}a$. This resultant force is shown as the dotted vector R in the figure. We note particularly that the resultant shear force does *not* pass through the center of gravity of the cross

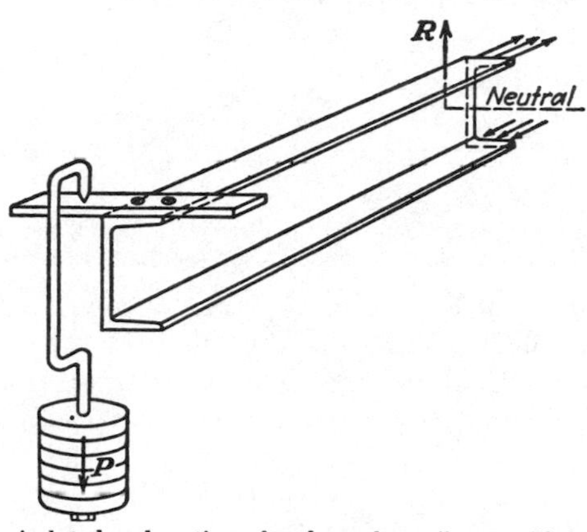

FIG. 108. The isolated end section of a channel cantilever. If the end load P is placed through the center of shear, there is equilibrium with a horizontal neutral line, and stresses, by Eqs. (11) and (14), at the far end. If the load P is made to wander across the section, an additional twisting couple is introduced.

section. Now let us arrange the cantilever of Fig. 108 so that the end load P can conveniently be shifted across the end section. If P is placed at $\frac{3}{8}a$ to the left of the web, the force P and the resultant shear force R form a pure couple in a vertical plane parallel to the axis of the beam; it is a pure bending couple and is held in equilibrium by the stresses [Eq. (11)] at the far end. If P is placed elsewhere, we introduce, in addition, a couple in a plane perpendicular to the beam axis, which is a twisting couple. In particular, if P is made to pass through the center of gravity, the additional twisting couple on the beam is $\frac{5}{8}Pa$ as shown in Fig. 109.

This completes our story for the channel placed in the position of Fig. 107 or 110*c*. Now let us look at a cantilever with the channel placed horizontally (Fig. 110*a* or *b*). Then, of course, the section is symmetrical about the line of action of the load, and hence the load must pass through the center of gravity. The lines of action of the loads in the two principal directions intersect in a point C. Now if we mount the cantilever channel obliquely (Fig. 110*d*) under a vertical load passing through that point C, we can resolve the load along the two principal directions of inertia and thus make Fig. 110*d* the superposition of Figs. 110*a* and 110*c*. Since neither 110*a* nor *c* has torsion, Fig. 110*d* will be free of it also.

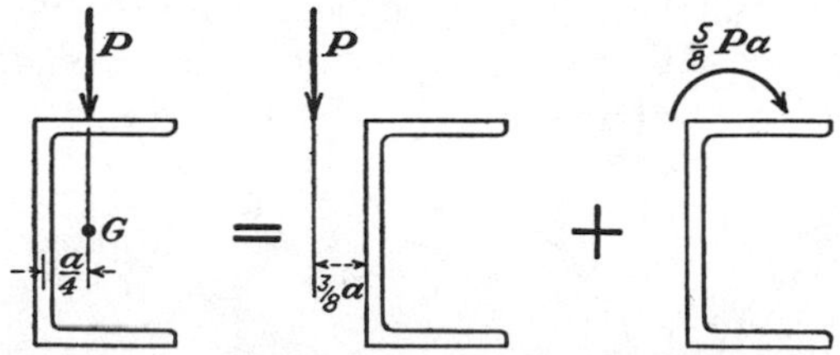

FIG. 109. An end load placed through the center of gravity can be considered the sum of a bending load through the center of shear and a twisting couple.

The point C, named the *center of shear* of the cross section, therefore has the property that, if an end cantilever load is made to pass through it in any skew direction, there will be no twist in the beam. For that reason C sometimes is also called the *center of twist*, and it is a very important point in a cross section. For

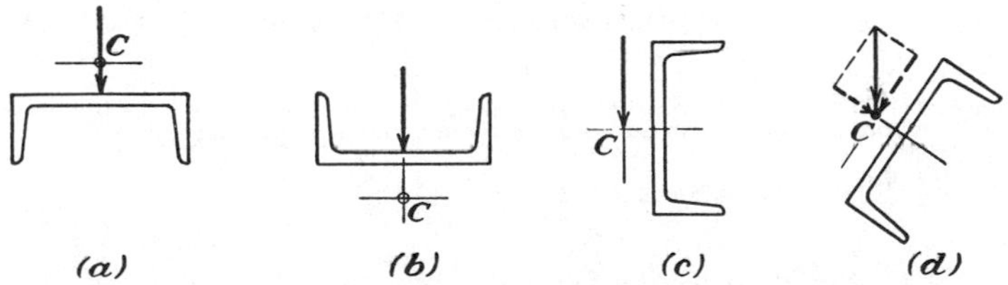

FIG. 110. Definition of the center of shear C of a cross section.

symmetrical sections, of course, it coincides with the center of gravity.

There is one more fact which will help to elucidate the center of twist and to which we alluded in Fig. 105. The beams of Figs. 110*a* and 110*c* both have bending-stress distributions with a horizontal neutral line, and the deformation associated with it (Fig. 72) is that of a horizontal, slightly curved neutral plane or ribbon. Consequently, the cross sections of the beam turn only about an axis perpendicular to the beam center line, but they do not rotate about the beam center line itself: they do not rotate in a twisting sense. Looking on the end face of a cantilever then, we see the section sag down parallel to itself without twisting. If we now turn to Fig. 110*d*, the end section will not rotate in a twisting sense either, because it is the sum of Figs. 110*a* and *c*. Thus, if the end load on a cantilever beam passes through the center of twist, the end section deflects (downward and sidewise) parallel to itself and if the end load does not pass through C the end section will deflect and rotate.

Now let us find the location of the shear center for the remaining three sections of Fig. 106, starting with the slit tube of Fig. 106*c*. If the load is applied horizontally, the section is symmetrical with respect to it; hence the x axis is one locus of point C. The other

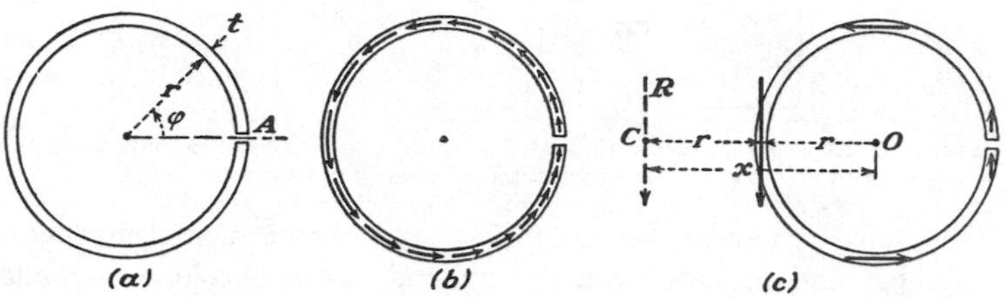

Fig. 111. Calculation of the shear center for a thin-walled slit tubular cross section.

locus belongs to the load along the y axis (Fig. 106*c*), for which the cross section is not symmetrical but which is a principal axis of inertia. Apply Eq. (14) to the case, and since we only want to find the location of the resultant and not its magnitude, we can disregard in Eq. (14) all factors which are constant for the whole cross section and which do not vary from point to point. Then Eq. (14) gives for the shear stress:

$$
\begin{aligned}
s_s &= \text{const} \int_y^{\frac{h}{2}} y \, dA = \text{const} \int_{\varphi=\varphi}^{\varphi=0} (r \sin \varphi)(tr \, d\varphi) \\
&= \text{const } r^2 t \int_\varphi^0 \sin \varphi \, d\varphi = - \text{const } r^2 t \cos \varphi \Big|_\varphi^0 \\
&= \text{const } r^2 t \, (1 - \cos \varphi).
\end{aligned}
$$

It may look strange at first sight that the upper limit of integration $h/2$ here becomes $\varphi = 0$ at the neutral line. The procedure becomes clearer when we look first at Fig. 30, then at Fig. 37. The symbol $h/2$ refers to Fig. 30, but it stands for the free end fiber where no shear stress can exist. If the flange of Fig. 37 is bent around downward, we arrive at the situation of Fig. 111*a*.

The result for the shear stress just obtained is illustrated in Fig. 111*b*, while Fig. 111*c* shows the effect roughly. It is seen that there is no horizontal resultant but that the horizontal shear forces form a counterclockwise couple similar to Fig. 107*c*, so that the resultant R must lie to the left at distance x from the center O. We calculate first the value of R, the resultant of all vertical components:

$$\begin{aligned} R &= \int_A (s_s\, dA) \cos \varphi = \int_{\varphi=0}^{\varphi=2\pi} [\text{const } r^2 t(1 - \cos \varphi)][tr\, d\varphi] \cos \varphi \\ &= \text{const } r^3 t^2 \int_0^{2\pi} (1 - \cos \varphi) \cos \varphi\, d\varphi \\ &= \text{const } r^3 t^2 \left(\int_0^{2\pi} \cos \varphi\, d\varphi - \int_0^{2\pi} \cos^2 \varphi\, d\varphi \right) \\ &= \text{const } r^3 t^2 [0 - \pi] = - \text{const } \pi r^3 t^2. \end{aligned}$$

The values of the two definite integrals appearing here should be looked up in a table or in a book on calculus, or the reader should calculate them for himself. The negative sign means that R is directed downward.

Next we calculate the moment of all these elemental shear forces about the center O, which is easy, because the moment arm of all the forces is simply r:

$$\begin{aligned} M &= \int (s_s\, dA) r = \int_0^{2\pi} [\text{const } r^2 t\, (1 - \cos \varphi)](tr\, d\varphi) \cdot r \\ &= \text{const } r^4 t^2 \int_0^{2\pi} (1 - \cos \varphi)\, d\varphi \\ &= \text{const } 2\pi r^4 t^2. \end{aligned}$$

This moment must be set equal to $- Rx$, which gives an equation in x. Its solution is $x = 2r$, or the center of shear is at distance $2r$ to the left of the center of the circle, as shown in Fig. 111*c*.

Thus if we had loaded the slit tube as in Fig. 106*c*, through its center of gravity, we should have applied a clockwise twisting couple $2Pr$ and as a consequence the end section would have deflected downward and rotated clockwise, the slit going down more than any other point of the section. If we require the end of the cantilever to sag down parallel to itself, without rotation, we must place the load at point C.

The next example we consider is Fig. 106*d*, in which we shall be able to find the center C without calculation, by reasoning only. Applying Eq. (14) we can see from the form of the integral that there is symmetry as shown in Fig. 112*a*, in which the shear stresses in the two flanges of the Z are identical. Then we can summarize in Fig. 112*b* that there are three forces F_1, F_1, and F_2 without having to know their magnitudes. The two F_1 forces have a resultant $2F_1$ through G, and compounding that with F_2 we get a resultant force through G, which must come out purely vertical, equal and opposite to the end load P. Hence the center of gravity and the center of shear coincide, which furnishes the post facto justification for our procedure on page 116, Figs. 101 and 103.

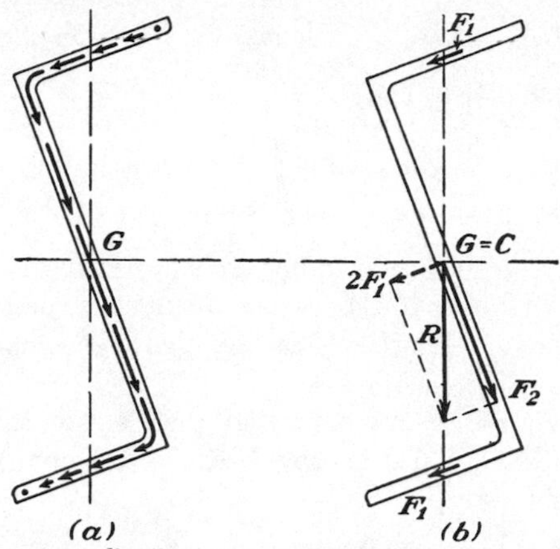

Fig. 112. Shear-stress distribution in a Z section bent in one of its principal planes. The center of twist coincides with the geometrical center and with the center of gravity of the figure.

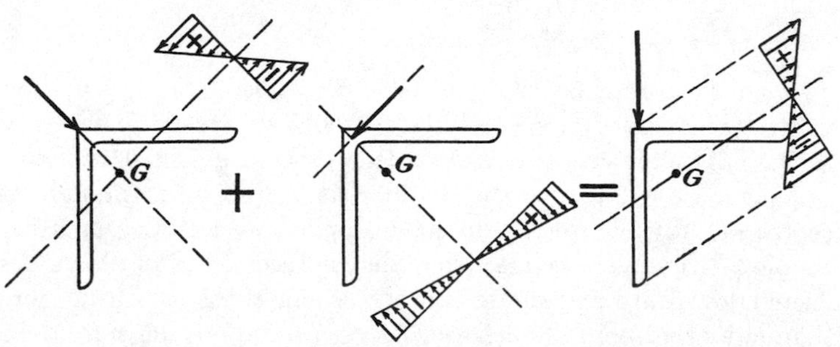

Fig. 113. Bending-stress distribution in a thin-walled angle cantilever, calculated by the method of page 114, showing compressive stress in the right-hand corner.

The last example we look at is Fig. 106b, where again we shall reason without calculation. Whatever form Eq. (14) may have, the shear stresses in a "thin" section must be along that section and have no crosswise component (Fig. 35, page 45). Hence in Fig. 106b the shear stresses are along the two legs, and all pass through the corner of the angle. Thus the resultant must pass through the corner of the angle, which therefore is the center of shear. The result of the analysis is shown in Fig. 113 for the bending stresses and in Fig. 114 for the shear stresses. The shear-stress distributions in Figs. 114*a* and *b* are "parabolic," as in Fig. 32; they are shown in Fig. 114*d*, in which the two legs have been straightened out into a single line. The distribution *c* in that figure is the sum of *a* and *b* and shows a neutral point of shear stress

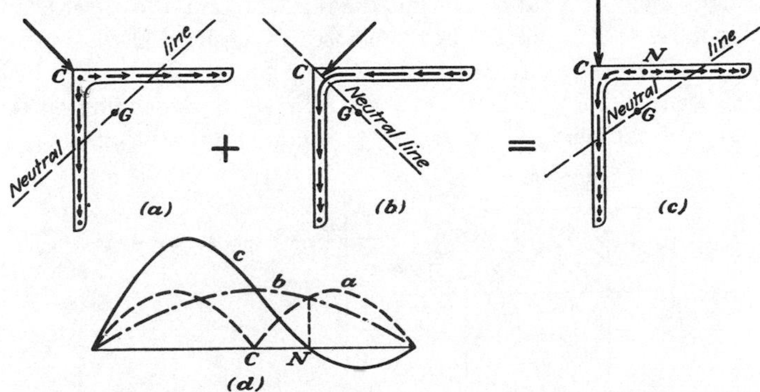

FIG. 114. Shear-stress distribution across the angle section, showing a point *N* of zero shear stress in the horizontal upper leg.

N in the upper leg. The numerical computation of the quantities in Figs. 113 and 114 is left as a problem to the reader.

Problems 214 *to* 218.

26. Reinforced Concrete. As a building material, concrete has the great advantages of being cheap, completely fireproof, and reasonably waterproof. It does not rust like iron or rot like wood, but while it can take compressive stresses of the same order of magnitude as soft wood, it suffers from the disadvantage of having practically zero strength in tension. Structural steel has an ultimate strength of about 60,000 lb/sq in. and a yield point of about 30,000 lb/sq in., and for static loads in building construction it is given a working or design stress up to 20,000 lb/sq in., tension or compression. Concrete has an ultimate strength in compression up to 4,000 lb/sq in., does not yield but breaks in a brittle fashion, and is given a design stress of about 1,000 lb/sq in. *in compression*. Its actual *tensile* strength is of the order of 200 lb/sq in. and for design purposes is usually considered to be zero. Looking at Mohr's circle for compression and tension (Figs. 57 and 58, page 66), we see that the maximum safe shear stress equals half the compressive stress. But if the concrete is *not* under compression, and subjected to pure shear, Mohr's circle is concentric with the origin (Fig. 60) and can assume no size without developing tensile stress, which the concrete cannot stand. Hence the shearing strength of uncompressed concrete is small.

The stress-strain diagram of concrete, when compressed, is not a straight line but a curve, so that the term "modulus of elasticity" loses its exact meaning. For purposes of design calculation we assume Hooke's law to be valid with a definite value of E, which runs between 2.0×10^6 and 3.75×10^6 lb/sq in. depending on the quality of the material, or from $\frac{1}{15}$ to about $\frac{1}{8}$ of that of steel.

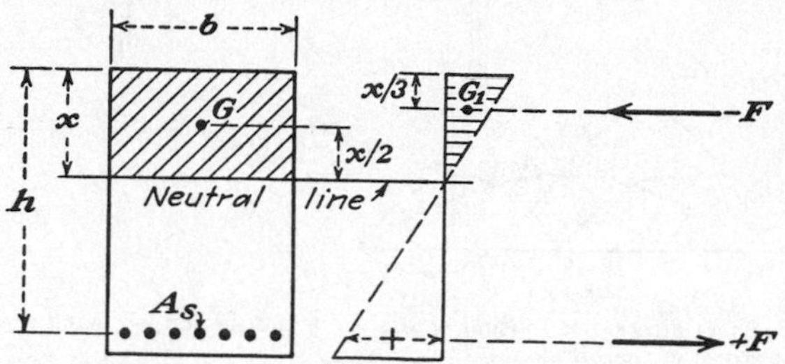

FIG. 115. A reinforced-concrete beam of rectangular section in bending.

Thus plain concrete can be used only for members under compression, such as columns. When applied in beams, subjected to bending moments, the concrete must be reinforced with iron bars to take the tensile stresses. These iron bars should be placed wherever tension occurs, *i.e.*, strictly speaking, along the tensile trajectories of the beam (Fig. 63). For long beams in bending this, however, is too complicated a procedure, and the bars are usually placed simply in one layer on the tensile side, as far from the center of the beam as possible. The bars have sufficient friction against the enveloping concrete so that they do not slip when strain occurs, and in extreme cases of short bars their ends are bent over to anchor in the concrete.

Figure 115 shows a cross section of a simple rectangular concrete beam with steel reinforcement of total cross-sectional area A_s in the bottom. The height h of the beam, as indicated, is the "effective" height; the extra one inch of concrete under the steel serves only as fire, rust, and buckling protection to the steel and has no structural function, since that concrete is in tension and is supposed to have lost its tensile stresses by cracking. The part of the concrete in compression above the neutral line is shaded.

If the angular deformation per unit length of the beam is called

$d\varphi$, then the average strain of the compressed concrete, *i.e.*, the strain at the center of gravity G, is $x\,d\varphi/2$; the compressive stress at G is $E_c x\,d\varphi/2$, and the total compressive force over the shaded area is $bxE_c x\,d\varphi/2$.

The tensile force in the steel bars by the same reasoning is $A_s E_s(h - x)d\varphi$, and, from equilibrium, this force must equal the compressive force. Hence:

$$\frac{E_s}{E_c} \cdot A_s(h - x) = \frac{x^2 b}{2} \tag{a}$$

is an equation from which the location x of the neutral line can be calculated if the dimensions of the beam and the ratio E_s/E_c are known.

The bending moment M of the beam is usually expressed in terms of the stresses in the materials. Let s_s be the tensile stress in the steel and s_c the (maximum) compressive stress in the uppermost layer of concrete. Then the force F is $A_s s_s$, and its distance from the compensating compressive force is $h - x/3$, so that the bending moment is:

$$M = A_s s_s(h - x/3). \tag{b}$$

A third relation, which the reader should verify for himself, is:

$$\frac{s_c}{s_s} = \frac{E_c}{E_s} \cdot \frac{x}{h - x}. \tag{c}$$

These three formulae suffice for most problems in the design of reinforced-concrete beams. For practical work in this field, even these formulae are too elaborate, and tables and graphs have been worked out whereby numerical answers are almost immediately obtained. Such tables can be found, for example, in the "Reinforced Concrete Design Handbook" of the American Concrete Institute, Detroit, Mich., 1948. The notations used by designers in reinforced concrete are entirely different from those in any other branch of engineering and have not been adopted in this text.

As a first illustration of the theory, we consider example 1 of the handbook, just mentioned, which asks for the "effective" height h of a beam of rectangular section (Fig. 115), in which $b = 7\frac{1}{2}$ in., $E_s/E_c = 12$, $s_s = 20{,}000$ lb/sq in., $s_c = 1{,}000$ lb/sq in., and $M = 15.8$ ft-kips. Looking at Eqs. (a), (b), and (c) we notice that all quantities are given except h, x, and A_s. Substituting numbers into Eq. (c) and solving for the ratio x/h we find:

$$x = \tfrac{3}{8}h.$$

Substituting numerical values in Eqs. (*a*) and (*b*) and not forgetting to convert the moment M from foot-pounds to inch-pounds, these equations become:

$$14.2A_s = h \qquad (a)$$

$$10.8 = hA_s \qquad (b)$$

from which we find $h = 12.5$ in. and $A_s = 0.88$ sq in.

The last result is important as it tells us how much steel is needed. If more steel were put in, the stress in the steel would become less than 20,000 lb/sq in.; the beam certainly would be

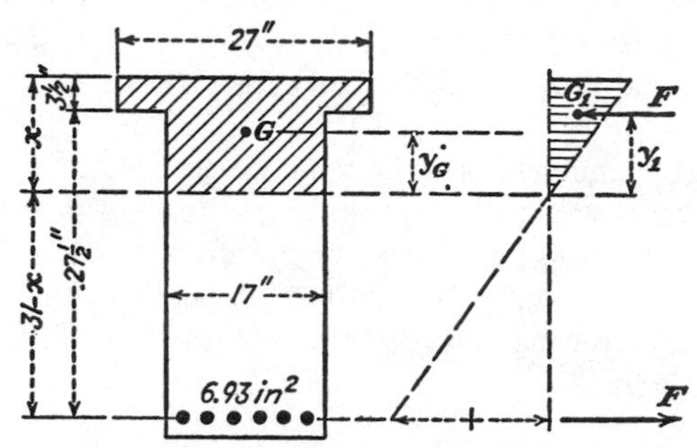

FIG. 116. T-shaped reinforced-concrete beam in which all dimensions are given; $E_s/E_c = 12$; $M = 320$ ft-kips. Calculate the stresses in both materials.

safe, but the design would not be economical. The solution for A_s is sometimes called a solution for "balanced design," and with 0.88 sq in. of steel the beam is said to have a "balanced" amount of steel in it.

For our second numerical case we take example 12 of the aforementioned handbook, which calls for the calculation of the stresses in a beam of T-shaped cross section in which all dimensions are given (Fig. 116). Furthermore, the bending moment is given as 320 ft-kips = 320,000 ft-lb = 3,840,000 in.-lb.

It is obvious why a designer likes to work with a T section rather than with a rectangular one. The concrete below the neutral line serves no purpose other than to hold the steel rods in place and keep them from rusting; hence material cut away below that neutral line represents a saving in weight as well as in cost.

Of the three general equations (*a*), (*b*), and (*c*), only (*c*) remains in force; the other two must be modified because the centers of gravity of the shaded compressive areas of Fig. 116 are no longer at $x/2$ and $x/3$ as before in Fig. 115. The right-hand side of Eq. (*a*) changes from $xb(x/2)$ to $A_c \cdot y_G$, and the parenthesis in Eq. (*b*) changes from $(h - x/3)$ to $(h - x + y_1)$; both statements should be checked by the reader by a fresh derivation of the formulae. Thus we have:

$$E_sA_s(h - x) = E_cA_cy_G, \tag{a_1}$$

$$M = A_ss_s(h - x + y_1), \tag{b_1}$$

$$\frac{s_c}{s_s} = \frac{E_c}{E_s}\frac{x}{h - x}, \tag{c_1}$$

three equations in the three unknowns x, s_s, and s_c, because y_G and y_1 are expressible in terms of x. We read from the figure:

$$A_c = 17x + 10 \times 3\tfrac{1}{2} = 17x + 35,$$

$$y_GA_c = 17x \cdot \frac{x}{2} + 35\left(x - 1\frac{3}{4}\right) = 8\frac{1}{2}x^2 + 35x - 61.$$

Substituting this into (a_1) leads to:

$$x^2 + 13.92x - 311 = 0$$

and

$$x = -6.96 + \sqrt{359.5} = 12 \text{ in.}$$

Next we must calculate y_1, of the point of application of the resultant of the compressive stresses:

$$y_1 = \frac{27\dfrac{x^3}{3} - 10\dfrac{(x - 3\frac{1}{2})^3}{3}}{27\dfrac{x^2}{2} - 10\dfrac{(x - 3\frac{1}{2})^2}{2}}.$$

Since we already know $x = 12$ in., this can be computed directly, giving $y_1 = 8.45$ in. Entering with this into Eq. (b_1) leads to $s_s = 21{,}600$ lb/sq in., and finally, from Eq. (c_1), $s_c = 1{,}140$ lb/sq in., which solves the problem.

Problems 219 *to* 224.

27. Plastic Deformations. All formulae discussed so far were derived on the basis of Hooke's law and therefore are valid only in the elastic range, *i.e.*, for stresses smaller than the yield stress. In practice, however, beams and other structures occasionally are subjected to loads that would give rise to stresses higher than the yield stress by the existing formulae, and such beams often do not fail but patiently keep on carrying their loads indefinitely. To investigate what happens in such a beam we must return to the stress-strain diagram of the material (Fig. 1, page 3) and formulate a suitable mathematical idealization of that diagram beyond the yield point A. An assumption not far from the actual facts and giving decent results is the one of Fig. 117, in which the stress remains constant, equal to the yield stress, while the strain grows indefinitely. In a test on mild steel the strain can be observed to grow to twenty times the largest elastic strain without

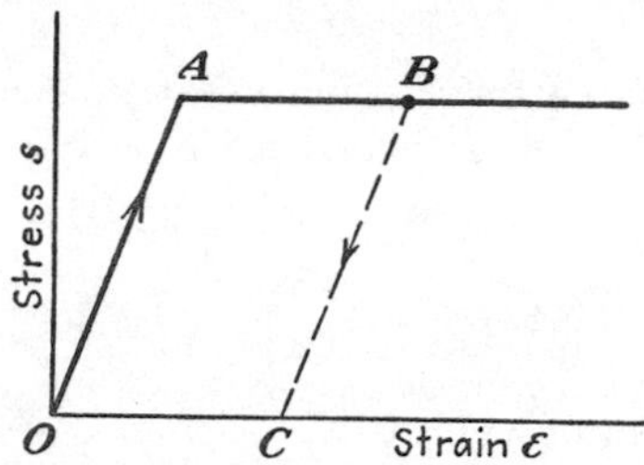

FIG. 117. Idealized stress-strain diagram for ductile materials, on which the theory of plastic deformations is based.

appreciable increase in stress (point B of Fig. 1), which justifies the assumption of Fig. 117.

First we look at a beam with a section symmetrical about its neutral line, such as a rectangle or an I beam, and we gradually increase the bending moment on it, as in Fig. 118. The stress

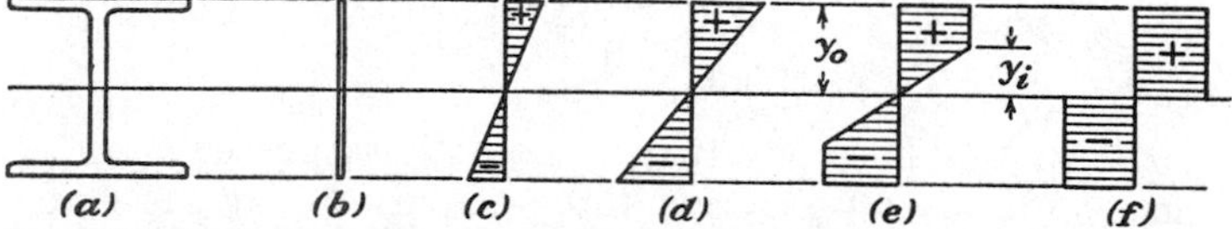

FIG. 118. Stress distributions for a growing bending moment; at d the yield point is reached; at f the beam carries its largest possible moment.

distribution of Fig. 118b occurs for zero bending moment; then c and d develop according to Eq. (11). At d the extreme fiber stress just equals the yield stress, and the bending moment carried is the largest possible one within the elastic limit. If the moment is increased beyond that, the outer fibers would develop stresses higher than the yield stress if this were possible. But Hooke's law no longer holds, and by Fig. 117 the stress just remains constant while the strain continues to grow. The outer fibers come into the plastic stage, and as the bending moment grows, more and more fibers become plastic until finally at Fig. 118f the whole mass is plastic. If we attempt to increase the bending moment beyond this value, we shall not succeed: the fibers will just elongate further, and the beam will bend like putty. Of course, the law of Fig. 117 holds only for strains up to about 20 times the elastic strain; or to about 20×0.001, which is just barely visible to the naked eye. The real cold-forging deformation of steel is much more complicated than even that of Fig. 117, involving cold working and lateral deformation by Poisson's ratio. But Figs. 117 and 118 do describe quite well what happens for strains up to 2 per cent. Now we have the picture (Fig. 118), and we also have good mathematical expression for the stages *b*, *c*, and *d* in Eq. (11) (page 39). What remains to be done is to translate the general picture of Fig. 118*e* (with *d* and *f* as limiting special cases) into mathematical language. That is shown in the three formulae below, where the subscripts *i* and *o* stand for inner and outer and refer to the elastic and plastic regions, respectively. The reader should derive these expressions and be sure that he understands them.

$$M_i = s_{\text{yield}} \cdot \frac{I_i}{y_i},$$

$$M_o = s_{\text{yield}} \cdot \int_0 y \, dA,$$

$$M = M_i + M_o = s_{\text{yield}} \left(\frac{I_i}{y_i} + \int_0 y \, dA \right).$$

In particular, for a rectangular cross section in the condition 118*f* of complete plasticity the first term in the parentheses of the last equation disappears, and we have:

$$\begin{aligned} M_{\max} &= s_{\text{yield}} \int y \, dA = s_{\text{yield}} \cdot 2A_{1/2} \cdot y_{G1/2} \\ &= s_{\text{yield}} \cdot 2 \cdot \frac{bh}{2} \cdot \frac{h}{4} = s_{\text{yield}} \cdot \frac{bh^2}{4}. \end{aligned}$$

On the other hand the maximum bending moment that can be carried elastically (condition 118*d*) is found from Eqs. (11*a*) and (12*b*):

$$M_{\max\ \text{elas}} = s_{\text{yield}} \frac{bh^2}{6}.$$

Thus we see that a rectangular beam can be subjected to a bending moment 50 per cent in excess over the maximum elastic one before the beam as a whole starts to yield under the load. This, of course, is caused by the inner fibers taking more load (Figs. 118*e* and *f*). In the case of an I beam, where most bending resisting fibers are in the flanges and very few in the web, this excess percentage is much smaller.

If in Fig. 117 an element is strained plastically as at point B and the stress is then let off, the element contracts as if it were elastic to point C, the line BC being parallel to AO. Applying this experimental fact to the condition of Fig. 118*f*, we arrive at Fig. 119. In order to find the stress in the beam for a bending moment less than the maximum one, we simply subtract the elastic-stress distribution from the state shown in Fig. 118*f*. This is done in three equal stages for a beam of rectangular cross section in Fig. 119, the upper row of figures being the state shown in Fig. 118*f*

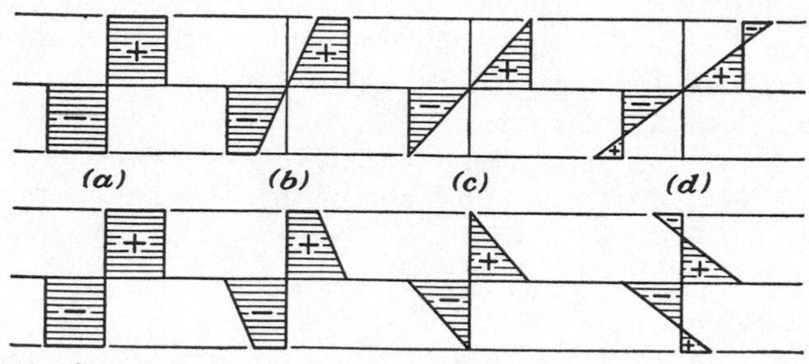

Fig. 119. Continuation of Fig. 118, for gradually diminishing bending moment; at d the external bending moment is zero, and internal residual stresses are left, leaving the interior fibers of the beam with a bending moment in the original direction, balanced by an equal and opposite bending moment in the outer fibers.

less a linear distribution, and the lower row of figures being the same as the upper ones but plotted from a vertical line as base. The bending moment in Fig. 119*a* is 150 per cent of the maximum possible elastic moment, as we have seen above. In Fig. 119*b*, 50 per cent has been let off, so that the figure carries just the maximum possible elastic moment, although the stress distribution is entirely different from Fig. 118*d*. The next stage (Fig. 119*c*) carries only 50 per cent of the maximum elastic moment, and finally in Fig. 119*d* the bending moment has been released altogether and we are left with considerably residual stresses in the beam. Such residual, or "locked-up," stresses always must form a statically balanced force system. (Why?)

Finally we consider what happens if bending beyond the yield stress is imposed on a beam, such as shown in Fig. 120, where the two extreme fibers are at different distances from the center of gravity. Obviously the yield point is reached first on one side only (Fig. 120*b*). If the bending moment is increased, plastic deformation at the constant yield stress occurs in the bottom fibers. If we should try in Fig. 120*c* to draw the diagram for the elastic part still through the center of gravity, we should find that we have violated the push-pull balance, and in order to ensure such balance the neutral line must be shifted up, as shown. In

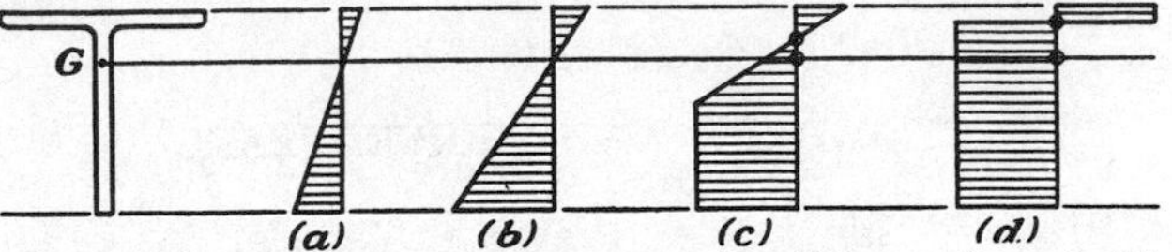

FIG. 120. Plastic bending of a T section, where the yield stress is reached first on one side only (stage *b*). As a consequence the neutral fiber in the plastic stages *c* and *d* no longer passes through the center of gravity.

the last stage of completely developed plasticity (Fig. 120*d*) the neutral line must be so located that the cross-sectional areas below and above it are equal (why?), and this offsets the neutral line a considerable distance from the center of gravity.

Problems 225 *to* 230.

CHAPTER VII

CYLINDERS AND CURVED BARS

28. Riveted Thin-walled Pressure Vessels. A cylindrical pressure vessel or tank is said to be "thin-walled" if the ratio t/r of the wall thickness to the radius is sufficiently small; in practice, t/r should be less than 10 per cent. This occurs in almost all tanks in practical use, and if t/r is small, no distinction is made between an inner radius r_i and outer radius r_o. The symbol r denotes the average radius, $r = (r_o + r_i)/2$. Then the stresses can be calculated

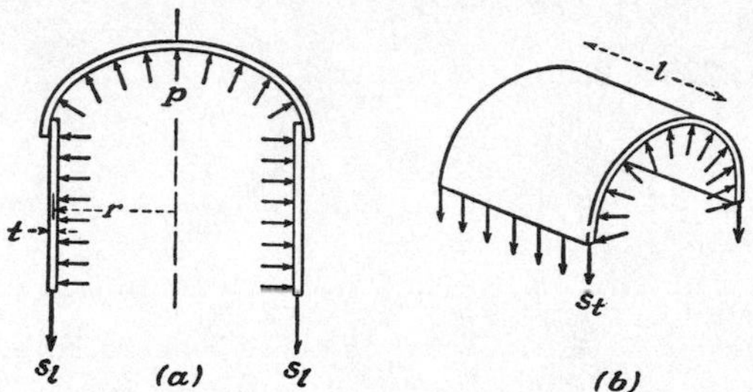

FIG. 121. Stresses in a thin-walled cylindrical tank.

by an extremely simple analysis, while for the case of thick-walled cylinders the calculation becomes much more complicated (pages 140 to 146).

Consider a thin-walled cylindrical tank of dimensions l, r, and t, subjected to an internal gas pressure p, which is constant everywhere. The stress conditions in the ends of the tank are quite complicated and cannot be discussed in this book, so that we limit our analysis to the cylindrical part at some distance from the ends. There, from symmetry, we can conclude that the principal directions of the stress must be longitudinal and tangential. Now we cut the tank in two along a circular section, as shown in Fig. 121*a*. The internal pressure pushes sidewise against the cylindrical inner part, but the resultant of those forces is zero since they are equally distributed in all directions. On the other hand the pressure against the inside of the head has an upward resultant $\pi r^2 p$, which must be held in equilibrium by tensile stresses s_l in the longitudinal direction of the steel of the cylindrical part. This stress acts on an area $t \times 2\pi r$, so that:

$$\pi r^2 p = s_l \times 2\pi r t$$

and

$$s_l = \frac{pr}{2t}. \tag{23a}$$

It has been tacitly assumed here that the stress s_l is constant across the section, *i.e.*, that it does not vary with r across the thickness. This assumption is more or less obvious in the present case, but it is valid only for thin-walled cylinders, while for thick-walled ones the stress *does* vary with the radius.

Next we cut the tank in halves along two longitudinal lines of the cylinder, 180 deg apart (Fig. 121*b*), and consider the equilibrium of a piece of length l. Again assuming the stress to be constant across the section, we write immediately:

$$p \times 2rl = s_t(lt + lt),$$

and consequently the tangential stress is:

$$s_t = \frac{pr}{t}, \tag{23b}$$

which is twice as large as the longitudinal stress. This explains why gas cylinders or steam boilers fail along a longitudinal line and never show a tear along the circle of Fig. 121*a*.

If the vessel is made of a single piece, as gas cylinders usually are, Eqs. (23*a*) and (23*b*) tell the whole story. But steam boilers and vessels in the chemical industry are usually made of a rolled-up sheet, riveted or welded along a longitudinal seam, and again riveted or welded to the end covers. A welded joint, if properly executed, is as strong as or stronger than the plate itself. The design and manufacture of such welded connections is a matter of "machine design" and "welding technique" and will not be discussed here any further.

The stresses occurring in a riveted connection are very complicated indeed, and in calculating them we idealize the actual phenomena even more than is usual with other calculations in strength of materials. In the assembly of an actual joint the two plates are held together, and a red-hot rivet is hammered in place without clearance. When the rivet cools off, it shrinks in length and thus presses the two plates together with considerable pressure. If these plates are pulled apart in the direction of their own planes, they tend to slide over each other, which motion is impeded by the friction between them. Owing to the great pressure of the plates on each other this friction force in some cases is as great as the strength of the full plate itself, so that the joint almost can carry its full load by friction and tension in the rivets only, without relying on the shear strength of the rivets. This is the actual fact, but in calculating the riveted connection all friction effects are neglected, and only the shear strength of the rivets is considered. Figure 122 shows a butt joint with a single cover plate. Another

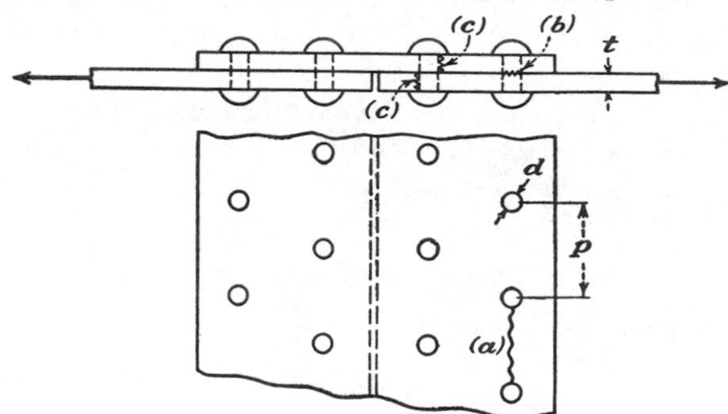

Fig. 122. Riveted joint; p is the pitch, and d is the rivet diameter. The three possibilities of failure as assumed by the A.S.M.E. Boiler Code are indicated by (a), (b), and (c) in the sketch.

frequently occurring type is a butt joint with two cover plates, one on each side of the main plates. For simple constructions no cover plate at all is used: the plates overlap each other, and such a joint is called a "lap joint." The calculation of all these joints has been standardized by the A.S.M.E. Boiler Code as follows. Three types of failure are considered possible:

a. Failure of the main plate in tension between rivet holes, as indicated by (a) in Fig. 122. By the presence of holes the main plate itself is weakened. The assumption is made that the tensile stress is distributed uniformly over the material between the holes, so that the tensile stress of the plate between the holes is $p/(p - d)$ times the stress in the main body of the plate.

b. Failure of the rivet shanks in shear. The assumption is made that the tensile force of the plates divides evenly, not only among all the rivets, but also across the cross section of each individual rivet. The shear stress in a rivet thus is $P/(n\pi d^2/4)$, where P is the tensile force per pitch and n is the number of rivet cross sections per pitch. In Fig. 122 we have $n = 2$; if there had been a second cover plate, n would have been 4 with the same rivet arrangement.

c. Failure by crushing of the main plate, cover plate, or rivet at the spot where the shear force of the rivet is transmitted to the plate by a compressive force. Again this compressive force is assumed to be evenly distributed over the projected area, which is td for a single rivet. Thus, with the above notation for P and n the crushing compressive stress is P/ntd, but this time n is not doubled for a double cover plate as it was under *b*.

The A.S.M.E. code states the following ultimate strengths:

In the plate, case *a*: $s = 55{,}000$ lb/sq in.,
In the rivet, case *b*: $s_s = 44{,}000$ lb/sq in.,
Crushing stress, case *c*: $-s = 95{,}000$ lb/sq in.,

and it further recommends to design joints with a factor of safety 5, *i.e.*, with working stresses one-fifth as large as those listed above.

As an example let us determine the allowable pull per pitch for the joint of Fig. 122 with $p = 5$ in., $d = 1$ in., and $t = 1$ in., for a cover plate of equal thickness as the main plates. The plate itself per pitch length of 5 in. can carry:

$$5 \times 1 \times 55{,}000 \text{ lb} = 275{,}000 \text{ lb}.$$

The holed plate (case *a*) can carry:

$$4 \times 1 \times 55{,}000 \text{ lb} = 220{,}000 \text{ lb, or } 80\% \text{ of the above.}$$

The rivets in shear (case *b*) are good for:

$$2 \times \frac{\pi}{4} \times 1^2 \times 44{,}000 = 69{,}000 \text{ lb, or } 25\% \text{ of the full plate.}$$

The crushing strength is:

$$2 \times 1 \times 95{,}000 = 190{,}000 \text{ lb, or } 69\% \text{ of the full plate.}$$

Of the three percentages that for the shear in the rivets is smallest; the joint is said to be 25 per cent efficient and will presumably fail by shear of the rivets at a load of 69,000 lb per pitch. The recommended working load is five times smaller, or 13,800 lb per pitch.

In looking over the three "efficiencies" *a*, *b*, *c*, we conclude that Fig. 122 is a badly designed joint; the rivets are too small. A better design is obtained by using a top and bottom cover plate as well as by stepping up d from 1 to $1\frac{1}{4}$ in. Then we have:

Case *a*: $3\frac{3}{4} \times 55{,}000 = 206{,}000$, or 75% of 275,000,

Case *b*: $2 \times 2 \times \frac{\pi}{4} \times \left(1\frac{1}{4}\right)^2 \times 44{,}000 = 216{,}000$, or 78%,

Case *c*: $2 \times 1\frac{1}{4} \times 95{,}000 = 238{,}000$, or 86%.

This time the working load of the joint is one-fifth of 206,000 lb per pitch, and the joint will presumably fail by tearing of the main plate between the holes of the outer row of rivets.

Problems 231 *to* 238.

29. Thick-walled Cylinders. When the outer and inner diameters of a cylinder under internal pressure differ by more than about 10 per cent, Eqs. (23) cease to be applicable, because they rest on the assumption of constant stress across the wall thickness, which is no longer tenable. The more complete theory, considering the non-uniform distribution of the stresses across the wall thickness, was worked out by Lamé (1795–1870) in France and is the subject of this article.

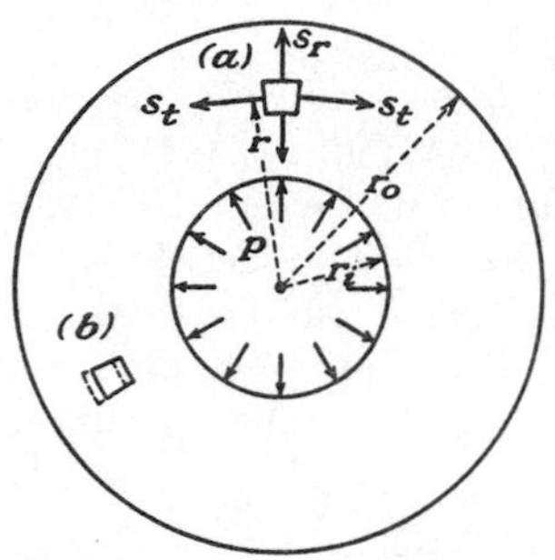

FIG. 123. Thick-walled cylinder under internal pressure, showing the stresses (*a*) and the deformation (*b*).

Consider in Fig. 123 a cylinder of inner radius r_i and outer radius r_o, subjected to internal pressure p. The radius of an in-between point is r, which is variable between r_i and r_o. Owing to the internal pressure p, the cylinder deforms by expanding, and we denote by u (measured in inches) the outward displacement of any point r. Thus u is a function of r, although u numerically is very small in comparison with r. A circle of radius r in the unstressed state thus becomes a circle of radius $r + u$ when stressed; all lengths of this circle are increased by a ratio $(r + u)/r$ or by $1 + (u/r)$. A piece ds of the circle becomes $ds + u\,ds/r$, which means that the tangential strain is:

$$\epsilon_t = \frac{u}{r}.$$

To find the strain in the radial direction we look at the detail *b* of Fig. 123, showing an element in the undistorted state in full lines, which after the loading goes to the dotted-line picture. The location of the inner point is r, and it goes to $r + u$, because of the pressure. The location of the outer point is $r + dr$, and it goes to $(r + dr) + (u + du)$, because u varies with r. Thus the original radial length of the piece was dr, and its distorted radial length is $dr + du = dr\left(1 + \frac{du}{dr}\right)\cdot$ Hence the radial strain is:

$$\epsilon_r = \frac{du}{dr}.$$

Let us consider the case where the pressure p acts on the cylindrical part only (Fig. 123), but not on the ends of the vessel, so that the longitudinal stress s_l is zero. This is the case in a shrink-fit problem. Then the two strains can be expressed in terms of the radial and tangential stresses s_r, s_t by means of Eq. (18) (page 75):

$$\frac{u}{r} = \frac{1}{E}(s_t - \mu s_r), \qquad (a)$$

$$\frac{du}{dr} = \frac{1}{E}(s_r - \mu s_t). \qquad (b)$$

These are two equations in the three unknowns u, s_t, and s_r. A third equation is found from the equilibrium of an element $dr \cdot r\,d\theta$, as shown in Fig. 124. The radial force 1 is $s_r \cdot r\,d\theta$ per unit length perpendicular to the paper. The force 2 on top listens to the same name, but both s_r and r are slightly different there. Thus that force is:

$$s_r \cdot r\,d\theta + d(s_r \cdot r\,d\theta),$$

and the resultant of 1 and 2 outward is:

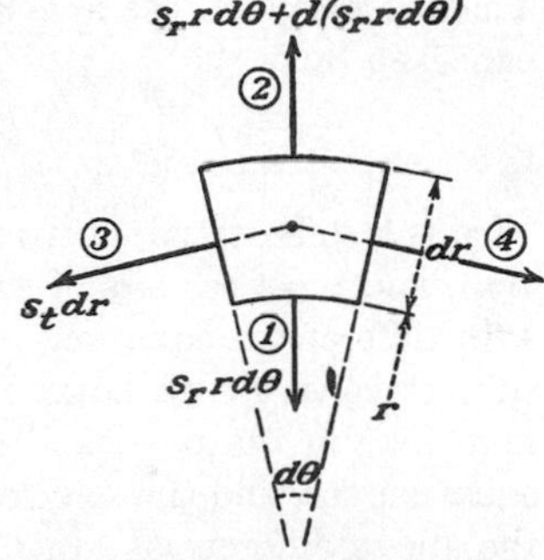

FIG. 124. Forces acting on an element.

$$d(s_r \cdot r\,d\theta) = d\theta\,d(s_r \cdot r) = d\theta\,\frac{d(rs_r)}{dr}\,dr = \frac{d(rs_r)}{dr}\,dr\,d\theta.$$

The forces 3 and 4 are numerically equal, but not quite in the same direction; they include the small angle $d\theta$. Hence their resultant, by the parallelogram of forces, is directed inward, and its magnitude is:

$$F_3\,d\theta = s_t\,dr\,d\theta.$$

The element is in equilibrium, so that the outward force from faces 1 and 2 must balance the inward one caused by faces 3 and 4:

$$s_t = \frac{d}{dr}(rs_r), \qquad (c)$$

which is the third equation we looked for. The set of three equations (a), (b), (c) can be solved algebraically in several different ways. We note that (c) allows us to find s_t if we know only s_r and that (a) gives us u if s_r and s_t are known. Therefore s_r seems to be the most involved variable, and we aim to find it first. Substitute (c) into (a) and (b) consecutively:

$$u = \frac{1}{E}\left[r\frac{d}{dr}(rs_r) - \mu(rs_r)\right], \tag{a^1}$$

$$\frac{du}{dr} = \frac{1}{E}\left[s_r - \mu\frac{d}{dr}(rs_r)\right]. \tag{b^1}$$

From these u can be eliminated by differentiating (a^1) and equating it to (b^1); the common factor $1/E$ can be left out:

$$\frac{d}{dr}(rs_r) + r\frac{d^2}{dr^2}(rs_r) - \mu\frac{d}{dr}(rs_r) = s_r - \mu\frac{d}{dr}(rs_r).$$

The last terms on the left- and right-hand sides are the same and cancel so that:

$$r^2\frac{d^2}{dr^2}(rs_r) + r\frac{d}{dr}(rs_r) - (rs_r) = 0. \tag{d}$$

This is a differential equation of the second order in the variable (rs_r), and most readers of this text are not yet on speaking terms with differential equations. The standard books on that general subject consist of a large number of solutions found by mathematicians in the past, and all we have to do is to recognize the equation and find the solution in the books. These books say that the above differential equation for (rs_r) is of a type such that its solutions are powers of the independent variable r. We thank the ancient mathematician for that information and try:

$$rs_r = r^n,$$

where n is an unknown exponent. Substituting this into the equation we get:

$$r^2 \cdot n(n-1)r^{n-2} + rnr^{n-1} - r \cdot r^n = 0.$$

We see that the r's drop out and find:

$$n(n-1) + n - 1 = 0,$$

a quadratic in n with the roots $n = \pm 1$, so that $rs_r = r^1$ and $rs_r = 1/r$ are solutions of Eq. (d), which can easily be verified by substitution. But we know that a process of two integrations [which is what the solution of (d) amounts to] is associated with two arbitrary constants of integration. The books on differential equations say and we can verify by substitution that the general solution of Eq. (*d*) is:

$$rs_r = C_1 r + \frac{C_2}{r}. \tag{e}$$

The values of these integration constants depend on the boundary conditions. For the case of Fig. 123 we know that the stress s_r at the inner radius is $-p$ and at the outer radius is zero. Substituting this into Eq. (*e*) gives:

$$-r_i p = C_1 r_i + \frac{C_2}{r_i},$$

$$0 = C_1 r_o + \frac{C_2}{r_o},$$

or, solved for C_1 and C_2:

$$C_1 = p\,\frac{r_i^2}{r_o^2 - r_i^2},$$

$$C_2 = -p\,\frac{r_o^2 r_i^2}{r_o^2 - r_i^2}.$$

Substituting these back into the expression (*e*) we have:

$$s_r = p\,\frac{r_i^2}{r_o^2 - r_i^2}\left(1 - \frac{r_o^2}{r^2}\right). \tag{f}$$

From Eq. (*c*) we then obtain:

$$s_t = p\,\frac{r_i^2}{r_o^2 - r_i^2}\left(1 + \frac{r_o^2}{r^2}\right). \tag{g}$$

In Fig. 125 these stresses are plotted against the radius r for the special case that $r_o = 2r_i$, and we see that the maximum stress occurs at the inside, as was to be expected. The tangential stress s_t, sometimes called "hoop" stress, is tensile, while the radial stress is compressive, diminishing from p at the inside to zero at the outside. We design usually by the maximum-shear theory of failure, so that we now calculate the shear stress at the inside, where it is a maximum:

$$s_s = \frac{1}{2}(s_t - s_r) = p\,\frac{r_i^2}{r_o^2 - r_i^2}\left(\frac{r_o^2}{r^2}\right)_{r=r_i} = p\,\frac{r_o^2}{r_o^2 - r_i^2}. \tag{h}$$

The reader should now reduce this result to the special cases of a thin ring and of an infinitely thick cylinder and see that the answers make sense in connection with Eq. (23*b*) (page 137) and Eqs. (*f*) and (*g*) above.

Now we solve for the radial displacement u, from Eq. (a) by substitution of the results (f) and (g):

$$u = \frac{pr}{E} \frac{r_i^2}{r_o^2 - r_i^2} \left[(1 - \mu) + \frac{r_o^2}{r^2} (1 + \mu) \right]. \qquad (i)$$

For the special case $r = r_i$, at the inside, this reduces to:

$$u_{r_i} = \frac{pr_i}{E(r_o^2 - r_i^2)} [(1 - \mu)r_i^2 + (1 + \mu)r_o^2] \qquad (j)$$

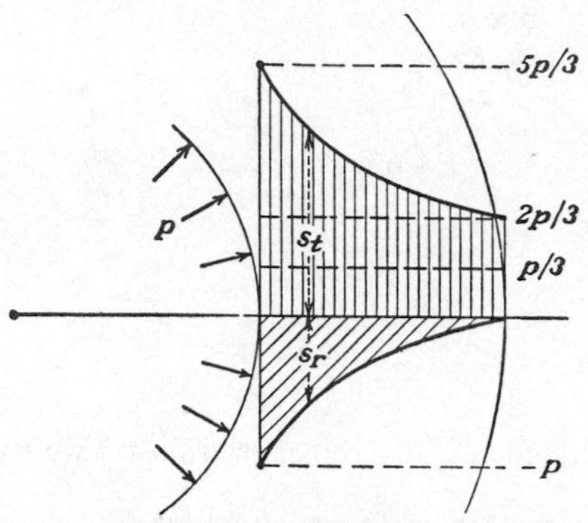

FIG. 125. Tangential- and radial-stress distribution in a cylinder with $r_o/r_i = 2$ under internal pressure p with zero radial load at the outer periphery.

The principal application of these formulae is in connection with shrink, or press, fits, where a cylinder is shrunk over a solid shaft of which the outer diameter is slightly larger than the bore diameter of the cylinder. Obviously the two then press on each other radially, and by Newton's third law the radial pressures p on the two elements are the same. By this pressure the inner radius of the cylinder is increased by the amount u_{r_i} of Eq. (j), denoted by δ_1 in Fig. 126. The shaft radius is decreased by an amount δ_2, to be calculated presently. Then the total radial "allowance" δ, or the difference between the unstrained radii of cylinder and shaft, is:

$$\delta = \delta_1 + \delta_2.$$

The amount δ_2 can be calculated from Eqs. (a), (b), (c), and (e) by taking $r_i = 0$; $s_{ro} = -p$, because the shaft with external pressure p is a special case of the general theory just discussed. This takes time and a few sheets of paper; it is left as a problem to the reader. However, we can think of the shaft as immersed in a liquid under pressure and thus being in a state of two-dimensional hydraulic pressure with $s_r = s_t = -p$ everywhere (Fig. 59, page 67). Then the strain is the same at every point and in every direction at any point, and it is, by Eq. (18):

$$\epsilon = -\frac{p}{E}(1 - \mu).$$

The radial inward deflection then is:

$$\delta_2 = \frac{pr}{E}(1 - \mu),$$

where r is the radius of the shaft, equal to r_i of Fig. 126 (except for

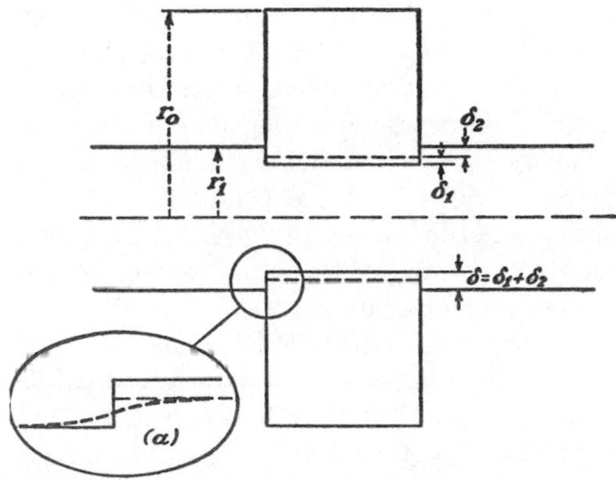

FIG. 126. Shrink, or press, fit with a radial tolerance δ.

a difference of order δ, which is considered infinitesimally small). Thus we find

$$\delta = \delta_1 + \delta_2 = \frac{pr_i}{E(r_o^2 - r_i^2)}[(1 - \mu)(r_i^2 + r_o^2 - r_i^2) + (1 + \mu)r_o^2],$$

or

$$\frac{\delta}{r_i} = \frac{2p}{E} \cdot \frac{r_o^2}{r_o^2 - r_i^2}. \tag{24}$$

This formula gives the relation between the allowance ratio δ/r_i and the shrinkage pressure p.

The maximum shear stress in the cylinder [Eq. (h)] can be expressed in terms of the allowance ratio by substitution of (24) into (h):

$$(s_s)_{\max} = \frac{E}{2(\delta/r_i)}, \tag{25}$$

a surprisingly simple result, independent of the dimensions of the cylinder, and hence the same for thick as well as for thin cylinders. For mild steel the yield stress in shear is 15,000 lb/sq in. and $E = 30{,}000{,}000$ lb/sq in., so that we draw the conclusion that a shrink allowance of 0.001 in./in. will just cause yielding at the inside of the cylinder. This is a good practical figure, known to any mechanic who has experience with such connections. For other materials with a different yield point or a different Young's modulus, Eq. (25) will give a correspondingly different shrink allowance.

Two remarks are in order before we leave this subject. The first one is indicated in the detail *a* of Fig. 126, where the actual deformed shape of the shaft is shown together with the one assumed in this analysis. The analysis is a two-dimensional one and is strictly true only for shafts which are just as long as the cylinder into which they are shrunk. When the shaft protrudes from the cylinder, as it must in practice, our theory demands that the shaft diameter suddenly jump from the elastically compressed one to the free unstressed one. This, of course, is not possible. Our theory is good only in the interior of the shrink fit, at some distance from the ends. No simple satisfactory theory for the effect of Fig. 126*a* has been worked out so far.

The second concluding remark is in connection with the third principal stress s_l, in the longitudinal direction, which we have studiously ignored so far. We are lucky, because by Eqs. (18*a*) (page 75) and (*f*) and (*g*) on page 143 it appears that for zero longitudinal stress the longitudinal strain is independent of r and is therefore the same over the entire cross section. Thus if there are no ends to our cylinder to hold the pressure and if the cylinder can freely expand longitudinally, its plane cross sections remain plane. When we then fit ends to the cylinder and hold the pressure in the manner of Fig. 121, a longitudinal stress has to be superposed which should be so that each fiber of the cylinder expands by the same amount longitudinally. Hence the longitudinal stress distributes uniformly over the cross section of the thick cylinder, and it can be calculated from (Eq. 23*a*) (page 137) for the thin-walled one.

Problems 239 *to* 249.

30. Thin Curved Bars. A curved bar or beam is said to be "thin" or the beam is said to be "slightly curved" if the height h of the beam is small compared with the radius of curvature R, say not more than 10 per cent of it. The definition of a "thin" curved bar therefore is the same as that of the "thin"-walled cylinder of page 136. In this article we shall discuss the bending of thin, initially curved beams. If we consider an element ds of such a beam, where the length ds is of the same order of magnitude as the height h, then the end sections, being perpendicular to the curved center line, are not quite parallel but the angle they include between them is small of the order h/R. Also, the length of the outside fiber of that element is greater than the length of the inside fiber, but again the percentage difference is small of the same order. If we neglect all these differences, we say that the beam element ds is substantially straight and then we can apply to it the complete theory of pages 37 to 41 for straight beams. In particular we conclude that the neutral fiber passes through the center of gravity of the cross section, that the bending stress distribution is linear (Fig. 26, page 38), that the stress is expressed by Eq. (11) (page 39), and that the deformation is given by Eq. (b) of page 39. That last equation was derived for a truly straight beam; hence $d\alpha$ is the actual angle between two perpendicular sections dx apart, the angle in the unstressed state being zero. In our present case the unstressed angle is not zero, so that we have to interpret $d\alpha$ as the *change* in angle between the bent and unbent states, and the symbol dx is replaced by ds, the usual convention of mathematical notation.

The calculation of the stress and deflection in a curved beam then reduces to a determination of the bending moment and to an integration of Eq. (b), which we shall now illustrate on the example of a cantilever beam of arbitrary initial curved shape in one plane, shown in Fig. 127. The loading is a concentrated end load P lying in the same plane. First we choose a coordinate system with the origin at the free end and with the y axis along the load as shown. Then the bending moment at an arbitrary point s of the beam can be found by isolating the free end of the beam from s outward, and it is $M = Px$. The bending stress can then be found immediately from Eq. (11) (page 39), in which the reader should satisfy himself as to the meaning of the symbol y, which differs from the y of Fig. 127.

In order to find the deflections at the free end B of the cantilever we integrate Eq. (b) in a peculiar manner. First we permit an element ds at C of Fig. 127 to deform according to the bending moment it carries, but we imagine all the rest of the beam to remain stiff, not changing its shape. Then the beam gets a slight local bend, or change in angle, at ds only, with the result that the left portion AC remains where it is and the right portion CB rotates as a rigid body through a slight angle. This causes at point C certain deflections: angular $d\varphi$, along the load $d\delta_y$ and across the load $d\delta_x$. The total deflections φ, δ_x, δ_y at C are then thought of as the sums of all the small contributions caused by consecutively

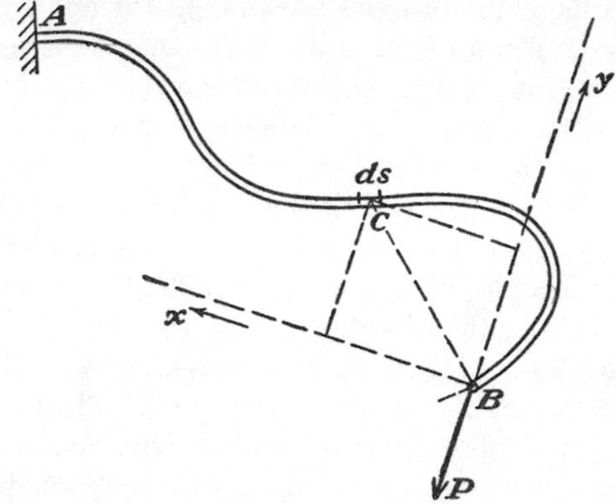

FIG. 127. Cantilever beam of arbitrary curved plane shape, loaded by a concentrated end load P in the plane of the beam.

permitting all the elements of the beam to flex, starting at A and ending at B. By Eq. (*b*) (page 39), we have:

$$d\varphi = \frac{M}{EI} ds,$$

where $d\varphi$ is positive clockwise. The negative sign in Eq. (*b*) is disregarded because all the sign conventions of Fig. 19 (page 32), adopted for straight beams, lose their meaning for a curved one. This small angular change $d\varphi$ at C makes point B move in a direction perpendicular to BC, by an amount $BC\ d\varphi$. Its components in the x and y directions are:

$$d\delta_x = y\ d\varphi = \frac{My}{EI} ds,$$

$$d\delta_y = -\ x\ d\varphi = -\ \frac{Mx}{EI} ds,$$

and, consequently, by integration:

$$\left.\begin{aligned} \varphi &= \int_s \frac{M}{EI} ds, \\ \delta_x &= \int_s \frac{My}{EI} ds, \\ \delta_y &= -\int_s \frac{Mx}{EI} ds. \end{aligned}\right\} \qquad (26)$$

The reader may have noticed that this process of integration is quite similar to that used in connection with Fig. 17 (page 28) and with Fig. 93 (page 102).

If the beam is of constant cross section, the stiffness EI can be brought in front of the integral sign; but, as written, the expressions (26) are true even for beams of variable cross section, provided always that at each point the height of the beam is small compared with the local radius of curvature. For straight beams, where $y = 0$, the expressions (26) give $\delta_x = 0$: a known result. Also we remark that in the derivation of Eqs. (26) only the symbol M was used for the bending moment at any point, so that the expressions are not limited to the loading of Fig. 127 but are good for any kind of loading, when only the proper expression for M is substituted.

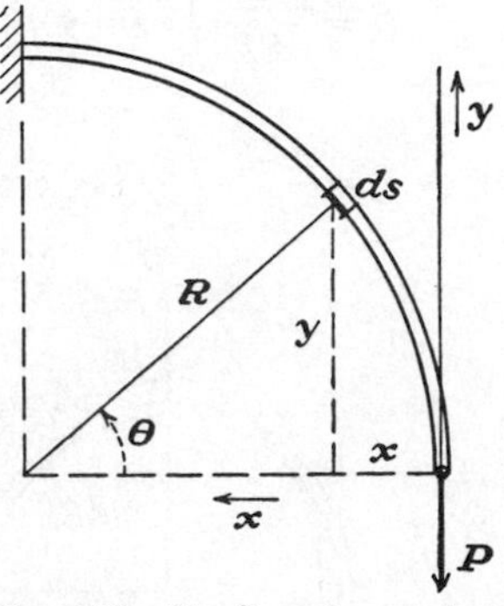

FIG. 128. Bending of a quarter-circle cantilever beam.

Now we apply the general theory to the specific case of a quarter-circle cantilever of radius R and of constant cross section EI (Fig. 128). We have:

$$M = Px, \qquad ds = R\,d\theta,$$
$$x = R(1 - \cos\theta), \qquad y = R\sin\theta,$$

so that:

$$\varphi = \frac{PR^2}{EI}\int_0^{\frac{\pi}{2}} (1 - \cos\theta)\,d\theta = \frac{PR^2}{EI}\left(\frac{\pi}{2} - 1\right) = 0.57\,\frac{PR^2}{EI},$$

$$\delta_x = \frac{PR^3}{EI}\int_0^{\frac{\pi}{2}} (1 - \cos\theta)\sin\theta\,d\theta = 0.50\,\frac{PR^3}{EI},$$

$$-\,\delta_y = \frac{PR^3}{EI}\int_0^{\frac{\pi}{2}} (1 - \cos\theta)^2\,d\theta = \frac{3\pi - 8}{4}\frac{PR^3}{EI} = 0.356\,\frac{PR^3}{EI}.$$

The actual working out of the above integrals is left as an exercise for the reader. The result shows, somewhat surprisingly, that the horizontal deflection to the left is substantially greater than the downward deflection along the load P.

Another example of the application of Eqs. (26) to a special case is when the load P in Fig. 128 is replaced by a clockwise bending moment M_0. Then the bending moment is M_0, constant all along the beam, and for that case we find:

$$\varphi = \frac{M_0}{EI}\int_0^{\frac{\pi}{2}} R\,d\theta = \frac{\pi}{2}\frac{M_0R}{EI} = 1.57\,\frac{M_0R}{EI},$$

$$\delta_x = \frac{M_0R^2}{EI}\int_0^{\frac{\pi}{2}} \sin\theta\,d\theta = 1.00\,\frac{M_0R^2}{EI},$$

$$-\delta_y = \frac{M_0R^2}{EI}\int_0^{\frac{\pi}{2}} (1-\cos\theta)\,d\theta = \left(\frac{\pi}{2}-1\right)\frac{M_0R^2}{EI} = 0.57\,\frac{M_0R^2}{EI}.$$

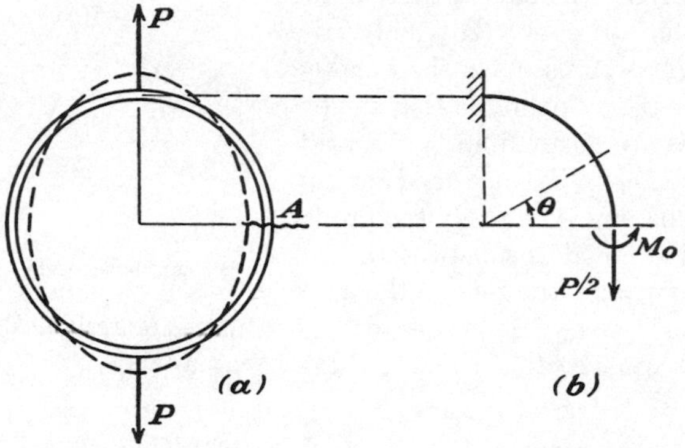

FIG. 129. A "thin" ring subjected to a pair of forces P.

Again the horizontal displacement is very much larger than the vertical one.

The two sets of results just obtained enable us to solve the statically indeterminate problem of the ring (Fig. 129), where we ask for the bending-moment distribution, for the increment in the diameter along P, and for the decrement in the crosswise diameter. The problem is statically indeterminate because we find it possible to cut the structure without having it collapse. If we cut it open at A, it will only deform elastically but it will still hang together. Between the two faces at A there will be acting a tensile force, a bending moment, and possibly a shear force. The magnitudes of these forces and moments can be found only from the condition that under their influence there must be no open gap at A, in other words, from a condition of deformation. No argument of static equilibrium can determine the magnitude of these forces at

A; in fact any arbitrary values assigned to them are all right as far as statics goes.

However, we can reach certain conclusions on these reaction forces at A from symmetry considerations, which are considerations of deformation rather than of statics. This is done with the help of Fig. 130, where the ring is cut open in two spots, 180 deg apart. No assumptions about the forces have been made in

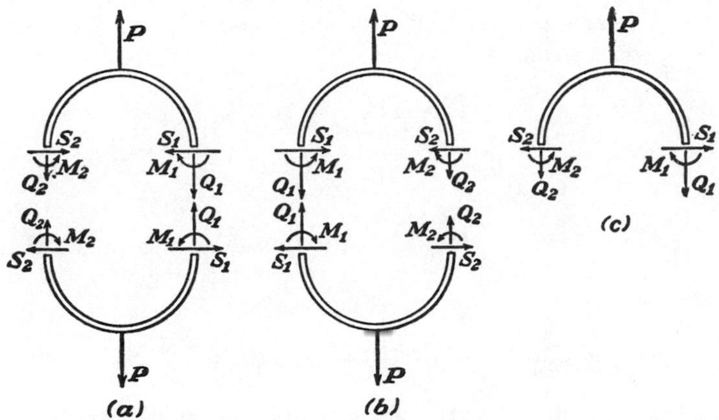

FIG. 130. Symmetry considerations applied to the ring of Fig. 129. Case b is derived from case a by a half turn about the PP axis. Case c is the lower half of case b redrawn.

Fig. 130a, except that by action equals reaction the quantities in the lower half of Fig. 130a are equal and opposite to those of the upper half. Now we take Fig. 130a and turn it 180 deg about the axis PP and redraw it as Fig. 130b. The two figures a and b represent the same ring under the same loading and hence must have the same reactions. Comparing a with b we can conclude that $S_1 = S_2$, $Q_1 = Q_2$, and $M_1 = M_2$. Moreover from vertical equilibrium of either half ring we find that Q_1 or $Q_2 = P/2$. Now we take the bottom half of Fig. 130b, turn it through 180 deg in the plane of the drawing, and slide it up to the position of Fig. 130c. The new figure represents the same half ring as the upper half of Fig. 130b, with identical loading P, so that the reactions must be the same as well. From it we conclude that $S_2 = -S_1$, but since S_2 also equals $+S_1$, the only possibility is that $S_1 = S_2 = 0$. Thus there is no shear force. This should not surprise us because the upper half of Fig. 130a has its ends pushed together by S, while the lower half has them pulled apart, and the two half rings could never fit together again if this were the case.

Now we return to Fig. 129*b*, showing one-quarter of the ring, subjected to a force $P/2$ and to an unknown bending moment M_0. From symmetry the free end of the quarter ring must be just vertical; the load $P/2$ tends to bend it elastically away from the vertical in one direction, and M_0 tends to produce an angle in the opposite direction. The condition for M_0 then is that it must be of such a value as to neutralize the angle produced by $P/2$. From the previous formulae we have:

$$\varphi = \left(\frac{\pi}{2} - 1\right)\frac{PR^2}{2EI} = \frac{\pi}{2}\frac{M_0R}{EI},$$

or

$$M_0 = \left(\frac{1}{2} - \frac{1}{\pi}\right)PR = 0.182PR.$$

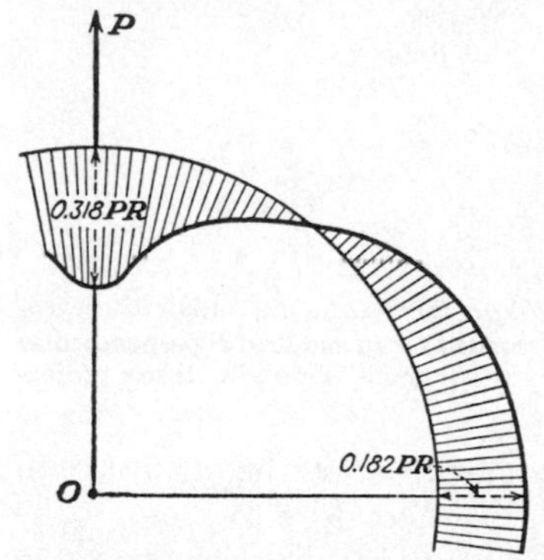

FIG. 131. Bending-moment distribution in a thin circular ring loaded by two radial forces P, 180 deg apart.

Also from Fig. 129*b* we see that at an arbitrary angle θ the bending moment is:

$$M_\theta = M_0 - \frac{PR}{2}(1 - \cos\theta),$$

a relation that is plotted in a radial direction in Fig. 131. Near the loads P the ring is bent so as to make its radius of curvature less than R; between the loads P the radius of curvature becomes greater than R. Thus Fig. 131 answers our first question completely.

Applying the previous results for the displacements to Fig. 129*b*. we find:

$$-\delta_y = \frac{3\pi - 8}{4}\frac{PR^3}{2EI} - \left(\frac{\pi}{2} - 1\right)\frac{M_0R^2}{EI}$$

$$= \frac{PR^3}{EI}\left[\frac{3\pi - 8}{8} - \left(\frac{\pi}{2} - 1\right)\left(\frac{1}{2} - \frac{1}{\pi}\right)\right]$$

$$= \frac{PR^3}{EI}\left(\frac{\pi}{8} - \frac{1}{\pi}\right),$$

$$\delta_x = \frac{1}{2}\frac{PR^3}{2EI} - \frac{M_0R^2}{EI} = \frac{PR^3}{EI}\left(\frac{1}{\pi} - \frac{1}{4}\right).$$

The increase in the vertical diameter ΔD_v of the ring (Fig. 129*a*) is twice $-\delta_y$, or

$$\Delta D_v = \frac{PR^3}{EI}\left(\frac{\pi}{4} - \frac{2}{\pi}\right) = 0.150\,\frac{PR^3}{EI}.$$

The decrease in the horizontal diameter $\Delta D_h = 2\delta_x$ is:

$$\Delta D_h = \frac{PR^3}{EI}\left(\frac{2}{\pi} - \frac{1}{2}\right) = 0.135\,\frac{PR^3}{EI}.$$

This completely solves the problem of Fig. 129.

The next and last example of this article is the quarter-circle cantilever of Fig. 132 loaded perpendicularly to its own plane. As we soon shall see, this involves not only bending but also twist of the cross section, so that we can solve the problem only for a circular cross section (page 25), which we shall assume in this case. At an arbitrary point B, determined by the angle θ, the force P bends the beam with a moment arm PD and twists it with the moment arm DB. Hence:

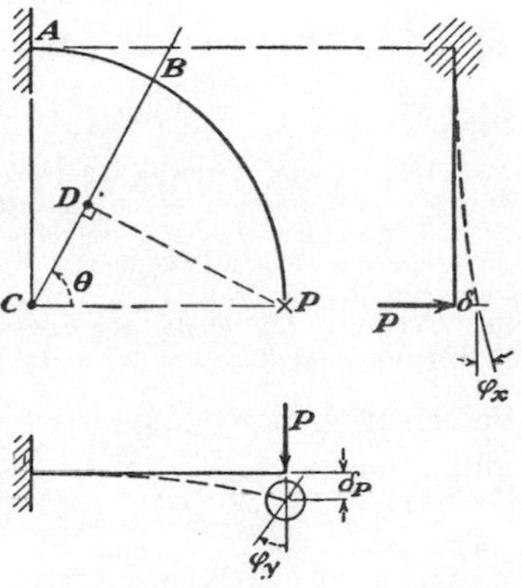

FIG. 132. Quarter-circle cantilever loaded by an end load P perpendicular to its plane, shown in three projections.

$$M_b = PR \sin \theta,$$
$$M_t = PR(1 - \cos \theta).$$

As in the previous example we assume the beam to be rigid with the exception of a small piece $ds = R\,d\theta$ at angle θ. This little piece will bend, by Eq. (*b*) (page 39), and it will twist, by Eq. (4) (page 19). Both angles will cause a downward displacement of the end point P, so that:

$$d\delta_P = \frac{M_b\,ds}{EI} R \sin \theta + \frac{M_t\,ds}{GI_p} R(1 - \cos \theta)$$
$$= \frac{PR^3}{EI} \sin^2 \theta\,d\theta + \frac{PR^3}{GI_p}(1 - \cos \theta)^2\,d\theta.$$

The downward displacement at P caused by all the elements of the beam is found by integrating the above expression from $\theta = 0$ to $\theta = 90$ deg, or

$$\delta_P = \frac{PR^3}{EI} \cdot \frac{\pi}{4} + \frac{PR^3}{GI_p}\left(\frac{3\pi}{4} - 2\right),$$

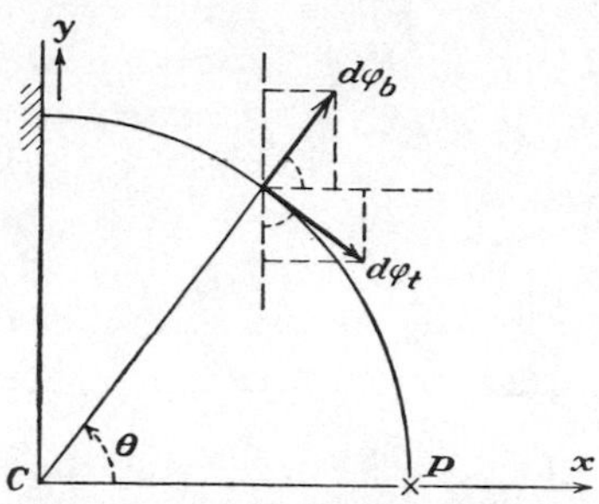

FIG. 133. Vectors showing the local angles of deformation of an element $ds = R\,d\theta$. Each vector is perpendicular to the plane in which it acts, and its direction corresponds to a right-handed screw. The angle $d\varphi_b$ is caused by bending, and $d\varphi_t$ is due to twist.

where again the arithmetic of integration is left as an exercise to the reader. No sidewise displacement takes place in this case. The question of the angular displacement of the end point P is somewhat more complicated, because at the point B the angles caused by bending and twist lie in perpendicular planes and hence cannot be added algebraically. Since these angles are infinitesimally small, they can be treated as vectors, as shown in Fig. 133. The local angles of deformation at B can be resolved into components parallel to the x and y axes:

$$d\varphi_x = d\varphi_b \cos\theta + d\varphi_t \sin\theta$$

$$= \frac{PR^2 \sin\theta}{EI}\,d\theta \cos\theta + \frac{PR^2(1 - \cos\theta)}{GI_p}\,d\theta \sin\theta,$$

$$d\varphi_y = \frac{PR^2 \sin\theta}{EI}\,d\theta \sin\theta - \frac{PR^2(1 - \cos\theta)}{GI_p}\,d\theta \cos\theta.$$

These small local angles of deformation are transmitted to the end P without change. Hence the total angle at P is found by integrating the above expressions from $\theta = 0$ to $\theta = 90$ deg $= \pi/2$. The reader should do this and find that:

$$\varphi_{xP} = \frac{PR^2}{2EI} + \frac{PR^2}{2GI_p},$$

$$\varphi_{yP} = \frac{\pi}{4}\frac{PR^2}{EI} - \frac{PR^2}{GI_p}\left(1 - \frac{\pi}{4}\right).$$

These angles as well as the deflection δ_P are indicated in the projections of Fig. 132.

Problems 250 *to* 259.

31. Thick Curved Bars. When the height of a curved bar becomes a considerable fraction of the radius of curvature, then the length of the unstressed outer fiber for a given subtended angle becomes considerably larger than the length of the inside fiber. In the derivation of Eq. (11) for the straight or slightly curved beam these fiber lengths between two cross sections perpendicular to the beam center line were assumed equal. As a consequence the bending-stress distribution over the cross section of a thick curved bar will no longer be linear, and a new theory has to be built up. In a sense this new theory of the bending of greatly curved beams compares with the bending theory of slightly curved

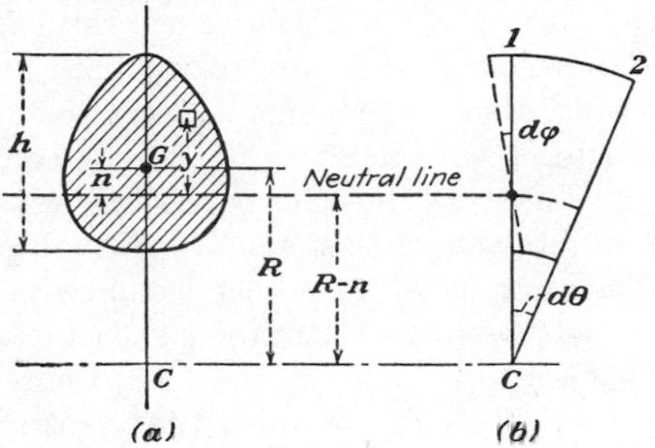

Fig. 134. Cross section of a highly curved beam, R being the unstressed radius of curvature of the center of gravity; n being the distance between the center of gravity and the neutral line of bending.

beams as the theory of thick cylinders (article 29) compares with that of the thin-walled tubes (article 28). Highly curved beams do not occur often in practice: crane hooks and the frames of punch presses are some examples. The theory to be derived is of much less importance than that of the straight and slightly curved beams; however, if a beam appears of which the height is about equal to the radius of curvature of the center of gravity of the cross section, the old theory of thin beams gives totally erroneous results, just as the stress distribution of Fig. 125 for the thick cylinder is very poorly approximated by the horizontal line of the thin-cylinder diagram.

Let Fig. 134a represent a cross section of the beam with its center of gravity marked G and the center of curvature marked C. Figure 134b shows the other projection: two such normal cross sections at angle $d\theta$ apart. We make the assumption that, as a result of the bending, plane cross sections remain plane, which is the same assumption made on page 37 for the straight beam. There exists no logical justification for this assumption, but the analytical results derived from it are fairly accurately confirmed by actual strain measurements. If plane sections remain plane and thus merely turn through a small angle, there must be a neutral line somewhere, as shown in the figure. For the straight beam this neutral line passes through the center of gravity; here it cannot be in the center of gravity, as we shall see soon. Assuming that the cross section 2 in Fig. 134*b* remains in place, the cross section 1 turns through the small angle $d\varphi$ to the dashed-line position. Now imagine for an instant that this turning would take place about point G. Then the fiber elongations would be linearly distributed as in the straight beam. To find the strains (which are the stresses divided by E) we must divide these elongations by the original fiber lengths, which were all the same in the straight beam. Here, however, the outer fibers are longer than the inner ones, so that the outer strains (and consequently stresses) are less than with the straight beam and the inner stresses, close to C, are greater. Thus the tensile stresses are smaller and the compressive ones greater than for the straight beam, and since these two balanced each other in the straight beam, they are no longer in balance here. They can be made to balance only by shifting the neutral fiber inward from G toward C, by an amount n, as yet unknown. (The letter n suggests *n*eutral fiber distance.) Let y denote the coordinate of an element dA with respect to the neutral line, as shown. Then the elongation of a fiber at y is $y\ d\varphi$; its original length is $(R - n + y)\ d\theta$; consequently the strain or unit elongation is $y\ d\varphi/(R - n + y)\ d\theta$, and the stress is:

$$s = E\frac{d\varphi}{d\theta}\frac{y}{R - n + y}. \tag{a}$$

The force of that fiber is $s\ dA$, and the total force over the entire cross section is the above expression integrated over the entire cross section. This force is zero for pure bending (where a tensile force along the bar is absent), so that, when the constants $E\ d\varphi/d\theta$ are divided out, we have:

$$\int_A \frac{y\ dA}{R - n + y} = 0. \tag{27}$$

In this expression the quantity n is the only unknown, so that we can solve for n from Eq. (27) for any given cross section and radius of curvature R.

The bending-moment contribution of the element dA is $sy\ dA$, so that the total bending moment is:

$$M = \int sy\,dA = E\frac{d\varphi}{d\theta}\int_A \frac{y^2\,dA}{R - n + y}.$$

This integral can be transformed by writing for one of the two y's of the numerator:

$$y = (R - n + y) - (R - n),$$

so that:

$$\int_A \frac{y^2\,dA}{R - n + y} = \int_A y\,dA - (R - n)\int_A \frac{y\,dA}{R - n + y}.$$

The last expression is zero by virtue of the force-balance equation (27), and hence the integral is:

$$\int_A y\,dA = y_G A = nA,$$

so that:

$$M = E\frac{d\varphi}{d\theta}\,nA. \tag{28}$$

This is the relation between the bending moment and the deformation $d\varphi$, since n is known from Eq. (27). To complete the analysis we need an expression for the stress at an arbitrary point y, not in terms of deformation, as in Eq. (a), but in terms of bending moment. This we find by substituting (28) into (a):

$$s_y = \frac{M}{nA}\,\frac{y}{R - n + y}. \tag{29}$$

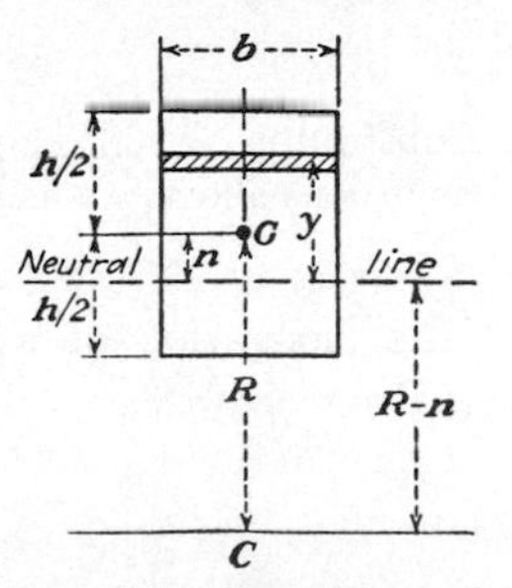

FIG. 135. Curved bar of rectangular cross section in bending.

Our problem is now formally solved, but in applying Eqs. (27) to (29) to a specific case it is not at all difficult to get hopelessly lost in a maze of algebra. We shall therefore restrict ourselves here to the simplest cross section: a rectangle of height h, width b, and radius of curvature R of its center G (Fig. 135). Here dA simply is $b\,dy$, and Eq. (27), by using the same trick that led to Eq. (28), becomes:

$$0 = \int_A \frac{y\,dy}{R - n + y} = \int_A dy - (R - n)\int \frac{dy}{R - n + y},$$

$$0 = y|_A - (R - n)\log_e (R - n + y)|_A,$$

$$0 = h - (R - n)\log_e \frac{R + (h/2)}{R - (h/2)}.$$

Hence:

$$\frac{h/R}{1 - (n/R)} = \log_e \frac{1 + (h/2R)}{1 - (h/2R)}. \tag{b}$$

In any numerical case where h and R are given, we can calculate n/R numerically, although an ordinary slide rule is a poor tool for it, as the reader soon will find out when he attempts it. The result of these calculations is shown in the table below:

h/R	0	$1/10$	$1/5$	$1/2$	1
n/h	0	0.008	0.017	0.042	0.090

We see from it that the shift n of the neutral line is negligibly small for small curvatures.

For very small values of h/R, Eq. (b) becomes unpleasant for numerical calculations, and we can do better by means of a series development. The series for the logarithm (b) can be found in any integeral table or in Marks' "Mechanical Engineers' Handbook," from which we quote:

$$\log_e \frac{1+x}{1-x} = 2\left(x + \frac{x^3}{3} + \cdots\right.$$

$$\frac{1}{1-x} = 1 + x + \cdots.$$

Substituting this into Eq. (b) and neglecting terms small of higher orders, as is done above, we find:

$$\frac{h}{R}\left(1 + \frac{n}{R}\right) = \frac{h}{R} + \frac{h^3}{12R^3},$$

or

$$\frac{n}{h} = \frac{1}{12}\frac{h}{R},$$

by which all the values of the above table for small h/R can be easily found.

In order to get a picture of the stress distribution across the section we calculate the variable part of Eq. (29) for the case of $h/R = 1$, representing a large curvature:

$$s_y = \text{const}\,\frac{y/h}{1 - 0.090 + (y/h)},$$

and for the top and bottom fibers:

$$s_{\text{top}} = \text{constant}\,\frac{0.590}{0.910 + 0.590} = 0.393,$$

$$s_{\text{bottom}} = \text{constant}\,\frac{0.410}{0.910 - 0.410} = 0.820.$$

The stress in the bottom fiber is more than twice as large as in the top one, while for the straight-beam theory these two stresses would have been equal (= 0.500). This is illustrated in Fig. 136.

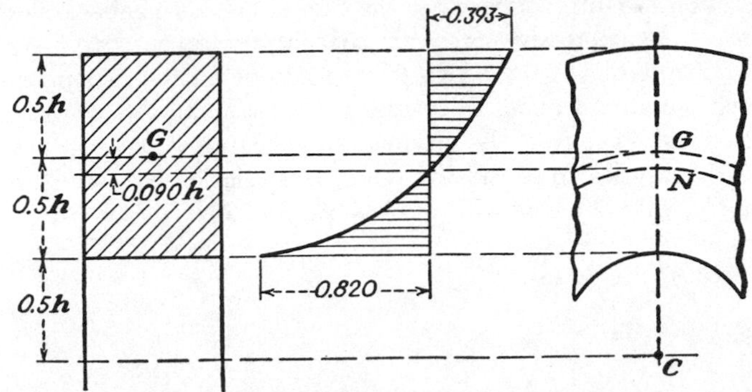

FIG. 136. Bending-stress distribution in a beam of rectangular cross section with $h = R_G$. The stress concentration at the inside of the curve may be either tensile or compressive, depending on the sign of the bending moment.

It is natural that the appearance of the large stress concentration at the inside of the curve has set designers busy making cross sections of which the center of gravity is shifted toward the inside (Fig. 137), so that by the straight-beam theory the outside fiber stress should be greater than the inside fiber stress, which effect then is to be neutralized by the curved-beam stress concentration

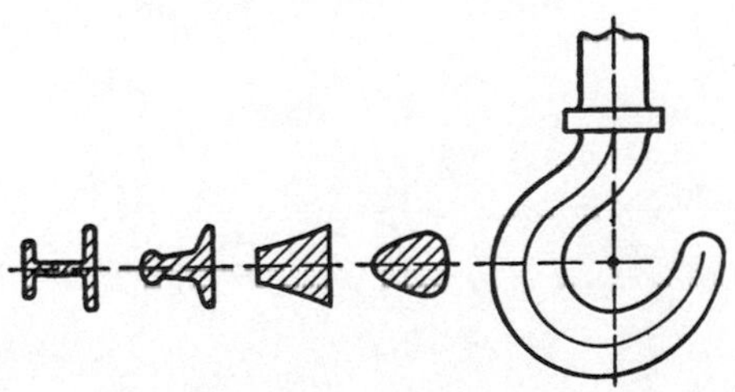

FIG. 137. Various practical cross sections of a crane hook with the center of gravity closer to the inner edge than to the outer one in order to counteract the stress-concentration effect of Fig. 136, and thus to achieve approximately equal stress in the extreme outer and inner fibers.

of Fig. 136. The actual calculation of these cross sections is laborious and time consuming, and we should not jump to conclusions too quickly. For example, if in the I section at the extreme left of Fig. 137 the material in the web is considered negligible and only the two flanges are counted, one would be tempted to make the inner flange larger than the outer one. However, this would not be of any utility, which we can recognize by looking at Eq. (27), expressing the fact that the tensile and compressive forces in the section must balance each other. In the cross section under consideration the integral becomes the sum of two terms, because the tensile and compressive forces are concentrated in the two flanges, and for a zero sum these two *forces* must be equal and opposite. To get equal and opposite *stresses* we must make the two flange areas equal, in spite of the pronounced curvature of the bar. The reader is advised to satisfy himself of this relation by a numerical example, with the help of Problems 263 and 264.

Problems 260 *to* 264.

CHAPTER VIII

THE ENERGY METHOD

32. Stored Elastic Energy. We remember that "energy" and "work" are practically synonyms; when outside forces do work on a system, the energy of the system is said to have increased by the amount of the work done. Sometimes this work done (or the energy gained) is recoverable, as, for example, when a rigid body is raised against its own weight (potential energy of gravitation) or when the force acting on a rigid body has given it speed (kinetic energy). In other cases the work is not recoverable; it is then said to have been "dissipated," usually in friction, but still the system has gained an amount of energy equal to the work done upon it, although that energy appears only in the form of heat.

If an elastic body is deformed by outside forces, these forces move through small distances, their displacements, and hence do some work. Thus the energy of the elastic system has been increased by the work done, and this work is recoverable; it is stored in the body in the form of "elastic energy," sometimes also called "resilience." The most common example of this is the mainspring of a clock; the work done by us in winding the spring is stored in it and is gradually used up during the day in overcoming friction in the mechanism.

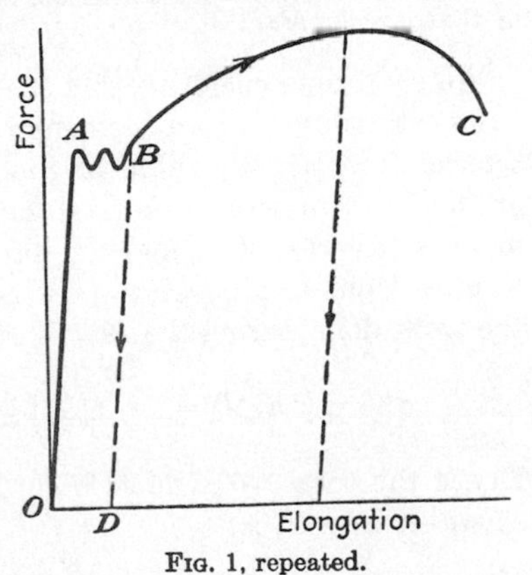

FIG. 1, repeated.

The amount of work done on an elastic element can be visualized in Fig. 1 above by the area under the curve from O up to the point to which the body is stressed. If this stress is "elastic," *i.e.*, below point A, then upon release the system returns to O, giving back as much work as was originally put into it. If the element is stressed above the elastic limit, say to point B, the work done on it still is the area under the curve OAB but this work is not all recoverable. Upon release of the stress, the system goes from B to D, giving back an amount of work shown by the area vertically under BD, which is but a small fraction of the original work put into the system. The rest has been dissipated in heat, by friction between the microscopic layers sliding over each other along the shear-slip lines. The closed curve $OABDO$ is called the "hysteresis loop." But, if we stay within the elastic range, *i.e.*, within the limits of Hooke's law, no hysteresis occurs and all work done on the body is stored in it in the form of resilience.

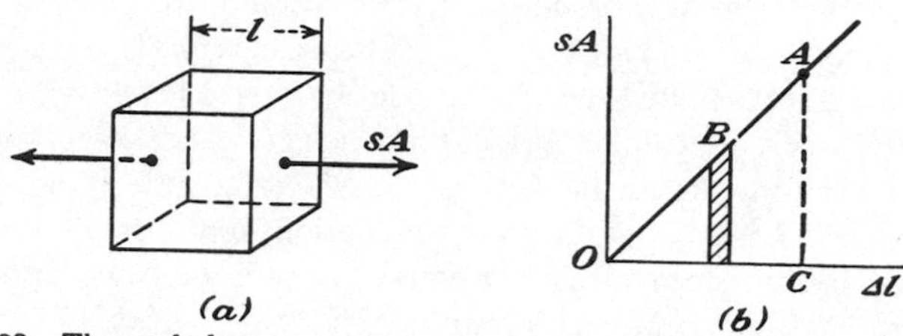

FIG. 138. The work done on an element by a single tensile stress s is represented by the triangular area OAC.

We shall now calculate this energy for a number of important cases, starting with an element lA (Fig. 138*a*) under a simple tensile stress s. We think of the stress s as a variable, growing gradually from zero to its final value. The elongation of the element is $\Delta l = \epsilon l = sl/E$, and the force is sA. If the force is allowed to grow from a value sA to $(s + ds)A$, as at point B in Fig. 138*b*, the work done during that small growth is:

$$dW = F\,d(\Delta l) = sA\,d\left(\frac{sl}{E}\right) = \frac{Al}{E}\,s\,ds = \frac{\text{volume}}{2E}\,d(s^2).$$

Hence the total work done from O to A in the figure, stored as elastic energy U, is:

$$U = \frac{s^2}{2E} \cdot \text{volume}. \qquad (30a)$$

By Hooke's law [Eq. (1)] this result can be written in several other forms:

$$\frac{U}{\text{Volume}} = \frac{s^2}{2E} = \frac{s\epsilon}{2} = \frac{E\epsilon^2}{2},$$

but the form (30*a*) is the most important one.

Figure 139 shows the analogous case of an element subjected to simple shear. By the same reasoning as on the tensile case, the work stored per unit volume is half the product of the maximum shear force s_sA and the displacement of that force $\gamma h = s_sh/G$, so that:

$$U = \frac{s_s^2}{2G} \cdot \text{volume}. \tag{30b}$$

Equations (30*a*) and (30*b*), by their derivation, hold for bodies on which the stress is equally distributed over the area A; these bodies may be large or small. In a case when the stress varies over the area, we can always cut out a sufficiently small cube on which the

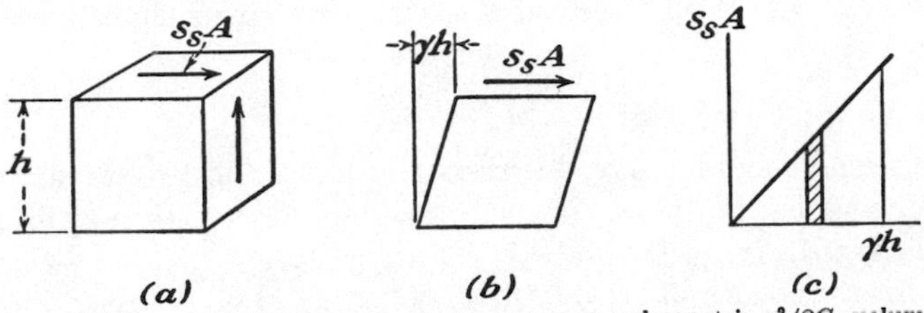

Fig. 139. The work done by a shear stress s_s on an element is $s_s^2/2G \cdot$ volume.

stress is locally constant. Thus Eqs. (30*a*) and (30*b*) can be regarded as differential formulae for a small element d vol.

Bars or beams of variable cross section under a tensile force P (Fig. 2, page 5) or loaded by a bending moment M (central part of Fig. 22, page 35) are made up of elements in pure tension, and their stored energies can be found by integration of Eq. (30*a*). For the tensile case this is obvious and leads to the result:

$$U = \int \frac{P^2}{2EA}\, dl. \tag{30c}$$

For the case of bending, the stress is given by Eq. (11) (page 39), and the integration of Eq. (30*a*) gives:

$$U = \int \frac{s^2}{2E}\, d\text{ vol} = \int\int \left(\frac{My}{I}\right)^2 \cdot \frac{1}{2E}\, dA\, dl$$

$$= \int_l dl \int_A \frac{M^2}{I^2 2E} y^2\, dA = \int_l \frac{M^2\, dl}{I^2 2E} \int_A y^2\, dA,$$

$$U = \int_l \frac{M^2}{2EI}\, dl. \tag{30d}$$

The latter formula is very important so that we shall derive it once more in a somewhat different manner. Consider an element dl of a beam under the influence of a pair of bending moments M.

The two faces of that element will turn with respect to each other through the angle

$$d\varphi = \frac{M\,dl}{EI},$$

and the work done by the moments M is half the product of M and $d\varphi$, because M grows proportionally with $d\varphi$, as in Fig. 138*b* or 139*c*. Thus:

$$dU = \frac{M^2}{2EI}\,dl. \tag{30d}$$

For the case of pure torsion of a shaft, we have a similar result:

$$dU = \frac{M_t^2}{2GI_p}\,dl \tag{30e}$$

The reader should derive this result for himself in two ways: first by integrating Eq. (30*b*) over the cross section, and then again by using Fig. 13 (page 18) and Eq. (4) (page 19).

The expressions (30*c*), (30*d*), and (30*e*) all have the same form: in the numerator appears the "load" squared; in the denominator, twice the "stiffness," EA, or EI, or GI_p as the case may be. The fact that the numerator is squared is important; from it follows that the stored energies due to two loads cannot be simply added to obtain the total energy, because in the expression

$$(M_1 + M_2)^2 = M_1^2 + M_2^2 + 2M_1M_2$$

the last term $2M_1M_2$ appears. The only occasion when the energies due to two loads may be added occurs when the deformation caused by one of the loads does not do work in conjunction with the other load. This occurs comparatively often; for example, if a shaft is subjected to simultaneous bending and twist, then the twist angle does no work with the bending moment, nor does the twisting torque do work when the beam is allowed to flex in bending.

If we superpose a tension P and a bending M on a beam, the force P does no work when the beam is bent through a small angle (the center of gravity of the cross section does not move in the direction of the force P), nor does the bending moment perform work on the parallel elongation of the bar caused by P.

In general the energy in a beam element dl, subjected to bending, twist, and tension simultaneously, thus is equal to the sum of the energies of bending, twist, and tension calculated separately by Eqs. (30*c*), (30*d*), and (30*e*).

A case where two component energies are *not* additive occurs when there are two principal stresses s_1 and s_2 in a plane (Fig. 140). If the two stresses are thought of as growing from zero to their full value simultaneously at the same rate, then the strains also grow at that same rate. The strain in the s_1 direction, by Eq. (18), is:

$$\epsilon_1 = \frac{1}{E}(s_1 - \mu s_2);$$

the elongation of the piece is $b\epsilon_1$, and the work done by the pair of forces $s_1 a$ is:

$$\frac{1}{2} \cdot s_1 a \cdot b\epsilon_1 = \frac{ab}{2E}(s_1^2 - \mu s_1 s_2).$$

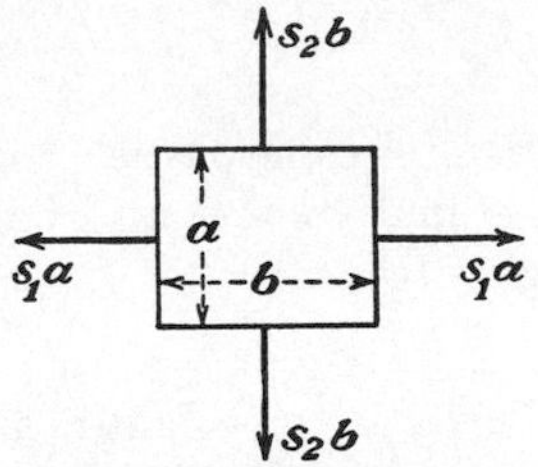

Fig. 140. A block of dimensions a and b, of unit thickness perpendicular to the paper, subjected to two principal stresses. The resilience stored in this block is expressed by Eq. (31).

The work done by the forces $s_2 b$ is derived similarly and, of course, comes out as the same expression in which the subscripts 1 and 2 are reversed. Hence the total energy stored in the block is:

$$dU = \frac{ab}{2E}(s_1^2 + s_2^2 - 2\mu s_1 s_2),$$

$$dU = (s_1^2 + s_2^2 - 2\mu s_1 s_2)\frac{d \text{ vol}}{2E}. \tag{31}$$

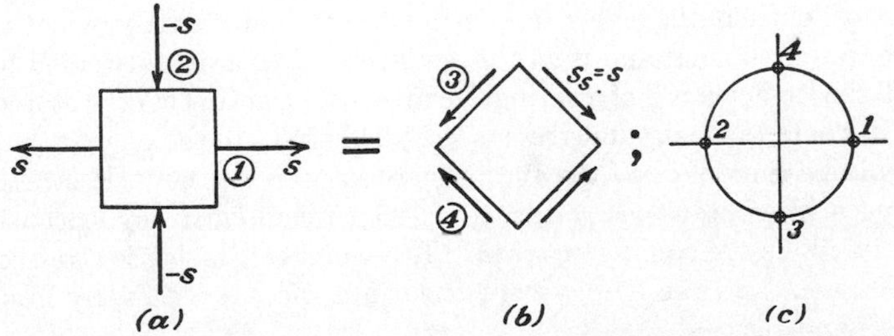

Fig. 141. An element in pure shear (a) or (b), together with its Mohr's circle (c). Writing the expressions for the energy content per unit volume in the two (identical) cases a and b leads to the relation of Eq. (32) between the elastic constants.

An interesting special case of this is the condition of "pure shear," where $s_1 = -s_2$, as illustrated in Fig. 141. The energy per unit volume in case a follows from Eq. (31):

$$dU = (s^2 + s^2 + 2\mu s^2)\frac{d \text{ vol}}{2E} = (1 + \mu)\frac{s^2}{E} d \text{ vol}.$$

The energy in case b is given by Eq. (30b), but $s_s = s$, by the Mohr-circle diagram. Hence:

$$dU = \frac{s^2}{2G}\, d \text{ vol.}$$

It is the same particle considered in two different ways; the energy per unit volume must be the same. Equating the two expressions gives:

$$E = 2G(1 + \mu). \tag{32}$$

This is an important relation: the three elastic constants E, G, and μ are not independent of each other; if two of the three have been determined by experiment for a certain material, then the third one can be calculated without further testing. For example, if a tensile test on steel gives $E = 29.0 \times 10^6$ lb/sq in. and a torsion test leads to $G = 11.9 \times 10^6$ lb/sq in., then we can calculate $\mu = 0.218$, a result that can be verified by subsequent experiment.

The most direct practical use that is made of elastic energy is in connection with springs, which are devices built for the purpose of storing such energy. It is clear that a given piece of material can store more energy when it is more highly stressed, and it is also clear that a spring, to be useful, cannot be stressed higher than the yield stress. Therefore, a spring contains the maximum possible amount of reversible energy if *all* of its material is stressed to the yield point. Very few practical springs are made that way; in most of them the high stress occurs in certain fibers only, whereas the rest of the material is at a lower stress. We are accustomed to call the "efficiency" of a spring the ratio of the actual energy stored in it when its maximum stress is the yield stress to the energy that would be there if *all* of the spring material were at the yield stress. Thus a wire in tension is a spring of 100 per cent efficiency, because all its fibers are equally stressed. The usual coil spring is stressed in torsion, and the energy contained in a piece $\pi r^2\, dl$ is, by Eqs. (30e) and (6):

$$dU = \frac{M_t^2}{2GI_p}\, dl = \frac{M_t^2}{\pi G r^4}\, dl.$$

The maximum stress, by Eq. (5), is:

$$s_s = \frac{M_t r}{I_p} = \frac{2M_t}{\pi r^3}.$$

If the entire volume $\pi r^2\, dl$ were stressed at this maximum stress, the stored energy would be, by Eq. (30b):

$$dU = \left(\frac{2M_t}{\pi r^3}\right)^2 \cdot \frac{\pi r^2\, dl}{2G} = \frac{2}{\pi}\frac{M_t^2}{Gr^4}\, dl,$$

or twice the amount actually stored. Hence the usual coil spring is 50 per cent efficient as an energy storer.

A similar calculation can be made for a flat leaf spring of rectangular cross section bh in pure bending. The actual energy, by Eq. (30d), is:

$$dU = \frac{M^2}{2EI}\, dl = \frac{6M^2}{Ebh^3}\, dl.$$

The stress in the outer fibers is:

$$s = \frac{M}{Z} = \frac{6M}{bh^2}.$$

If the total volume $bh\, dl$ were thus stressed, Eq. (30a) would give for the energy:

$$dU = \left(\frac{6M}{bh^2}\right)^2 \cdot \frac{bh\, dl}{2E} = \frac{18M^2}{Ebh^3}\, dl.$$

Thus a leaf spring in pure bending is only 33 per cent efficient. Of course, this low efficiency is caused by the presence of "dead" or stressless material near the neutral line. If this material is eliminated (as for instance in an I cross section with large flanges and an insignificant web), the efficiency of a spring in bending can be brought up close to 100 per cent.

Problems 265 *to* 273.

33. The Theorem of Castigliano. Carlo Alberto Castigliano (1847–1884) of Milan, Italy, found a general theorem whereby deflections in elastic systems can be calculated in a manner entirely different from that of Chap. V, often leading to the answer with less algebra than the Myosotis method. Before we state and derive the theorem some preliminary remarks are in order.

Let a simple cantilever beam (Fig. 142a) be subjected to an end force P under which it sags through distance δ. Within the elastic region P is proportional to δ, which we can express as:

$$P = k\delta,$$

where $k = P/\delta$, a constant expressed in pounds per inch, is called the "stiffness" of the cantilever as a spring. The work done by P

is $\frac{1}{2}P\delta$, as we have seen before in connection with Fig. 138*b* or 139*c*. Therefore, the energy stored in the beam can be written:

$$U = \frac{1}{2} P\delta = \frac{1}{2} \frac{P^2}{k} = \frac{1}{2} k\delta^2,$$

in three ways, the first "mixed" form containing both the load P and the deflection δ, the second form in terms of the load only, and a third form in terms of the deflection only. The last two forms are quadratic and are shown graphically in Figs. 142*b* and *c*. Now we increase the energy by increasing the force to $P + dP$ and as a

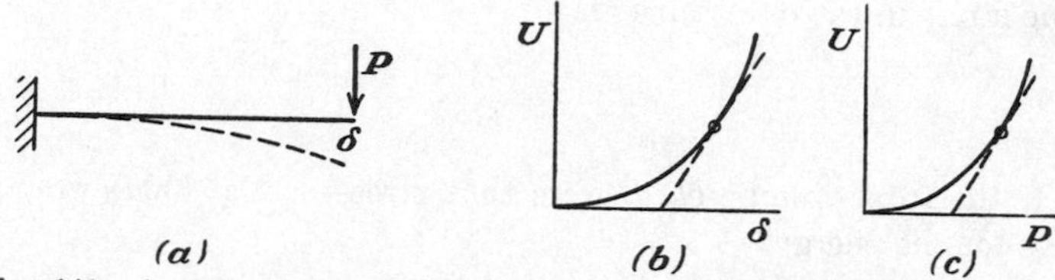

Fig. 142. A cantilever beam with its stored energy plotted as a function of the end deflection (*b*) and of the end load (*c*). The slope of the *b* diagram is P, by the definition of work, while the slope of the *c* diagram is δ, by Castigliano's theorem.

consequence by increasing the deflection to $\delta + d\delta$. During this small growth the value of the force is substantially P, and this force goes down through distance $d\delta$, so that the work it does is $P\,d\delta = dU$. Hence:

$$\frac{dU}{d\delta} = P.$$

If we use for U the form $\frac{1}{2}k\delta^2$ and differentiate it with respect to δ, we find $k\delta = P$. This is a consequence of the usual definition of work. Now Castigliano does something quite different. He takes for U the expression in terms of the load and differentiates it with respect to the load:

$$\frac{dU}{dP} = \frac{d}{dP}\left(\frac{1}{2}\frac{P^2}{k}\right) = \frac{P}{k} = \delta.$$

In this particular case we find the deflection of the beam due to the load.

As another example we take a cantilever shaft twisted by a torque M_t at its end, causing an angle of twist φ at the torque. Again we can write $M_t = k\varphi$, where k is the "torsional stiffness" and

$$U = \frac{1}{2} M_t\varphi = \frac{1}{2} \frac{M_t^2}{k} = \frac{1}{2} k\varphi^2.$$

If, with Castigliano, we use the form for U in terms of the "load" M_t and differentiate, we find:

$$\frac{dU}{dM_t} = \frac{d}{dM_t}\left(\frac{1}{2}\frac{M_t^2}{k}\right) = \frac{M_t}{k} = \varphi,$$

getting the angle of twist as an answer. Now when we use the word "load" in a narrow sense, we mean a force, but the word "load" can be meant in a more general sense to mean a force or a bending moment or a twisting torque. In the same way we can generalize the meaning of "deflection" and understand by it not

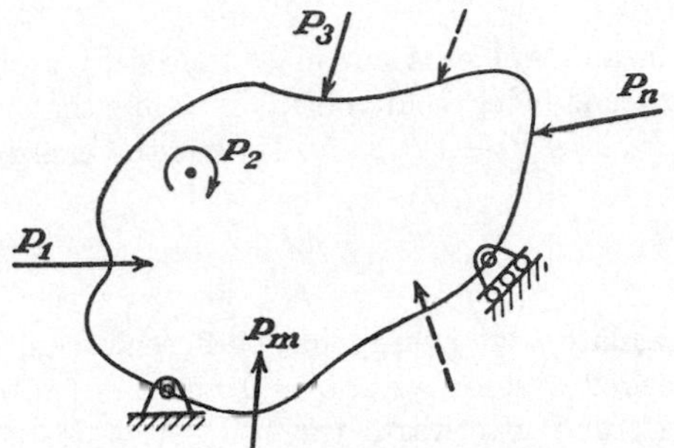

FIG. 143. A general elastic system, for the proof of Castigliano's theorem.

only a linear displacement but also an angle of bending or an angle of twist. Then the results of the two previous examples can be pronounced in one sentence as follows:

If the energy U stored in an elastic structure is expressed in terms of the loads acting on it, then the derivative of U with respect to one of the loads (all other loads being held constant) is the work-absorbing component of the deflection under that load.

This is the general statement of Castigliano's theorem. In the two examples discussed so far there was only one load on the structure. When that load is a force, the "work-absorbing component of deflection" is a displacement in the direction of the force; when the "load" is a couple, either bending or twist, the corresponding deflection is an angular displacement in the plane of the couple. The statement as given above can be understood without a knowledge of "partial derivatives," but the reader who is familiar with such derivatives will have recognized them in the statement.

Now we are ready for the general proof of the theorem. Consider the "elastic system" of Fig. 143, under the influence of m

"loads" $P_1, P_2, \ldots, P_m$, some of which may be forces, other moments, as indicated. The system is supported in a statically determinate manner, *i.e.*, just sufficiently to prevent collapse. As a result of all these "loads," the system will have some energy stored in it, which we call U. Now we increase *one* of the loads, say P_n, to $P_n + dP_n$, and as a result the energy increases to $U + dU$. To prevent misunderstanding we write the dU somewhat more elaborately:

$$dU = \frac{\partial U}{\partial P_n} dP_n,$$

by which we mean the increment in U caused by a change in P_n only, all other loads being held constant. Thus the system under the loads $P_1, P_2, \ldots, P_n + dP_n, \ldots, P_m$ has the energy

$$U + \frac{\partial U}{\partial P_n} dP_n.$$

Now we calculate that energy in a different way, by starting from the unloaded system with $U = 0$ and first putting on the small load dP_n and afterward the full set of loads $P_1 \ldots P_m$. Ending up with the same loads, the system must have the same energy. When the load dP_n is put on first, it causes an infinitesimally small displacement, and the work done by it is $\frac{1}{2} dP_n$ times that displacement and hence small of the second order. We shall neglect quantities of the second order, retaining only those of the first order. Then we put on the full set of loads. They by themselves will do an amount of work U as before. But while this is going on the small force dP_n sits there all the time, and its point of application deflects through δ_n; hence it performs work to an amount $\delta_n \, dP_n$. (There is no factor $\frac{1}{2}$ this time; why not?) Hence the total stored energy is:

$$U + \delta_n \, dP_n.$$

Equating this to the previous result for the energy we find:

$$\frac{\partial U}{\partial P_n} = \delta_n, \qquad (33)$$

which is Castigliano's statement. From the nature of the proof we see that δ_n must be the "work-absorbing component of the deflection under the load P."

As a first example illustrating the use of this powerful theorem, we take in Fig. 144 a cantilever loaded at its end by a load P_0 and a bending moment M_0. Taking the coordinate x from the free end to the left, we have for the bending moment at a general section x:

$$M_x = M_0 + P_0 x,$$

and by Eq. (30*d*):

$$U = \frac{1}{2EI}\int_0^l M_x^2\,dx = \frac{1}{2EI}\left(M_0^2 l + M_0 P_0 l^2 + P_0^2 \frac{l^3}{3}\right).$$

Incidentally we notice that the energy is *not* the sum of the energies due to P_0 and M_0 separately; there is a third term proportional to

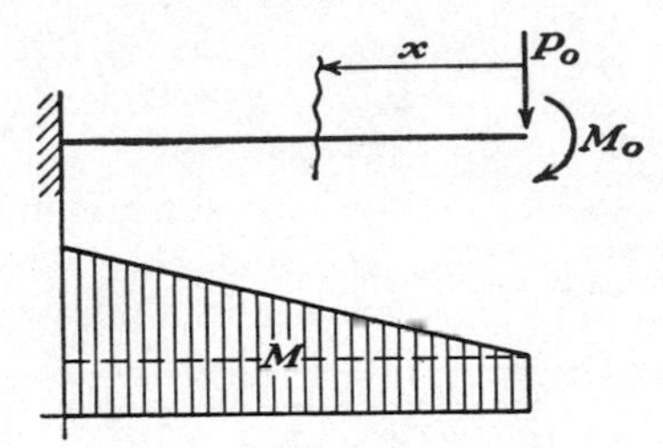

FIG. 144. Cantilever beam under two "loads" at its end.

$M_0 P_0$. This is due to the fact that the load P_0 not only causes a linear end deflection but also an angular one, on which the moment M_0 does work.

Now, by Castigliano, we first vary M_0 only, keeping P_0 constant:

$$\frac{\partial U}{\partial M_0} = \frac{1}{2EI}(2M_0 l + P_0 l^2) = \varphi_{\text{end}}.$$

This must be the component of the deflection by which M absorbs work, *i.e.*, an angular deflection φ. The diligent reader will recognize this answer to consist of two familiar Myosotis expressions.

Then, again by Castigliano, we vary P_0, keeping M_0 constant:

$$\frac{\partial U}{\partial P_0} = \frac{1}{2EI}\left(M_0 l^2 + 2P_0 \frac{l^3}{3}\right) = \delta.$$

This must be interpreted as the work-absorbing deflection of P_0, that is, the vertical downward displacement, which again checks two of our trusty Myosotis formulae.

As a second example we consider the cantilever of Fig. 145 with a 90-deg bend in it, under the influence of a single load P. We ask for the horizontal and vertical deflection of the free end. To obtain the horizontal deflection is simple enough: we calculate the energy and differentiate it with respect to P. The vertical displacement is harder to get. Here, then, we employ a standard procedure in Castigliano's method: when a deflection is asked for in a direction in which there is no force, we introduce an auxiliary force in that direction, go through all the calculations, and at the very end give the auxiliary force its true value, zero. Thus we draw in Fig. 145 the force Q.

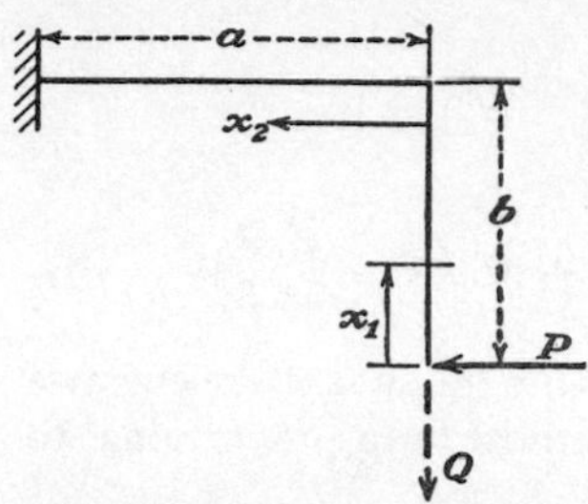

FIG. 145. A cantilever beam of constant cross section EI and EA, with a 90-deg bend, loaded by a force P. In order to find the vertical displacement under P, the fictitious force Q must be introduced, which at the end of the calculation is made equal to zero.

The bending moment in the vertical arm is Px_1, and hence the bending energy in the vertical arm is:

$$U_1 = \frac{1}{2EI}\int M^2\,dx = \frac{P^2}{2EI}\int_0^b x_1^2\,dx_1 = \frac{P^2b^3}{6EI}.$$

The tensile force in the vertical arm is Q, and the corresponding energy is:

$$U_2 = \frac{1}{2EA}\int_0^b Q^2\,dx_1 = \frac{Q^2b}{2EA},$$

which, by page 164, can be directly added to the bending energy. In the horizontal arm we have:

$$M_{x2} = Pb + Qx_2$$

and

$$U_3 = \frac{1}{2EI}\int_0^a (Pb + Qx_2)^2\,dx_2$$

$$= \frac{1}{2EI}\left(P^2b^2a + PQba^2 + Q^2\frac{a^3}{3}\right).$$

In addition there is a compressive force P with the energy:

$$U_4 = \frac{1}{2EA}\int_0^a P^2\,dx = \frac{P^2a}{2EA}.$$

The total energy of the system thus is:

$$U = U_1 + U_2 + U_3 + U_4$$

$$= \frac{1}{EI}\left[P^2\left(\frac{b^3}{6} + \frac{b^2a}{2}\right) + PQ\frac{ba^2}{2} + Q^2\frac{a^3}{6}\right] + \frac{1}{2EA}(P^2a + Q^2b),$$

and, by Castigliano:

$$\delta_{\text{hor}\leftarrow} = \frac{\partial U}{\partial P} = \frac{1}{EI}\left[2P\left(\frac{b^3}{6} + \frac{b^2a}{2}\right) + \frac{Qba^2}{2}\right] + \frac{Pa}{EA},$$

$$\delta_{\text{vert}\downarrow} = \frac{\partial U}{\partial Q} = \frac{1}{EI}\left(\frac{Pb^2a}{2} + 2Q\frac{a^3}{6}\right) + \frac{Qb}{EA}.$$

The answer to our specific question is found by setting $Q = 0$ in the above expressions. However, they can be interpreted as the answers to the more general question of the deflections of Fig. 145 under two arbitrary loads P and Q combined.

Before proceeding to the next example, an important concluding remark must be made. The above expressions contain terms proportional to $1/EI$, giving the deflections due to bending, and terms proportional to $1/EA$, being the contribution due to direct tension or compression. Comparing their relative magnitudes, we see that they are about as:

$$\frac{\text{Bending term}}{\text{Tension term}} = \frac{b^2/I}{1/A} = \frac{b^2/Ak^2}{1/A} = \frac{b^2}{k^2},$$

where k is the "radius of gyration" of the cross section of the beam, a length somewhat smaller than half its height. Now a beam is no beam if it is not some ten times as long as it is high, so that the above ratio is well over 100. Therefore, the deflection due to direct tension and compression is negligibly small compared with the bending deflection, and in all future cases we shall neglect the energy of tension-compression, when it occurs in conjunction with bending.

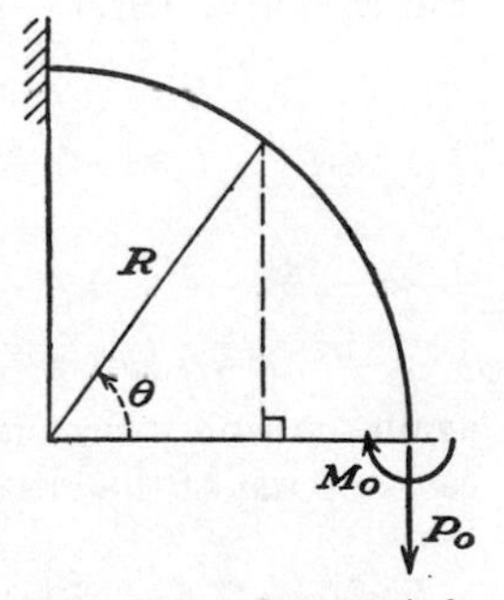

FIG. 146. Quarter-circle cantilever. Castigliano's method leads to the results of Eq. (26) (page 148).

Our last example is the quarter-circle cantilever of Fig. 146, loaded by a force P_0 and by a bending moment M_0. We ask for the downward deflection and for the angular deflection of the free end. The bending moment in a general section θ is:

$$M_\theta = M_0 + P_0R(1 - \cos\theta),$$

and the length of an element of beam is $R\,d\theta$. Hence, by Eq. (30*d*):

$$U = \frac{1}{2EI}\int_0^{\frac{\pi}{2}} M_\theta^2 \cdot R\,d\theta.$$

Now in order to find the deflections we can "differentiate under the integral sign," an expedient that often saves much work in Castigliano's calculations.

$$\varphi = \frac{\partial U}{\partial M_0} = \frac{\partial U}{\partial M_\theta}\frac{\partial M_\theta}{\partial M_0}$$

$$= \frac{1}{2EI}\int_0^{\frac{\pi}{2}} 2M_\theta \frac{\partial M_\theta}{\partial M_0} \cdot R\,d\theta.$$

But $\partial M_\theta/\partial M_0 = 1$, so that:

$$\varphi = \frac{1}{EI}\int M_\theta R\,d\theta.$$

Without calculating any further we recognize that this is the same as Eq. (26) (page 148), derived in an entirely different manner.

Similarly, we find for the vertical downward deflection:

$$\delta = \frac{\partial U}{\partial P_0} = \frac{\partial U}{\partial M_\theta}\frac{\partial M_\theta}{\partial P_0}.$$

But $\partial M_\theta/\partial P_0 = R(1 - \cos\theta)$, so that:

$$\delta = \frac{1}{2EI}\int_0^{\frac{\pi}{2}} 2M_\theta \frac{\partial M_\theta}{\partial P_0} \cdot R\,d\theta$$

$$= \frac{1}{EI}\int_0^{\frac{\pi}{2}} M_\theta \cdot R(1 - \cos\theta)\,R\,d\theta,$$

again the same expression as that for $-\delta_y$ in Eq. (26). The further carrying out of the integrations is left as an exercise to the reader.

Problems 274 *to* 281; *also* 178, 179, 183, *and* 250 *to* 255.

34. Statically Indeterminate Systems. The theorem of Castigliano is particularly useful in its applications to statically indeterminate structures. In such structures there are one or more superfluous, or "redundant," supports, at which the deflection usually must be zero. With the method of Chap. V, the redundant reaction is replaced by an unknown force (or moment), and the deflection of the system at the support is calculated in terms of this unknown. This part of the calculation can be performed either by the Myosotis formulae or by Castigliano's theorem. The last step is the same in both cases: the deflection is set equal to zero, and the support reaction is calculated from that equation.

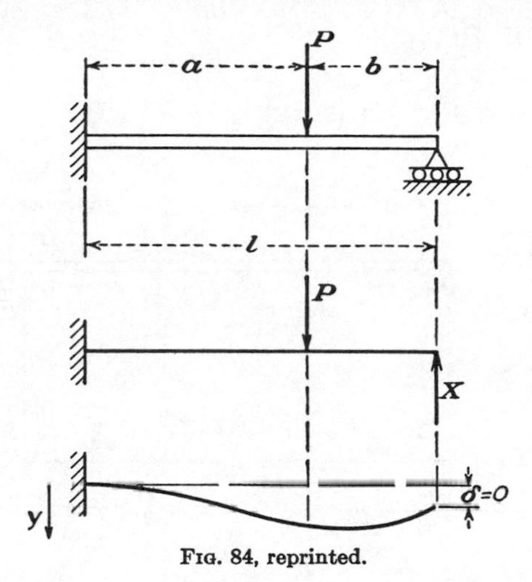

FIG. 84, reprinted.

This general method will now be illustrated in a number of examples.

First we take the case of Fig. 84, which was solved completely on page 93. With Castigliano's method we say that the bending moment in the right-hand part of the beam is Xx, and hence the stored energy in that part is:

$$U_1 = \frac{1}{2EI}\int_0^b X^2x^2\,dx = \frac{X^2}{2EI}\cdot\frac{b^3}{3}.$$

For calculating the bending moment in the left-hand part, we can simplify our expressions somewhat by taking a new coordinate x, starting at the load P, and counting positive to the left. Then:

$$M_2 = Xb + (X - P)x,$$

$$U_2 = \frac{1}{2EI}\int_0^a [Xb + (X - P)x]^2\,dx$$

$$= \frac{1}{2EI}\left[X^2b^2a + 2bX(X - P)\frac{a^2}{2} + (X - P)^2\frac{a^3}{3}\right].$$

The deflection at X is $\partial(U_1 + U_2)/\partial X$, and it must be zero. Hence, by dividing out the common factor $1/2EI$, we find:

$$0 = 2X\frac{b^3}{3} + 2Xb^2a + 2(X-P)\frac{ba^2}{2} + 2X\frac{ba^2}{2} + 2(X-P)\frac{a^3}{3},$$

$$0 = X\left(\frac{2b^3}{3} + 2b^2a + ba^2 + ba^2 + \frac{2a^3}{3}\right) - P\left(ba^2 + \frac{2a^3}{3}\right),$$

$$0 = X\frac{2}{3}(b+a)^3 - Pa^2\left(b + \frac{2}{3}a\right).$$

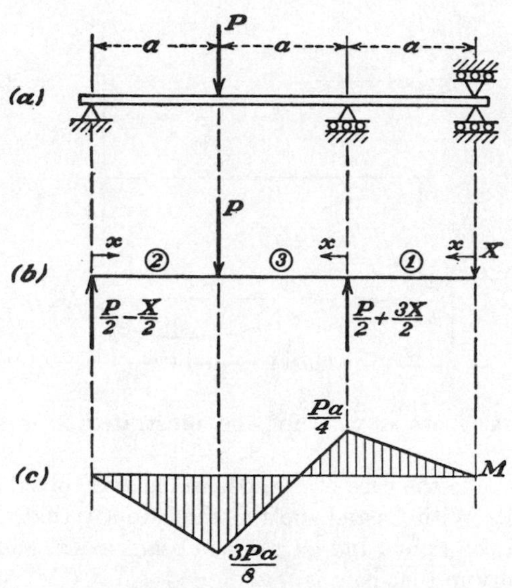

FIG. 147. Beam on three supports.

Remembering that $a + b = l$, this can be written as:

$$X = P\left(\frac{a}{l}\right)^2\left(\frac{3}{2} - \frac{1}{2}\frac{a}{l}\right),$$

which is the result of page 93, plotted in Fig. 85.

As our second example, we take the beam of Fig. 147, loaded by a single force P in the center of the long span. We choose the extreme right support as the one to be replaced by an unknown reaction X, and from the setup we suspect that that force acts downward, as shown. Then the other two reaction forces can be calculated from statics, with the results shown in the figure. The bending moment assumes different algebraic expressions in the three sections of the beam. For the two outer sections 1 and 2 these expressions are quite simple, while that for the center section 3 is somewhat more complicated. Taking the x coordinate as shown in the figure, it is:

$$M_3 = Xa - \left(\frac{P}{2} + \frac{X}{2}\right)x.$$

Then the energy is:

$$2EI \cdot U = \int_0^a (Xx)^2\,dx + \int_0^a \left(\frac{P}{2} - \frac{X}{2}\right)^2 x^2\,dx + \int_0^a \left[Xa - \left(\frac{P}{2} + \frac{X}{2}\right)x\right]^2 dx$$

$$= \frac{X^2a^3}{3} + \left(\frac{P}{2} - \frac{X}{2}\right)^2 \frac{a^3}{3} + X^2a^3 - 2Xa\left(\frac{P}{2} + \frac{X}{2}\right)\frac{a^2}{2} + \left(\frac{P}{2} + \frac{X}{2}\right)^2 \frac{a^3}{3}.$$

The deflection at the right extremity, which must be zero, is:

$$\delta = 0 = \frac{\partial U}{\partial X} = \frac{2Xa^3}{3} - \left(\frac{P}{2} - \frac{X}{2}\right)\frac{a^3}{3} + 2Xa^3 - 2\left(\frac{P}{2} + \frac{X}{2}\right)\frac{a^3}{2} - X\frac{a^3}{2} + \left(\frac{P}{2} + \frac{X}{2}\right)\frac{a^3}{3},$$

or, worked out, $X = P/4$.

With this the other reactions follow, and the bending-moment distribution can be calculated as shown in Fig. 147*c*. This problem can be solved by the Myosotis method with less calculation.

Our third example is the ring of Figs. 129 to 131. The argument involving symmetry of Fig. 130 (page 151), by which the shear force S was shown to be non-existent in the section midway between the loads, should always precede Castigliano's calculation. We thus start from Fig. 129*b*, where the system has been reduced to a single unknown quantity M_0. The energy in a quarter circle is:

$$U_{1/4} = \frac{1}{2EI}\int_0^{\frac{\pi}{2}} M_\theta^2 R\,d\theta = \frac{1}{2EI}\int_0^{\frac{\pi}{2}} \left[M_0 - \frac{PR}{2}(1 - \cos\theta)\right]^2 R\,d\theta.$$

The angle through which M_0 rotates in Fig. 129*b* must be zero, from symmetry. Hence:

$$0 = \frac{\partial U_{1/4}}{\partial M_0} = \frac{\partial U_{1/4}}{\partial M_\theta}\frac{\partial M_\theta}{\partial M_0} = \frac{\partial U_{1/4}}{\partial M_\theta} \times 1$$

$$= \frac{2R}{2EI}\int_0^{\frac{\pi}{2}} \left[M_0 - \frac{PR}{2}(1 - \cos\theta)\right] d\theta.$$

The integral is:

$$0 = M_0 \frac{\pi}{2} - \frac{PR}{2}\left(\frac{\pi}{2} - 1\right),$$

from which we solve for the unknown:

$$M_0 = PR\left(\frac{1}{2} - \frac{1}{\pi}\right),$$

a result found previously on page 152.

To find the extension of the ring between the two forces P, we must find the total energy of the ring, which, by symmetry, is four times that of the quarter ring. The result for M_0 is substituted, so that:

$$U = \frac{4}{2EI}\int_0^{\frac{\pi}{2}} \left[\frac{PR}{2}\left(\cos\theta - \frac{2}{\pi}\right)\right]^2 R\, d\theta$$

$$= \frac{P^2R^3}{2EI}\int_0^{\frac{\pi}{2}} \left(\cos\theta - \frac{2}{\pi}\right)^2 d\theta.$$

Now $\partial U/\partial P$ is the deflection δ of the system due to P. We can imagine one of the two points P as standing still; then the other point P must move through distance δ, which therefore is the diametral increase.

$$\delta = \frac{\partial U}{\partial P} = \frac{2PR^3}{2EI}\int_0^{\frac{\pi}{2}} \left(\cos^2\theta - \frac{4}{\pi}\cos\theta + \frac{4}{\pi^2}\right) d\theta$$

$$= \frac{PR^3}{EI}\left(\frac{\pi}{4} - \frac{4}{\pi} + \frac{4}{\pi^2}\cdot\frac{\pi}{2}\right) = \frac{PR^3}{EI}\left(\frac{\pi}{4} - \frac{2}{\pi}\right),$$

as found before on page 153.

As a fourth and last example, we consider the rectangular frame of Fig. 148*a* with all four sides of equal cross section EI. An argument of symmetry, completely analogous to that with Fig. 130 (page 151), shows us that there is no shear force in the mid-sections AA, so that we can consider the quarter frame of Fig. 148*b*. Then, writing only the bending energy and neglecting that of direct tension, we have:

$$U_{1/4} = \frac{1}{2EI}\int_0^b M_0^2\,dx + \frac{1}{2EI}\int_0^a \left(M_0 - \frac{P}{2}x\right)^2 dx$$

$$= \frac{1}{2EI}\left(M_0^2 b + M_0^2 a - M_0 P\frac{a^2}{2} + \frac{P^2a^3}{12}\right).$$

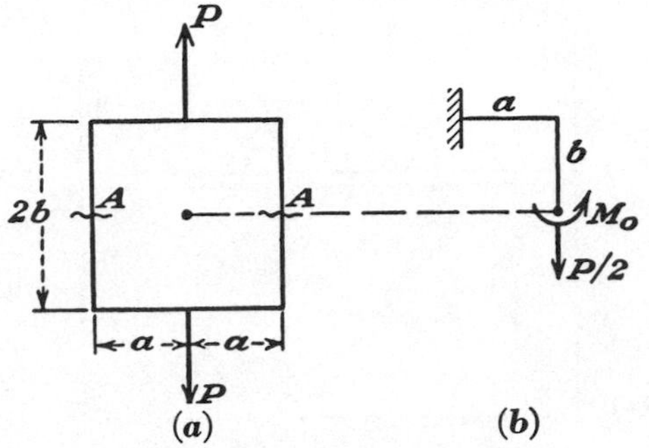

Fig. 148. Rectangular frame.

By Castigliano, $\partial U_{1/4}/\partial M_0$ (with P held constant) must be the "work-absorbing deflection of M" or the angle at that point, which must be zero for reasons of symmetry. Hence:

$$0 = \frac{\partial U_{1/4}}{\partial M_0} = \frac{1}{2EI}\left[2M_0(b+a) - P\frac{a^2}{2}\right],$$

so that:

$$M_0 = \frac{Pa^2}{4(a+b)}.$$

To find δ, the elastic increase in the distance between the two points of application P in Fig. 148a, we must write:

$$\delta = \frac{\partial}{\partial P}(4U_{1/4}) = \frac{\partial}{\partial P}\left\{\frac{4}{2EI}\left[\frac{P^2a^4}{16(a+b)^2}(a+b) - \frac{P^2a^4}{8(a+b)} + \frac{P^2a^3}{12}\right]\right\}$$

$$= \frac{\partial}{\partial P}\left[\frac{2P^2}{EI}\left(-\frac{a^4}{16(a+b)} + \frac{a^3}{12}\right)\right]$$

$$= \frac{4P}{EI}a^3\left(\frac{-a}{16(a+b)} + \frac{1}{12}\right) = \frac{Pa^3}{12EI}\,\frac{a+4b}{a+b}.$$

In this result we can recognize two special cases, for $b = 0$ and for $b = \infty$. In the first case, the two horizontal bars of Fig. 148 are "built in" or "clamped" at their ends; in the second case, they are "freely supported," since the very long bars $2b$ furnish no restraint to angular motion. The reader should verify the correctness of the answer for both cases.

Problems 282 *to* 293.

35. Maxwell's Reciprocal Theorem. Consider the beam of Fig. 149, subject to two forces P_1 and P_2. It is sometimes useful to

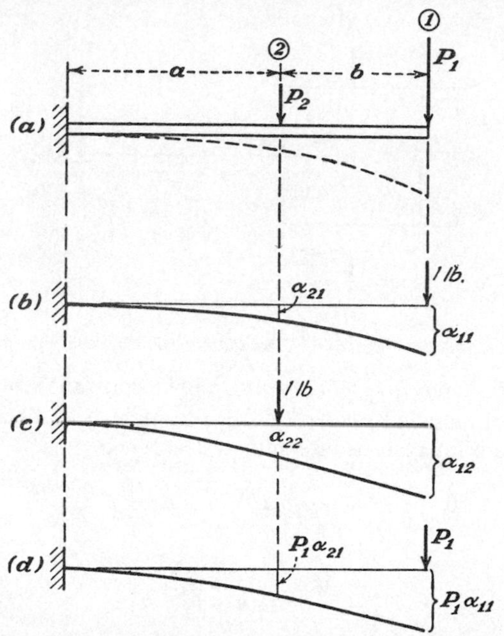

Fig. 149. The definition of an influence number α_{pq} as the deflection at location p caused by a unit force at location q.

analyze the deflections caused by these forces by means of "influence numbers." An influence number α_{12} is defined as the deflection at location 1 (the first subscript) caused by a unit force at location 2 (the second subscript). The two subscripts may be the same; then, of course, we mean the deflection under the unit load itself. The four influence numbers α_{11}, α_{12}, α_{21}, and α_{22} of our example are shown in Figs. 149*b* and *c*. In Fig. 149*d* the load of Fig. 149*b* has been multiplied by a factor P_1, and hence the deflections have been increased by the same factor. The total deflection δ_1 under P_1 in Fig. 149*a* can be written as:

$$\delta_1 = P_1\alpha_{11} + P_2\alpha_{12}.$$

Now we set out to calculate the work done by the forces P_1 and P_2, and hence the elastic energy stored in the beam, in terms of the influence numbers. We calculate that energy in two different ways:

a. We first apply P_1 and then P_2.

b. We first apply P_2 and then P_1.

By the first, or *a*, process, the deflection under P_1 is $P_1\alpha_{11}$, and the work done by P_1 is $\frac{1}{2}P_1 \cdot P_1\alpha_{11}$. There is some deflection at location 2 during this, but since there is no force at 2, no work is being done there. Now we apply P_2. At location 2 the deflection is $P_2\alpha_{22}$, and hence the work of P_2 is $\frac{1}{2}P_2 \cdot P_2\alpha_{22}$, because P_2 grows from zero to its full value gradually with the deflection. However, while P_2 was being applied, the load P_1 was sitting there at its full value, and its point of support went down by $P_2\alpha_{12}$ so that the work done by P_1 is $P_1 \cdot P_2\alpha_{12}$. This time no factor $\frac{1}{2}$ appears, because P_1 was not gradually growing. Thus the total energy in the beam is:

$$U = \tfrac{1}{2}P_1^2\alpha_{11} + \tfrac{1}{2}P_2^2\alpha_{22} + P_1P_2\alpha_{12}.$$

The calculation of the energy by the *b* process follows exactly the same story; only the numbers 1 and 2 are reversed, so that the result is:

$$U = \tfrac{1}{2}P_2^2\alpha_{22} + \tfrac{1}{2}P_1^2\alpha_{11} + P_2P_1\alpha_{21}.$$

These two expressions for the final energy must be the same, so that we conclude that:

$$\alpha_{12} = \alpha_{21}. \tag{34}$$

This is the reciprocal theorem of Clerk Maxwell (1831–1879), and it states that in an elastic system the deflection at location 1 due to a unit load at location 2 is the same as the deflection at location 2 due to a unit load at 1.

Although the "loads" in Fig. 149 are forces, and consequently the "deflections" are linear ones, the theorem is not restricted to forces but can be understood in the same general sense as Castigliano's theorem of page 169. By the word "loads" we mean forces or moments; by "deflection" we mean the "work-absorbing component of the deflection." This will become clearer from the examples to which we now proceed.

First we turn to Figs. 149*b* and *c* and calculate α_{12} and α_{21} by the Myosotis formulae. The reader should do this for himself and find that:

$$\alpha_{12} = \alpha_{21} = \frac{a^2}{EI}\left(\frac{a}{3} + \frac{b}{2}\right).$$

Next we look at Fig. 150. In Fig. 150*b* the angle at the end of the beam happens to be the same as at location 1. It is $Pa^2/2EI$; hence:

$$\alpha_{21} = \frac{a^2}{2EI}.$$

In Fig. 150*c* the deflection at 1 caused by the moment is $Ma^2/2EI$; and hence:

$$\alpha_{12} = \frac{a^2}{2EI}.$$

Next, in Fig. 146, Maxwell's theorem states that the angle at the free end caused by a unit load P alone equals the downward

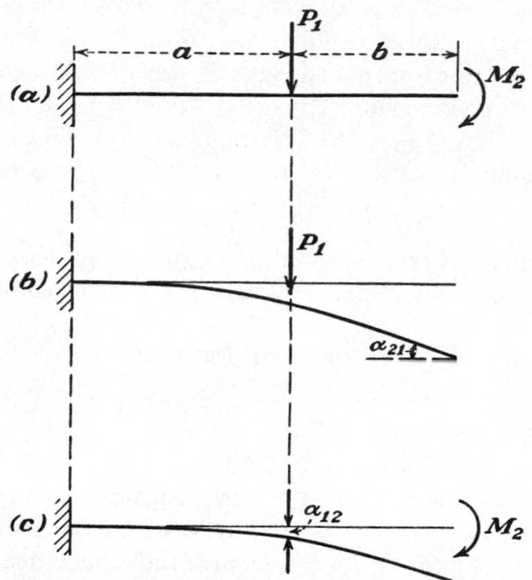

FIG. 150. Maxwell's theorem applied to a cantilever loaded by a force and a couple.

deflection of the free end caused by a unit moment M alone. This can be checked by Castigliano's theorem. We have:

$$M_\theta = M_0 + P_0R(1 - \cos\theta),$$

$$U = \frac{1}{2EI}\int M_\theta^2\,ds = \frac{1}{2EI}\int_0^{\frac{\pi}{2}} [M_0 + P_0R(1 - \cos\theta)]^2R\,d\theta,$$

$$\delta = \frac{\partial U}{\partial P_0} = \frac{\partial U}{\partial M_\theta}\frac{\partial M_\theta}{\partial P_0}$$

$$= \frac{2}{2EI}\int_0^{\frac{\pi}{2}} [M_0 + P_0R(1 - \cos\theta)]R(1 - \cos\theta)R\,d\theta,$$

$$\varphi = \frac{\partial U}{\partial M_0} = \frac{\partial U}{\partial M_\theta}\frac{\partial M_\theta}{\partial M_0} = \frac{2}{2EI}\int_0^{\frac{\pi}{2}} [M_0 + P_0R(1 - \cos\theta)]R\,d\theta.$$

Now if in the expression for δ we set $P = 0$ and $M_0 = 1$, we find:

$$\alpha_{P_0M_0} = \frac{2R^2}{2EI}\int_0^{\frac{\pi}{2}} (1 - \cos\theta)\,d\theta = \frac{R^2}{EI}\left(\frac{\pi}{2} - 1\right).$$

And when in the expression for φ we set $M_0 = 0$ and $P_0 = 1$, we find:

$$\alpha_{M_0P_0} = \frac{2R^2}{2EI}\int_0^{\frac{\pi}{2}} (1 - \cos\theta)\,d\theta,$$

so that indeed $\alpha_{P_0M_0} = \alpha_{M_0P_0}$.

The most interesting application of Maxwell's theorem is to the center of shear. On pages 120 to 127, we saw that the center of shear of a cross section is a point through which a load must be made to pass in order to deflect the end of a cantilever beam parallel to itself, without twisting. Now consider Fig. 151, a cantilever, loaded at its end with a force P (not necessarily through the center of shear) and with a torque M_t. Then we have four influence numbers:

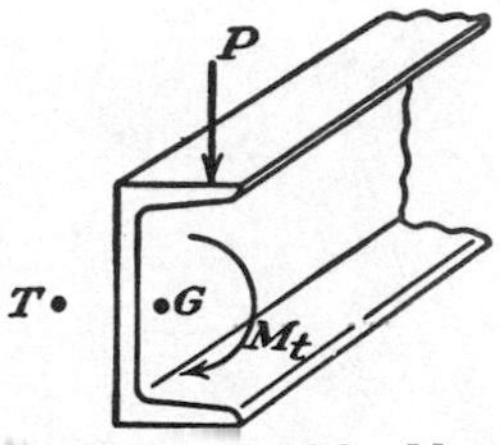

FIG. 151. Proof by Maxwell's theorem that the center of shear is the point about which a cross section of a cantilever rotates when it is loaded by a twisting torque only.

α_{11} = vertical deflection under load due to $P = 1$ lb; $M_t = 0$.
α_{22} = angle of twist of the end section, caused by $M_t = 1$ in.-lb and $P = 0$.
α_{21} = angle of twist due to $P = 1$ lb and $M_t = 0$.
α_{12} = deflection under the load, for $M_t = 1$ in.-lb; $P = 0$.

Now if, for zero M_t, the load P passes through the center of shear, then by our definition the cross section does not rotate so that $\alpha_{21} = 0$. Then by Maxwell's theorem α_{12} must be zero also, which means that the center of shear does not deflect vertically when the shaft is twisted by a torque. If we next apply a horizontal load and repeat the proof, the center of shear does not move horizontally either under the influence of a torque. Hence we see that the center of shear is the point of the section which stands still when the shaft is twisted, *i.e.*, it is the center about which the cross section rotates when the beam is loaded by a torque M_t only.

Problems 294 *to* 300.

CHAPTER IX

BUCKLING

36. Euler's Column Theory. When a short column or strut is put under a gradually increasing compressive load, it ultimately fails by crushing. The stress during the process simply is P/A when the load applies purely centrally through the center of gravity of the cross section, or it is found from Fig. 44 (page 55) when the load is off center. When the column is long, however, it will buckle out sidewise at a load P very much smaller than is required to crush the material. This sidewise deflection then increases very much while the load remains practically constant, and the column fails by bending somewhere near the middle of its length.

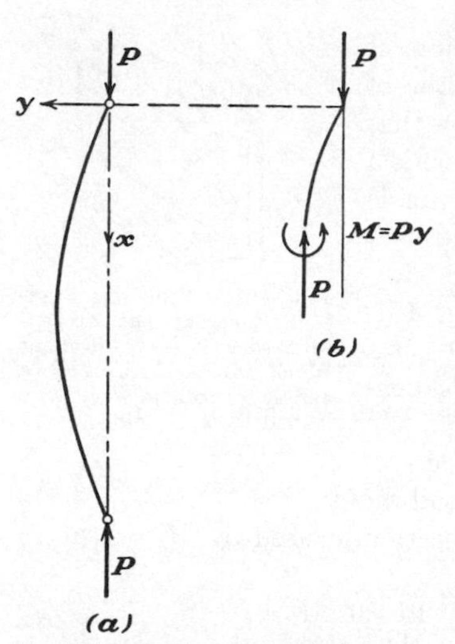

FIG. 152. A column with hinged ends, originally straight, in a state of equilibrium bent out under the influence of end loads P.

This can be seen by a simple experiment on a yardstick. The usual dimensions of such a stick are 36 by 1 by $\frac{3}{16}$ in.; it is made of white pine with $E = 10^6$ lb/sq in. and a compressive crushing strength of about 3,000 lb/sq in. If the stick were short, say 1 in. long, it could support an end load of $\frac{3}{16} \times 3{,}000 = 565$ lb before crushing. But when 36 in. long, and placed with one end on the ground and pushed down by hand from the top end, a very small force, of the order of 5 lb, makes the stick bend out sidewise in the middle, and we hardly feel any increase in our hand force when we increase that side deflection by pushing farther down.

A theory for this phenomenon was given by the great Leonhard Euler (1707–1783) of Switzerland. Consider in Fig. 152 the column of which only the center line is shown. It is supposed to be pin-jointed or hinged at its ends, which means that the ends are supposed not to be able to take a bending moment. The column is put under a pair of end loads P, which are along the line connecting all the centers of gravity of the cross sections of the straight bar. Our yardstick experiment makes us suspect that the bar may be in equilibrium when bent out sidewise, as shown, under end loads P of a certain value. Let that sidewise deflection be $y = f(x)$. The equilibrium of a cutoff piece (Fig. 152*b*) requires then that an arbitrary cross section transmit not only the force P but also a bending moment Py. We then have a beam in bending to which Eq. (19) (page 80) applies:

$$-Py = EI \frac{d^2y}{dx^2}.$$

The negative sign before Py can be understood either by comparing Fig. 152 with Fig. 21 or by observing that in the upper part of the bar of Fig. 152 the slope dy/dx is positive and that it *de*creases with increasing x, so that d^2y/dx^2 is negative. Hence we must have a negative sign before Py.

This differential equation cannot be integrated directly, as could all those of Chap. V, because here y'' is a function of y and not of x, as in the previous cases. The equation here is of the same form as that of a simple vibrating body in dynamics:

$$m \frac{d^2x}{dt^2} = -kx,$$

although there the deflection, called x, was a function of the time t, while here the deflection y is a function of x. In both cases the solution can be written immediately by readers familiar with "linear differential equations," while for those not so far advanced, we write the buckling equation as follows:

$$\frac{d^2y}{dx^2} = -\frac{P}{EI} y, \tag{a}$$

and pronounce it, in words: The deflection y is such a function of x that, when differentiated twice, the same expression y reappears, only multiplied by a negative constant. The reader may remember that sines and cosines behave just like that and by trial find the following two solutions:

$$y = C_1 \sin\left(x\sqrt{\frac{P}{EI}}\right) \quad \text{or} \quad y = C_2 \cos\left(x\sqrt{\frac{P}{EI}}\right).$$

Then we can build up the combined solution:

$$y = C_1 \sin\left(x\sqrt{\frac{P}{EI}}\right) + C_2 \cos\left(x\sqrt{\frac{P}{EI}}\right), \tag{b}$$

which is called the "general" solution, since it contains two arbitrary integration constants and since the solution of Eq. (*a*) is tantamount to a double integration. The values of C_1 and C_2 must be found from the boundary conditions of the problem, which are:

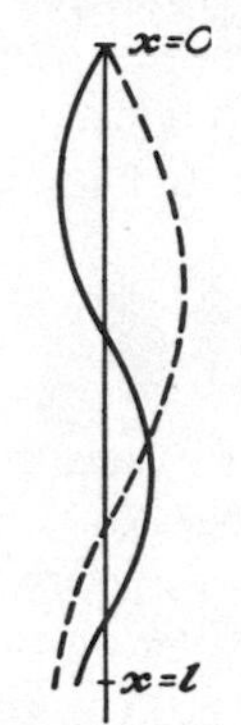

FIG. 153. Solutions of the Euler differential equation of buckling that satisfy the boundary condition at the top, but not the one at the bottom.

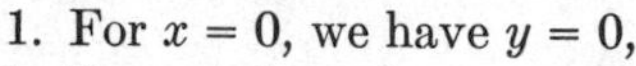

1. For $x = 0$, we have $y = 0$,
2. For $x = l$, we have $y = 0$.

Substitution of the first condition into (*b*) gives:

$$0 = C_1 \times 0 + C_2 \times 1 \qquad \text{or} \qquad C_2 = 0,$$

so that the equation reduces to

$$y = C_1 \sin\left(x\sqrt{\frac{P}{EI}}\right), \tag{c}$$

which means that the curve of Fig. 152 starts as a sine wave at the top and hence may have any of the shapes of Fig. 153. By manipulating the constants of Eq. (*c*), we can push the curves of Fig. 153 into various shapes, provided that they always remain sine waves. The constant C_1 only changes the amplitude of the sine wave, leaving its wave length unaffected, while by varying the load P we can change the wave length in Fig. 153, leaving the maximum amplitude the same.

Now we are ready to satisfy the second boundary condition at $x = l$, and we see that nothing can be done by changing C_1 but that we must manipulate the load P. It is possible to make $y = 0$ at $x = l$ by adjusting the curve to ½, 1, 1½ sine waves along the length l. This means that the angle $x\sqrt{P/EI}$ at $x = l$ is π, 2π, 3π, . . . radians, or

$$l\sqrt{\frac{P}{EI}} = n\pi, \qquad (n = 1, 2, 3, \ldots)$$

from which follows:

$$P_{\text{crit}} = n^2 \frac{\pi^2 EI}{l^2}. \tag{35}$$

The most important of these values is for $n = 1$, with a shape of half a sine wave, as shown in Fig. 154*a*. The significance of Eq. (35)

is that an equilibrium of the column in any of the curved positions of Fig. 154 is possible *only* if the end load assumes one of these "critical" values. For any other value of the end load, the only possible configuration of equilibrium is the straight line. With the first critical loading on, the bar is in equilibrium for any value of C_1 in Eq. (c), which, incidentally, for this case becomes very simply:

$$y = C_1 \sin \frac{\pi x}{l}.$$

One possibility is $C_1 = 0$, that is, even at the critical load the column can be straight. The only limitation placed on the value

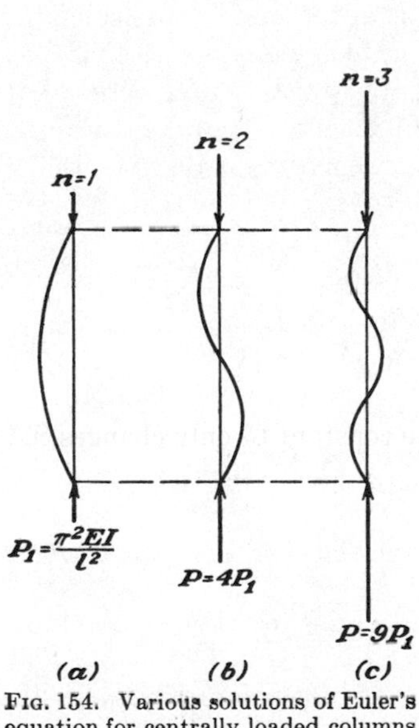

FIG. 154. Various solutions of Euler's equation for centrally loaded columns with pinned ends.

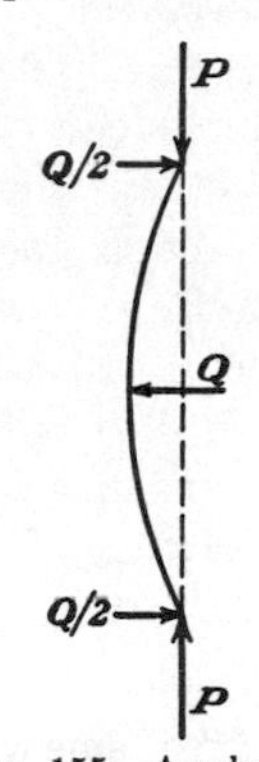

FIG. 155. A column loaded by forces P and Q. The (small) force Q is as shown for the case that P is smaller than the critical Euler load; $Q = 0$ for $P = P_{crit}$, and Q is negative, pushing to the left, for P greater than the Euler buckling load.

of C_1 harks back to the derivation, to Eq. (a) (page 38), and from there to Eq. (19) (page 80). That equation is good only for "small slopes," and hence the same limitation carries through to C_1. But if C_1 is kept sufficiently small, it can have *any* value; therefore, the equilibrium of the bar in the configuration Fig. 154a is said to be *indifferent*.

This can be understood better by imagining a side force Q applied to the center of the column, as in Fig. 155, and by adjusting this Q to such a value that the sidewise deflection in the middle, at Q, is constant, say 1 in. If the end loads P are zero, we have a case of ordinary bending and Q must have a certain value, which can easily be calculated. Then when we apply a small P and let P increase gradually, we notice that the required Q becomes smaller. When $P = P_{crit}$, $Q = 0$; no side force at all is required for equilibrium. When P becomes greater than P_{crit}, Q becomes negative; we must push the bar to the other side in order to stop the bulge from becoming greater.

Now returning to Fig. 154*a*, suppose a straight column is loaded with the critical P; will it buckle or not? The answer is that, within our theory so far, it will not buckle but will stay where it is. However, if a fly flies against it in the middle, the column will be pushed into a curved shape. Or, to be more practical, if the load P is not exactly in the center of gravity of the cross section, but 0.001 in. away from it, the column will buckle. The stability of a column under the Euler load is of the same nature as that of a billiard ball lying on a table; the slightest inclination of the table top will cause the ball to start rolling away.

In conclusion, let us apply Euler's equation (35) to the yardstick example of page 184. We have:

$$I = \frac{bh^3}{12} = \frac{1}{12}\left(\frac{3}{16}\right)^3 = 5.5 \times 10^{-4} \text{ in.}^4,$$

$$P_{crit} = \frac{\pi^2 \times 10^6 \times 5.5 \times 10^{-4}}{36^2} = 4.2 \text{ lb},$$

while, as we saw before, the crushing load on a bar of this cross section is 565 lb.

Problems 301 *to* 306.

37. Other End Conditions. The columns of Figs. 152 to 155, to which Eq. (35) applies, all have pinned ends and hence have no bending moment or curvature at those ends. Now we shall investigate a column with built-in ends, where the end condition is that of zero tangent. There may be a bending moment at the ends; in fact, if there is curvature at the end, there must be a bending moment. Consider the column in Fig. 156, and call the end bending moment M_0. Then the equilibrium of a piece (Fig. 156*b*) requires a bending moment at location x of

$$M_x = M_0 - Py,$$

and, by Eq. (19), the differential equation of the column becomes:

$$+EI\frac{d^2y}{dx^2} = M_0 - Py.$$

The reader should satisfy himself of the correctness of the sign by looking at Fig. 156. This equation can be rewritten as:

$$\frac{d^2y}{dx^2} = -\frac{P}{EI}y + \frac{M_0}{EI}$$

$$= -\frac{P}{EI}\left(y - \frac{M_0}{P}\right),$$

and it differs from Eq. (*a*) (page 185) only by the addition of the last term, which is constant, not depending on x. The general solution of it differs from Eq. (*b*) (page 186) again only by the addition of one extra constant term:

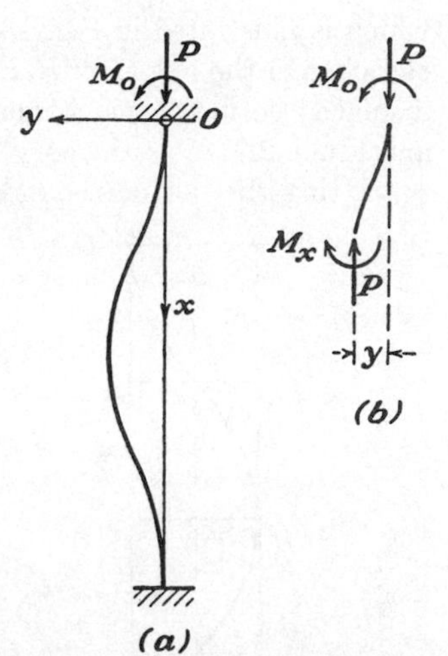

FIG. 156. Column with built-in ends.

$$y = C_1 \sin\left(x\sqrt{\frac{P}{EI}}\right) + C_2 \cos\left(x\sqrt{\frac{P}{EI}}\right) + \frac{M_0}{P}.$$

This should be verified by substitution into the differential equation. The constants C_1, C_2 and the unknown quantities P and M_0 must be found from the boundary conditions, which are:

For $x = 0$, we must have $y = 0$, $y' = 0$,

For $x = l$, we must have $y = 0$, $y' = 0$.

The first condition leads to:

$$0 = C_1 \times 0 + C_2 \times 1 + \frac{M_0}{P} \quad \text{or} \quad C_2 = -\frac{M_0}{P}.$$

For the second condition we first must differentiate the expression for y and afterward substitute $x = 0$:

$$0 = C_1\sqrt{\frac{P}{EI}} \times 1 - C_2\sqrt{\frac{P}{EI}} \times 0 + 0,$$

or

$$C_1 = 0.$$

With these two results the expression for the shape y reduces to:

$$y = \frac{M_0}{P}\left[1 - \cos\left(x\sqrt{\frac{P}{EI}}\right)\right],$$

which is illustrated in Fig. 157. By manipulating the load P, we can change the angle $x\sqrt{P/EI}$ and hence the wave length of the displaced cosine curve; by manipulating M_0, we can change the amplitude $2M_0/P$ of the curve. We see at once that it is useless to try fitting the bottom boundary conditions by changing M_0; we must get at the vital wave length, *i.e.*, we must choose P properly by forcing the point A of Fig. 157 to

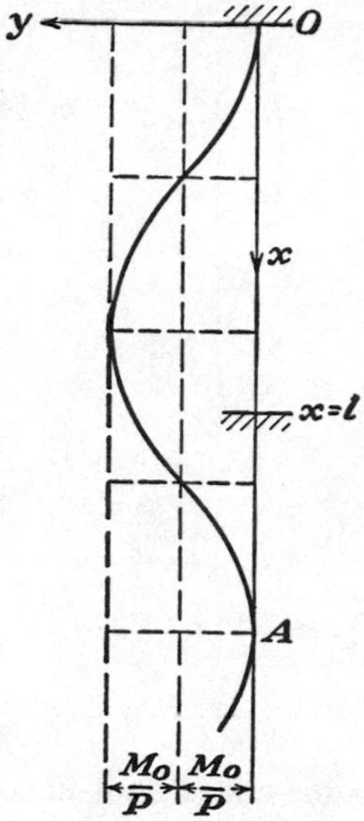

FIG. 157. Column with fixed ends. Solution of Euler's differential equation, satisfying the boundary conditions at $x = 0$, but not yet those at $x = l$.

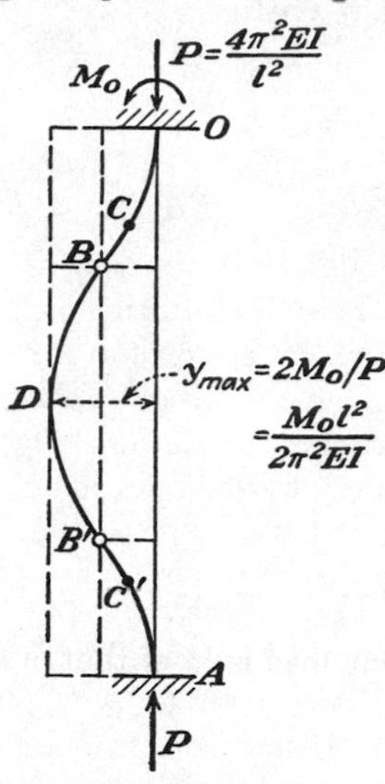

FIG. 158. Fundamental shape of buckling of a column built in at both ends.

the bottom end of the column. When we do that, we satisfy both boundary conditions at once, since at A both y and y' are zero. From O to A is 360 deg or 2π radians of cosine wave. Hence:

$$l\sqrt{\frac{P}{EI}} = 2\pi,$$

and

$$P_{\text{crit}} = \frac{4\pi^2 EI}{l^2}, \tag{35a}$$

which is illustrated in Fig. 158.

We see that the critical load for a column with built-in ends is four times greater than that for a column of the same length with hinged ends. Or we can say that a column with built-in ends can carry the same end load as a column with hinged ends of the same cross section and half the length. This becomes clear when we look at the piece BB' of Fig. 158. At those points the cosine wave has zero curvature; hence the column is without bending and transmits only the compressive forces P at B and B'. Thus the column BB' with hinged ends is in the same condition as the column OA with fixed ends.

Incidentally, Fig. 158 gives us two more buckling formulae for other somewhat artificial boundary conditions. If we look at the portion OB of the column, then B is a point of zero curvature. Hence if $OB = l$, the length of the column, we have the buckled shape of a bar of which one end O is built in, while the other end B is hinged, the hinge being permitted to slide sidewise, *i.e.*, a hinge on rollers. For this kind of column the critical load can be found by replacing the factor $(2\pi)^2$ of Eq. (35*a*) by $(\pi/2)^2$. Therefore, that kind of column will buckle at $\frac{1}{16}$ of the critical load required for a doubly built-in column of the same length and bending stiffness.

The other case appears when we take the piece OD of Fig. 158 as the length l. We have then a column built in at both ends, but one end is on rollers and can slide sidewise. Of course the shape then consists of 180 deg, or π radians, of cosine wave, so that the buckling load is $\frac{1}{4}$ of that of a truly double-ended built-in column.

In practice a column is neither truly hinged nor truly built in but finds itself somewhere in between. This can be expressed by a stiffness factor k; the bending moment at the ends is $k\varphi$, where φ is the angle of the center line there with respect to the equilibrium position. For hinged ends this k is zero (points B and B' of Fig. 158); for fixed ends k is infinity (points O and A). Now if we choose a pair of points CC' in that figure, there will be a certain angle $\varphi = y'$ and a certain curvature y'' at those points, so that there is a value of k between 0 and ∞. Hence the deflection curve for a column with intermediate end conditions will be the portion CC' of Fig. 158, and the critical load will be the Euler load of Eq. (35) multiplied by a factor between 1 and 4. For practical design we are accustomed to err on the safe side, so that we always assume hinged ends, although we know that the critical load is somewhat higher on account of the partial restraint against rotation of the ends.

Next we consider a column with hinged ends on which the end loads P are not placed centrally, but with a certain eccentricity e, measured in inches from the center of gravity of the cross section, Fig. 159. This causes a moment Pe at the hinged end, and therefore we might be tempted to think it to be the same case as that of the bar CC' of Fig. 158 with partially restrained ends. This, however, is not so, because the sign of the moment is opposite in the two cases. In Fig. 159 the couple Pe tends to increase the angle φ of the center line at the ends, while at C and C' in Fig. 158 the couple, being a restraining couple, of course tries to diminish the angle. But still we can use Fig. 158, by shifting the points CC' to the inside of BB'. Then the slopes still have the same sign, but the curvatures have reversed. Thus in Fig. 159 we draw a half sine wave on the line of action PP as base and consider only a portion of it, CC', as the deflected shape of the bar. The length l of the bar is known, but as yet we do not know the quarter wave length λ. Taking the origin in the center as shown, the equation of the curve is:

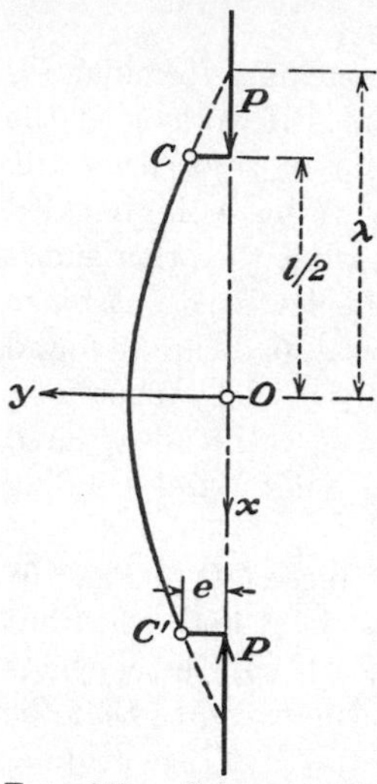

FIG. 159. Column with hinged ends, loaded by two loads P placed eccentrically at distance e from the center of gravity of the cross section.

$$y = y_{\max} \cos\left(\frac{\pi}{2}\frac{x}{\lambda}\right).$$

From the figure we then read:

$$e = y_{\max} \cos\left(\frac{\pi}{2}\frac{l/2}{\lambda}\right).$$

Substituting this into the above, the equation of the curve becomes:

$$y = e\,\frac{\cos\dfrac{\pi}{2}\dfrac{x}{\lambda}}{\cos\dfrac{\pi}{2}\dfrac{l}{2\lambda}}.$$

The column behaves as a hinged-end one of length 2λ with central loads P, which are the Euler critical loads for length 2λ. Thus, applying Eq. (35), we have:

$$P = \frac{\pi^2 EI}{(2\lambda)^2} \qquad \text{or} \qquad \lambda = \frac{\pi}{2}\sqrt{\frac{EI}{P}},$$

so that the curve is:

$$y = e\,\frac{\cos\,(x\sqrt{P/EI})}{\cos\,(\tfrac{1}{2}l\sqrt{P/EI})}.$$

It is well to point out that the load P is smaller than the critical Euler load for our bar of length l; it is the critical load for a longer column 2λ. The above equation is the curve of deformation under the influence of this pair of eccentric subcritical loads. We now set out to calculate the maximum stress, which of course occurs at the point of maximum bending in the center of the column. The curvature is found by differentiation:

$$y'' = -\frac{eP}{EI}\frac{\cos(x\sqrt{P/EI})}{\cos(\frac{1}{2}l\sqrt{P/EI})},$$

and the maximum bending moment is:

$$M_{\max} = EI(y'')_{x=0} = -\frac{eP}{\cos(\frac{1}{2}l\sqrt{P/EI})} = -eP\sec\left(\frac{l}{2}\sqrt{\frac{P}{EI}}\right).$$

The stress is made up of a direct compressive stress P/A and a bending stress, in the manner of Fig. 44 (page 55):

$$s_{\max} = \frac{P}{A} + \frac{M_{\max}}{I}\frac{h}{2},$$

where $h/2$ denotes the distance of the extreme fiber in the cross section. [We do not use the notation $y_{\max}$ from Eqs. (11) and (11a) because $y_{\max}$ in our present figure 159 has an entirely different meaning.] Then for I we can write:

$$I = Ak^2,$$

where k is the radius of gyration of the cross section, so that:

$$s_{\max} = \frac{P}{A}\left(1 + \frac{eh/2}{k^2}\sec\sqrt{\frac{Pl^2}{4EI}}\right). \tag{36}$$

This equation, which is of no particular fundamental importance and certainly not good enough to be memorized, has been given a number because it is often mentioned in design literature as the "secant formula."

It is represented graphically in Fig. 160. In any graphical representation of a general formula, we first try to reduce it to dimensionless ratios. Now the combination appearing under the square root suggests the Euler load $P_{\text{crit}} = \pi^2 EI/l^2$, and with it we rewrite the square root and the entire formula (36):

$$\frac{s_{\max}}{P/A} = 1 + \frac{eh/2}{k^2}\sec\left(\frac{\pi}{2}\sqrt{\frac{P}{P_{\text{crit}}}}\right),$$

in which three dimensionless ratios appear. The first is the actual stress measured in terms of the ideal stress that would be there if the column were short and centrally loaded. The second ratio is P/P_{crit}, of the actual load in terms of the Euler buckling load of a hinged-end centrally loaded column of length l. The third ratio is $eh/2k^2$, representing the eccentricity in terms of the dimensions of the cross section. When this ratio equals unity, it does not represent a very bad eccentricity.

Figure 160 tells us that for zero eccentricity the stress is just the compressive stress P/A for all loads less than the critical one, while

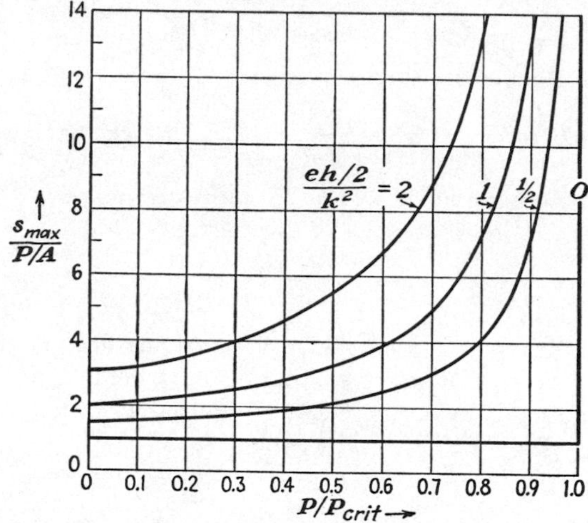

FIG. 160. Stress *vs.* load curves at various eccentricities for a hinged column. This illustrates the "secant formula," Eq. (36).

at the critical load the stress can be anything from P/A on up to infinity. This is the ideal case. But the figure also tells us that, for a column with an eccentricity $eh/2k^2 = 1$ (which is not much), the stress becomes greatly exaggerated when the end loads become more than 60 per cent of critical. This, then, is a good reason not to approach the critical load too closely in a practical case and justifies the customary factor of safety 3, which is applied to Euler's formula.

Problems 307 *to* 314.

38. Practical Column Design. Let us look once more at the simple experiment (page 184) with the yardstick. When we make it buckle, it obviously will bow out in the limber plane of the cross section, *i.e.*, the 1-in. sides will be on the outside and inside of the curve, while the $\frac{3}{16}$-in. faces will remain plane. Buckling in the stiff plane would theoretically be possible, but it would take a larger critical load; by Euler's equation (35) that load is found from the small one in ratio of the two I's, that is, $28\frac{1}{2} \times 4.2$ lb = 120 lb in this case. For the actual design of a column we are not interested in this higher value of the critical load; the lowest critical load corresponding to the smallest moment of inertia of the cross section governs. But then there is no virtue whatever in having a higher moment of inertia in the 90-deg direction. Thus columns to be loaded by compression only are always designed with equal

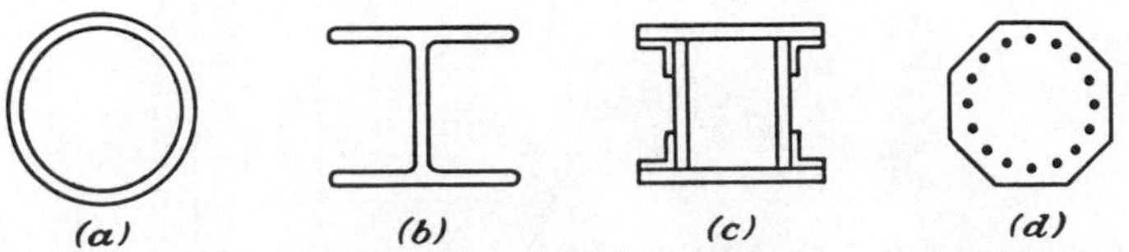

FIG. 161. Cross sections of columns loaded by compression only must have equal moments of inertia in all directions, such as shown in steel (*a*), (*b*), (*c*) or in reinforced concrete (*d*).

values of I in two principal directions and hence equal I's in all directions (Fig. 55, page 65). Column sections in steel are made tubular, H-shaped, or box-shaped, and in reinforced concrete they must have steel reinforcing bars all around the periphery (Fig. 161).

The most significant fact of Euler's equation (35) is that, if the column becomes sufficiently long, its carrying capacity becomes smaller very rapidly. This can be brought out more clearly by rewriting the equation in terms of the "radius of gyration" $k = \sqrt{I/A}$ of the cross section:

$$P_{\text{crit}} = \frac{\pi^2 EI}{l^2} = \pi^2 EA \left(\frac{k}{l}\right)^2,$$

or

$$s = \frac{P_{\text{crit}}}{A} = \pi^2 E \left(\frac{k}{l}\right)^2. \qquad (35b)$$

Here s is the compressive stress in the column in the unbuckled, straight state, and the ratio l/k is called the "slenderness ratio." In the important diagram of Fig. 162 this "stress" is plotted against the slenderness ratio, for steel of $E = 30 \times 10^6$ lb/sq in. We see that, if we were to design a column with a slenderness ratio of 200, it would buckle at a direct stress of about 7,400 lb/sq in., whereas the material can stand 30,000 lb/sq in. before it starts to yield. The column thus is poorly designed; we do not utilize our material efficiently. We do better, for a given length l, to spread out our cross-sectional material to a greater distance k, so that then we go up on Euler's curve (Fig. 162) and can put more stress on our material. Thus it would appear that the most efficient slenderness ratio for structural steel is about $l/k = 100$, because then the direct stress s is just equal to the yield stress of 30,000 lb/sq in. Above the yield stress, of course, Euler's formula is not

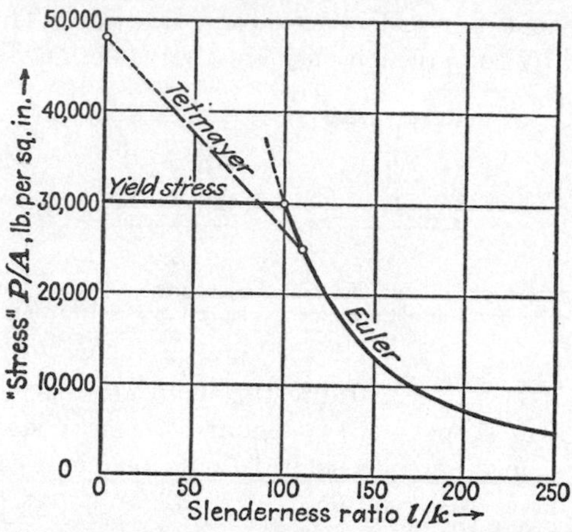

Fig. 162. Graphical representation of Euler's and Tetmajer's buckling formulas for structural steel columns with pinned ends.

valid, because it was derived on the basis of the bending differential equation (19), which again is based on the linear distribution of bending stresses and on Hooke's law. Thus, for columns of a slenderness ratio less than 100, either we must work out a new theory involving plasticity, or we must conduct experiments. The theory has indeed been evolved, but it is extremely complicated, and for practical design we rely on tests, of which many accurate ones were made extending over more than a century. Of course, experiments give scattered test data, and the individual investigator has to use his judgment to reduce these data to some empirical formula. One such formula for structural steel (and incidentally the simplest and hence the preferable one) is:

$$s = \frac{P}{A} = 48{,}000 - 210\,\frac{l}{k} \qquad \text{lb/sq in.,} \tag{37}$$

known as "Tetmajer's formula." It is shown in Fig. 162, and it intersects Euler's curve at a slenderness ratio of about 110.

Above that value, Euler's curve agrees completely with the experiments. At first we should expect Euler's analysis to agree with the facts also in the region l/k between 100 and 110. This would be the case for a completely homogeneous bar loaded completely centrally. A glance at Fig. 160, however, instructs us that slight deviations from these ideal conditions become serious if we are close to the critical load, which for l/k near 100 means also close to the yield point. Therefore, we should not be surprised at deviations from the theory in a practical test in that region.

For columns shorter than $l/k = 85$, the final breaking stress is greater than the yield stress. Tetmajer's expression holds for l/k between 30 and 110. For extremely short struts, say $l/k = 1$, the ultimate strength is the crushing stress, about 60,000 lb/sq in., well above the extrapolated Tetmajer value of 48,000 lb/sq in.

Thus Fig. 162 is all we need to design a steel column decently. First we avoid slenderness ratios of over 150, except in dire necessity. Then we calculate the ultimate load by Tetmajer's formula (37) for $l/k < 110$ or by Euler's equation (35) for $l/k > 110$ and divide this load by a suitable factor of safety, say 3. This is exactly what the building code of the city of Chicago prescribes; it writes Tetmajer's formula as $16{,}000 - 70\,l/k$, absorbing the safety factor in it.

However, here the complications begin. Tetmajer was just one investigator; there were many others. The test data were properly scattered and were interpreted by some as lying on a straight line like Tetmajer's, by others as lying on a parabolic curve, and by still others (Gordon and Rankine) as lying on an S-shaped curve expressed by:

$$s = \frac{P}{A} = \frac{C_1}{1 + C_2(l/k)^2}.$$

After the investigators came the engineer-lawmakers on the building-code committees of large cities and on standardization committees of engineering societies. Each committee chose a formula from among the rich available collection and applied its own safety factor to it. The result is that one poor innocent column may have twenty different allowable loads according to the building codes of New York, Chicago, Philadelphia, and Boston, or by the "standardization" formulae of the various engineering societies. All these formulae group themselves around the Chicago curve (which is Tetmajer divided by 3) with deviations of the order of 10 or 15 per cent. Therefore, it is perfectly safe to design with a single formula, Eq. (37).

For materials other than structural steel, such as alloy steel, aluminum, or magnesium, diagrams similar to Fig. 162, with different numerical values, apply. The companies manufacturing beams of these materials of sections fit for columns have made their own tests and have published suitable formulae. For such cases the reader is referred to engineering handbooks or to the catalogues of the manufacturers.

Problems 315 *to* 324.

CHAPTER X

EXPERIMENTAL ELASTICITY

39. Photoelasticity. It has been known for more than a century that, if a transparent elastic object, such as glass, is viewed in "polarized" light, colors appear in it when the object is stressed. This experimental fact has been worked out into a full-fledged accurate technique for measuring complicated stress distributions. Figure 163 shows the test setup, consisting of a source of monochromatic (single-colored) light, usually a bluish-green mercury-arc lamp, two ordinary glass lenses, two plates of "polaroid," a

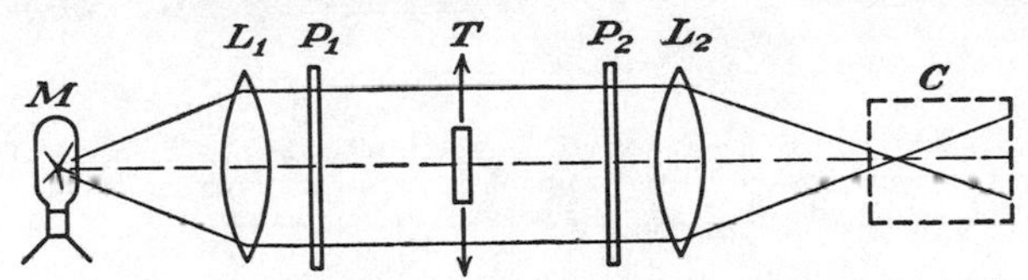

FIG. 163. Apparatus for photoelastic tests. M is a mercury-arc lamp; L_1 and L_2 are condensing lenses, P_1 and P_2 are disks of polaroid, T is the bakelite test piece, and C is a camera.

camera, and the test piece. The first lens serves to convert the diverging beam of light from the lamp into a parallel beam, which passes through the test piece, and the second lens changes that parallel beam into a converging one to the objective of the camera. The two circular disks of polaroid, of some 10 in. diameter and about 1/16 in. thickness, look like ordinary glass or celluloid, slightly tinted. They have the important property of "polarizing" the light, a process of which the details must be studied in books on optics and which we shall not discuss here. The test piece between the polaroid plates is made of special yellowish, transparent bakelite, usually about 1/4 or 1/2 in. thick, cut out from a flat plate to the desired contour and placed with its flat side perpendicular to the rays of light so that the light traverses the 1/4-in. thickness. It is mounted in a loading frame by means of which accurately measurable loads can be imposed on it. The camera is used to photograph the picture, but for preliminary visual observation it and the second lens L_2 can be omitted and replaced by a milk-glass screen.

We start with a test piece T in the form of a simple straight bar (about $\frac{1}{4}$ by 2 by 10 in.) in pure tension so that, when loaded, all points of the bar have the same stress. When the polaroid plates are properly adjusted, the entire milk-glass screen, including the picture of the bar, is dark when the bar is unstressed. Then when the stress is put on and gradually increased, the entire bar becomes uniformly light, then dark again, and so on, until the yield point of the bakelite test piece is reached. For good bakelite of $\frac{1}{4}$ in. thickness the piece can go through about 20 successive darkenings

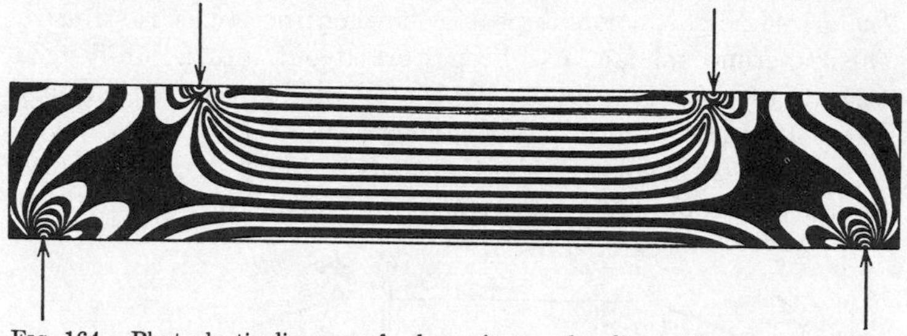

FIG. 164. Photoelastic diagram of a beam in pure bending. The upper and lower fibers in the center portion have 6 units of stress; in the four corners of the beam the stress is zero; under the loads it is greater than 10 units.

and lightings up before the yield point appears, and the load increments between successive cycles are found to be all equal.

As a second experiment we take a flat beam (rectangular cross section of $\frac{1}{4}$ by 2 in.) and load it in pure bending in the 2- by 10-in. plane, like the center part of the beam of Fig. 22 (page 35). When we gradually increase the load, each point of the beam (except the central neutral line) will alternately become light and dark, the outer fibers at a faster rate than the ones closer to the center, because the stress there rises more rapidly. The final picture is shown in Fig. 164, and the fact that the black and white lines (in the center portion) are equally spaced can be interpreted as an experimental proof for the linear stress distribution of Eq. (11) or Fig. 26 (page 38). Each individual black line in Fig. 164 represents a certain stress, and when we start from zero load (uniformly dark bar) and gradually increase the bending moment, we first see a light line appear at the upper and lower fibers. These light lines then move inward toward the neutral line, and the outside becomes black. Gradually more lines appear, starting at the outside and gradually moving inward. For example, the line in Fig. 164 that is midway between the neutral line and the outer fiber started to appear first at the outer fiber at an instant when the bending load was half the ultimate one. (Why?)

A third experiment can be made by taking a square plate and pulling it with equal stress in two perpendicular directions so that the bakelite is in a state of "two-dimensional hydrostatic tension," like Fig. 59 (page 67), with reversed sign. In this case we shall see that the test piece remains dark at all loads; no optical effect of the applied stress can be seen at all. From this third experiment we conclude that the photoelastic effect, as measured by the number of light alternations at a point between zero and full load, is proportional to the *shear* stress or to the difference between the two principal stresses at the point. This is fortunate, because from a practical standpoint the shear stress is even more important than a principal stress. But in most cases we can deduce even the maximum principal stress itself by the circumstance that this maximum principal stress usually occurs on the periphery of the test piece rather than in the interior. Now, the free edge of the periphery obviously carries no normal stress or shear stress at all, so that at a free edge the principal stress must be along the edge, while the second principal stress across the edge is zero. Hence (by Fig. 57) that principal stress is twice the local shear stress, which is measured photoelastically.

The three experiments described thus far tell us nothing new about the state of stress, and they serve us only as a means to establish the technique; they should be considered as *calibration experiments.* From them we deduce accurately the "calibration constant," or "sensitivity," of the material used, measured in pounds per square inch stress increment for one complete cycle of light change. This constant roughly is 1/20 of the yield stress of the bakelite, for 1/4 in. thickness, as was mentioned before. The sensitivity of a bakelite plate is proportional to its thickness, *i.e.*, proportional to the length of the path of light across the plate; hence a 1/2-in. plate will show twice as much detail as a plate of 1/4 in. thickness.

Now we can place between the polaroid plates a specimen of a shape and with a loading so complicated that we are unable to calculate the stress in advance. Then, by counting the number of black lines passing a point of our specimen while loading it, we determine the stress experimentally.

As a first example of this we take Fig. 165, which shows the central portion of the bar of Fig. 164 in which two semicircular notches have been cut with a diameter equal to one-quarter of the height of the beam. At the left and right of the picture, at some distance from the notches, the pattern is the same set of parallel lines as in Fig. 164, and, counting from the central black line (the zero line) toward the upper or lower edge, we cross six lines, in Fig. 164 as well as in Fig. 165, so that the bending stress at the edges of those beams is the same: 6 units. But in Fig. 165, when

FIG. 165. Central portion of the beam of Fig. 164 in which two semicircular notches have been cut. The extreme fibers of the beam away from the notches have 6 units of stress; in the bottom of the notch the stress is greater than 11 units.

we count from the center to the bottom of the notch, we cross at least 14 lines, the last ones being so close together that they become indistinct. (A more accurate way of counting consists of watching the picture grow from zero load to the maximum and counting the lines at the bottom of the notch as they are born.) Thus we see that the stress at the bottom of the notch is more than 14 units, or about 2½ times the stress at the edge away from the notch. We say that the notch causes a *stress concentration factor* of 2½.

In Fig. 166 the notch in our bent bar is somewhat deeper and of narrow rectangular shape with sharp square edges at the bottom. The stress concentration here is so great that, if the beam had been bent to 6 units as in the previous cases, the stress at the bottom of the notch would have been in excess of the 20 units allowable below the yield point of the bakelite. Therefore, this test was made by increasing the bending load up from zero, and counting the lines as they formed at the bottom corners of the notch until fourteen such lines had appeared. Then the photograph was taken as here reproduced. Away from the notch, calling the central black-line number zero, we count till 2 (or $2\frac{1}{2}$) at the outer edge. Starting from the center of the picture to the bottom corner of the notch, we count to 6 distinctly. From then on the lines are so

FIG. 166. Central portion of the beam of Fig. 164 in pure bending, with narrow rectangular notches. The stress at the bottom of these notches is so great that the beam itself could not be loaded to 6 units as were those in Figs. 165 and 164.

close together that they merge, but we know there are 14 of them from the way they grew. Thus the stress concentration factor is about $14/2.5 = 6$.

The next picture is Fig. 167, again a bar in bending with the same kind of notch, but this time with many of them closely spaced. From common sense we suspect that the pieces of beam between notches (call them "tongues") have little stress and are practically dead material; the experiment bears that out. If we start from the central neutral line, calling it zero, and proceed in a direction perpendicular to it and midway between two notches, we count a sequence as follows: 0-1-2-3-4-5-4-3-2-1-0, ending up with zero stress in the middle of the tongue. Now, how do we know that we count up to 5 and then down again, instead of continuing up from 0 to 10? The answer is that we do *not* know that from this photograph alone, but that we must watch the picture growing up from zero stress to the maximum, as shown. We then see that (disregarding the bottom of the notch) the first black line appears at 5, splits up into two lines, which move off in opposite directions; a second black line appears again at 5, splits up, etc. Looking at the bottom of the notch, we see two more lines beyond line 5, which are interpreted as numbers 6 and 7. If the beam were streamlined by cutting off all the tongues there would be five lines only in pure

FIG. 167. The same beam as Fig. 166 in bending but with many notches. The tongues between the notches are "dead" material with small stress, and the stress concentration at the bottom of the notches is reasonably small.

bending. This, with seven lines at the bottom of the notch, is interpreted as a stress concentration factor 1.4 due to the series of notches.

The last example is Fig. 168, which is a flat plate with a central hole in a state of pure tension (not bending this time). At some distance from the hole, the plate is in uniform tension and hence is either all black or all white, showing no lines at all. Near the hole, however, there are many lines, and the maximum stress appears at the sides of the hole on a horizontal diameter. The stress there is about three times as large as in the full plate, *i.e.*, the stress concentration factor is 3. This cannot be seen at all from Fig. 168 by itself, but is deduced by counting the number of light alternations at the periphery of the hole and dividing that number by the number of light alternations in the full plate far from the hole.

Now we return once more to Fig. 164 and look at the detail near the points of load application. Right under the loads we count a large number of lines, which means that the stress immediately under a load is very large (theoretically infinitely large for a load concentrated at a "point"). We see from the figure that these high stresses are limited to a very small region near the load. In actual practice there will be some plastic deformation locally, and

FIG. 168. A bar with a central hole in pure tension. There is a large stress concentration near the edges of the hole, but this illustration cannot be understood by itself; for its understanding it is necessary to see it grow up from zero stress.

as a result the load becomes distributed over a finite small area. At some distance from the load, the state of stress corresponding to "pure bending" appears.

Thus, summarizing, we see that a stress concentration factor of the order 3 appears at the edge of a round hole in a surrounding field of tension (Figs. 165 and 168), and that the stress concentration factor becomes materially greater than that in the bottom of a sharp, long crack (Fig. 166) or under a concentrated load (Fig. 164). On the other hand, little concentration of stress occurs when a number of cracks of equal depth are so closely spaced (Fig. 167) that the intermediate material is "dead" or stressless.

The photoelastic method is widely used to investigate two-dimensional states of stress that are too complicated for calculation in terms of equivalent "beams." The method has also been adapted to three-dimensional stress distributions, but then it is so extremely complicated experimentally that it remains a tool of research alone and is not yet fit for practical application.

Problems 325 *to* 328.

40. Strain Gages. The elastic strain in most materials (with the exception of rubber) is of the order of 1 part in 1,000 so that, in order to measure it, considerable magnification is required in the measuring instrument. Until quite recently this magnification was accomplished mechanically, by means of tiny lever systems, or optically, by means of mirrors or microscopes. These "strain gages" were large, cumbersome, and often none too accurate.

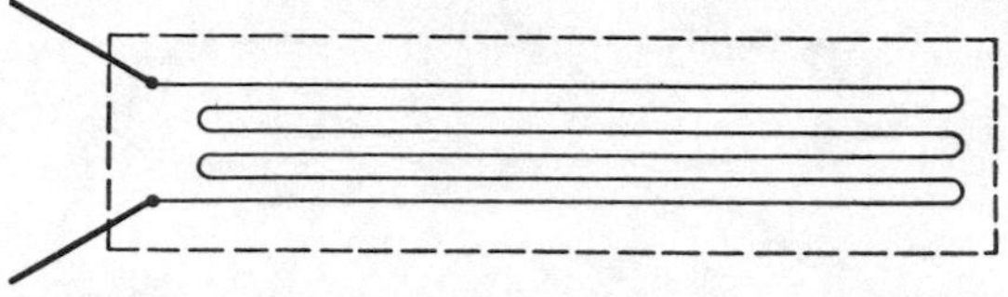

FIG. 169. Strain gage, actual size about 1 by ¼ in., consisting of michrome wire of 0.001 in. diameter, glued between two very thin sheets of paper.

They have now been completely overshadowed by strain gages with electrical magnification, in which the fundamental part, the "pickup," is a tiny piece of very thin electrical-resistance wire glued to the object of which the strain is to be measured. The gage is shown and described in Fig. 169. It is attached to the test specimen by a special, good-quality glue that will not come loose with reasonable moisture or with temperatures up to 180°F.

When the test piece experiences strain, the gage follows that strain and as a result changes its electrical resistance. If the wire of the gage is strained, it elongates and becomes thinner, by Poisson's ratio, both of which actions increase the electrical resistance, even if the molecular structure of the wire remains the same. If ϵ is the strain, *i.e.*, the percentage increase in length, then the percentage increase in electrical resistance due to this effect alone would be $(1 + 2\mu)\epsilon$, or roughly 1.6ϵ. In addition to this, the molecules are pulled farther away from each other, which increases the resistance still more. For nichrome wire, experiments show that the resistance increases at the rate of 3ϵ, which is usually expressed by saying that the "sensitivity factor" of the gage is 3. This sensitivity factor is defined as the percentage increase in electrical resistance R divided by the percentage increase in length l, or

$$\frac{\Delta R}{R} = 3\,\frac{\Delta l}{l} = 3\epsilon$$

The strain gage G is usually inserted as one of the arms of a Wheatstone bridge (Fig. 170) in which a second arm consists of another, identical gage D, not attached to the test piece but to a stressless piece which is made to follow any temperature fluctuations the main test piece may experience. This second gage then

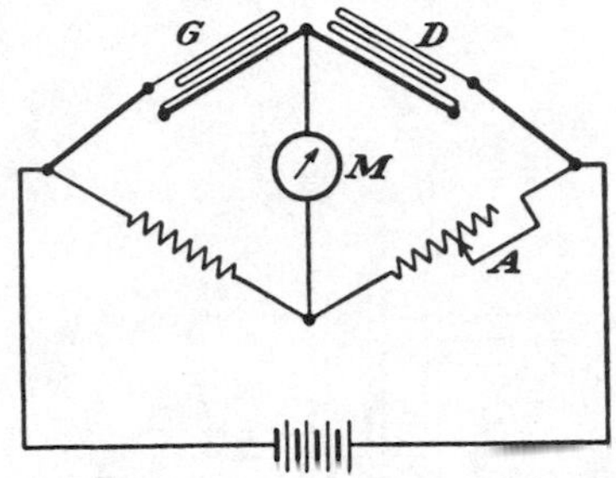

Fig. 170. Fundamental Wheatstone bridge circuit into which the strain gage G is inserted.

is called a "dummy"; hence the letter D. The bridge is energized either by a direct-current battery of the order of 45 volts as shown, or by an alternating-current source of a frequency of the order of 1,000 cycles/sec, generated by a vacuum-tube oscillator. The bridge is electrically balanced by the adjustment A so that the meter M shows zero reading for zero strain. Then, when the gage G is strained, the bridge goes out of balance and the meter M shows a reading. That meter may be a very sensitive galvanometer, but more likely it is an oscillograph, and then the electrical "signal" M must be amplified by an electronic amplifier before it can operate the oscillograph. With this device, strains can be measured as small as 10^{-6}, corresponding to a stress in steel of 30 lb/sq in., and this strain is not necessarily steady in time but can vary with a frequency of several hundred cycles per second. Whenever possible, the dummy gage D is replaced by an active one; for instance, when a bending moment in a beam is to be measured, the gage G is glued to the extreme tension fiber and the gage D is glued on the opposite side of the beam to the compression fiber. Then when G increases its resistance, D decreases its resistance by an equal amount, thus doubling the strength of the "signal" through M. The circuit then is designated by the name "push-pull." As an example, Fig. 171 shows a common test setup for measuring the torque in a shaft, which is used for determining the efficiency of the turbines of a ship drive. (When we measure the fuel consumption of the machinery and the torque and speed of the propeller shaft, the efficiency can be calculated.) The strain gages are glued on in the two principal directions of stress in torsion (Figs. 60 and 61, page 68), so that if one of the gages is elongated,

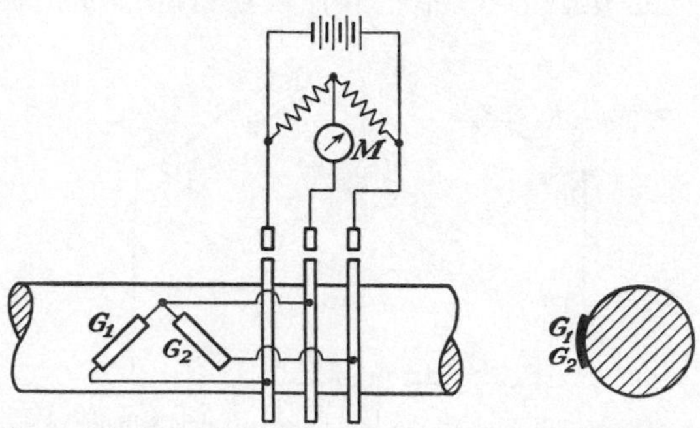

FIG. 171. Two strain gages glued at 45 deg on a ship's drive shaft in a push-pull circuit with three slip rings, for measuring the torque in the shaft.

the other one is shortened. If the shaft is in longitudinal compression (which it actually is on account of the propeller push), this compression will shorten both gages by the same amount, which will not unbalance the bridge circuit of Fig. 171, so that the instrument M reads the torsion effect only and disregards compression (or bending) of the shaft. The circuit can read the average or "steady" torque of the shaft if a slow reading amperemeter is inserted at M, but it can also be made to register fast variations of the torque (usually with the frequency of the rpm, multiplied by the number of propeller blades; why?) if the "meter" M is a fast oscillograph with photographic recording.

In the examples shown so far the stresses can be deduced from the measured strains in a very simple manner, but this is not always the case. If in a complicated piece, for example, near a point of stress concentration (Fig. 168), we wish to determine the maximum stress at a point experimentally, we cannot do it by just gluing on a single strain gage. We do not know the direction of the principal stress, and we might locate our gage in an unfortunate direction.

For example, in Fig. 171, if the gage were located along the shaft, it would show a zero reading no matter how high the torque were. In general, the state of stress at a surface point of a structure contains three unknowns: the values of the principal stresses s_{max},

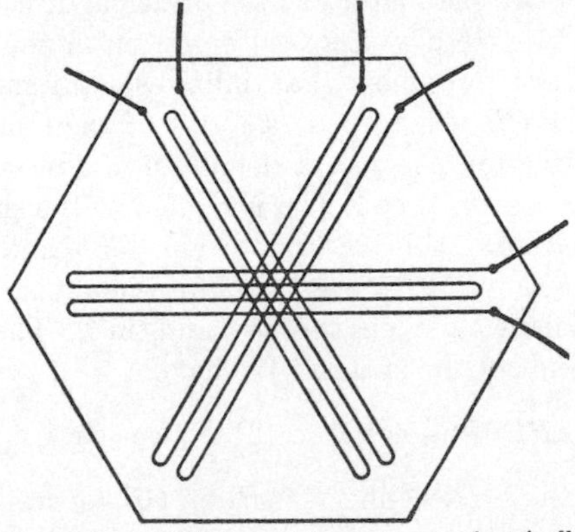

FIG. 172. Strain "rosette" consisting of three gages, electrically insulated from each other, for the purpose of determining the general state of stress at a surface point of the structure. Slightly larger than full size.

s_{min}, and the angle between the direction of s_{max} and a prescribed "zero" direction. To find three unknowns three experimental readings must be obtained. For this purpose "rosette gages" can be bought, consisting of three electrically separate gages with six

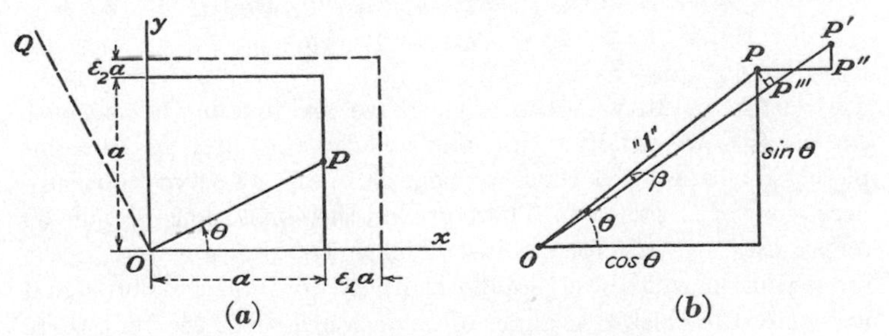

FIG. 173. Plane strains at a point.

terminals glued at 60 deg from each other between two sheets of paper (Fig. 172). By reading the three strains at the point, the stresses can be determined, and we now proceed to see how this is done by first studying the geometrical properties of the strains at a point.

Consider in Fig. 173*a* a small square of side *a*, which is strained to $a(1 + \epsilon_1)$ in the x direction and to $a(1 + \epsilon_2)$ in the y direction. In drawing the figure we have assumed that the x and y axes before and after the strain retain their original directions and that thus the angle xOy remains exactly 90 deg. (We have laid the x and y axes along what we shall later call the principal directions of strain at the point O.) We now ask for the strain in the direction OP at angle θ, and also we ask by what (infinitesimally small clockwise) angle β the line OP will turn as a result of the strain. (For the time being the letter β is just a definition of the angle shown in Fig. 173*b*; soon we shall see how β is related to the shear strain γ). For the calculation of this we turn to Fig. 173*b*, and we make OP of unit length (the "unit" must be infinitesimally small); then the sides of the triangle have lengths $\cos\theta$ and $\sin\theta$. The line OP goes to OP' as a result of the strain. We have:

$$PP'' = \epsilon_1 \cos\theta, \qquad P''P' = \epsilon_2 \sin\theta.$$

If we project the broken line $PP''P'$ on OP (or on OP', which is the same thing), we find for the elongation $P'''P'$ of the line OP:

$$\begin{aligned} P'''P' &= PP'' \cos\theta + P''P' \sin\theta \\ &= \epsilon_1 \cos^2\theta + \epsilon_2 \sin^2\theta = \epsilon_\theta \end{aligned} \qquad (a)$$

This quantity is ϵ_θ because OP is of unit length.

Projecting $PP''P'$ on PP''', we find:

$$\begin{aligned} PP''' &= PP'' \sin\theta - P'P'' \cos\theta \\ &= (\epsilon_1 - \epsilon_2) \sin\theta \cos\theta = \beta. \end{aligned} \qquad (b)$$

The latter quantity is the angle β we are looking for because $OP = 1$. Now we notice that our equations (*a*) and (*b*) are completely identical with those of page 59, only we have "strains" here, instead of stresses. Therefore, all the conclusions of page 60 for stresses are true for strains. Figure 174 shows Mohr's circle for strain, in which we plot horizontally the linear strain ϵ and vertically the clockwise angle of deviation β. We see that there are two and only two principal directions of strain in which the angle of deviation β is zero. We also see that for two perpendicular directions in Fig. 173 the values of β must be equal and opposite, because the corresponding points in Mohr's circle are $2 \times 90 = 180°$ apart. Thus an unstrained 90-deg angle in Fig. 173 will become $90 + 2\beta$ when strained, and the relation between the angle of deviation β and the angle of strain γ is that γ is twice as large as β.

When we inspect the signs involved, we shall see that it is more accurately:

$$\gamma = -2\beta.$$

The sign of $\gamma = s_s/G$ was defined in Fig. 51 as positive when the stress s_s tends to turn the element in a clockwise direction. This

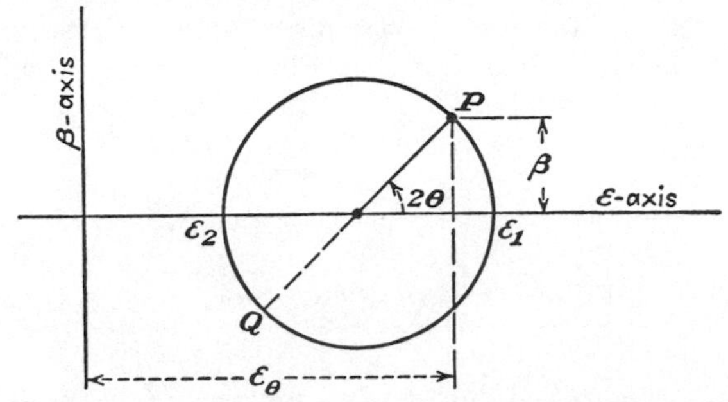

FIG. 174. Mohr's circle for strain at a point. The properties of the diagram are the same as those of Mohr's circle for stress (Fig. 50) or for moment of inertia (Fig. 55).

is repeated in Fig. 175*a*. When the element of Fig. 175*a* is allowed to strain, we can draw the distorted shape in several ways, as in Figs. 175*b*, *c*, and *d*. These three pictures have the same shape: they are only turned with respect to each other. Only the shape

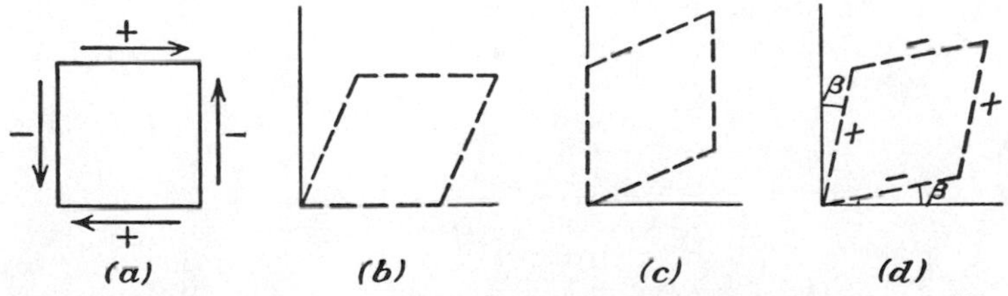

FIG. 175. A shear *stress* is positive when turning clockwise, by Fig. 51, page 61, and an angle of deviation β is positive clockwise by Fig. 173. Comparing Fig. 175*a* with 175*d* we see that a positive shear stress corresponds to a negative β. Figures 175*b* and *c* cannot be considered, because the principal axes rotate, in violation of the assumption of Fig. 173.

of Fig. 175*d* has equal and opposite angles β and thus is represented by Mohr's circle (Fig. 174). Another way of saying this is that it is only in Fig. 175*d* that the principal axes (which are the diagonals) have not rotated, while in Figs. 175*b* and *c* these principal axes have rotated. Now, on comparing Fig. 175*a* with Fig. 175*d*, we notice that a positive stress s_s (tending to turn the element clockwise) corresponds to a negative (*i.e.*, counterclockwise) angle β. Hence the relation $\beta = -\gamma/2$.

Now we are ready to apply Mohr's strain circle to the problem of the strain rosette (Fig. 172). When this rosette is glued on the test piece, we have no way of knowing the principal directions of strain so that the rosette is put on with an arbitrary unknown angle α between the first rosette direction ϵ_a and the principal direction ϵ_1. This is shown in Fig. 176. The three experimental

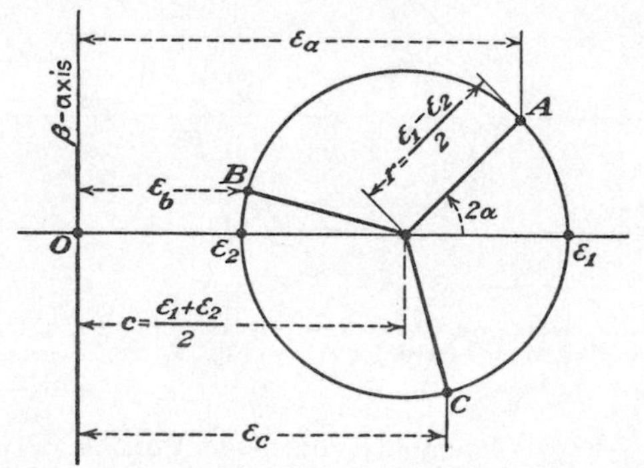

FIG. 176. The three strains ϵ_a, ϵ_b, and ϵ_c are measured by the rosette gage; then the principal strains ϵ_1, ϵ_2, and the angle α are calculated by Eqs. (*c*) and (*d*).

readings ϵ_a, ϵ_b, and ϵ_c are the horizontal projections of three points A, B, C on the circle, 120 deg apart.

The unknowns are the principal strains ϵ_1, ϵ_2, and the angle α. We read off the diagram immediately:

$$\begin{aligned}
\epsilon_a &= c + r \cos 2\alpha, \\
\epsilon_b &= c + r \cos (2\alpha + 120^\circ), \\
\epsilon_c &= c + r \cos (2\alpha + 240^\circ),
\end{aligned}$$

where $c = (\epsilon_1 + \epsilon_2)/2$ is the center distance and $r = (\epsilon_1 - \epsilon_2)/2$ is the radius of the circle, while ϵ_a is the *largest* of the three strains measured and 2α is less than 60 deg.

The solution of this set of equations is cumbersome. First we work out the cosines of the compound angles:

$$\begin{aligned}
\epsilon_a &= c + r \cos 2\alpha, \\
\epsilon_b &= c + r\left(-\frac{1}{2} \cos 2\alpha - \frac{\sqrt{3}}{2} \sin 2\alpha\right), \\
\epsilon_c &= c + r\left(-\frac{1}{2} \cos 2\alpha + \frac{\sqrt{3}}{2} \sin 2\alpha\right).
\end{aligned}$$

Adding these three together gives:

$$c = \frac{\epsilon_a + \epsilon_b + \epsilon_c}{3}, \tag{38a}$$

so that the location of the center of the circle of Fig. 176 can be calculated from the experimental data. Next, in the two first equations we bring c to the left side:

$$\epsilon_a - c = \frac{1}{3}(2\epsilon_a - \epsilon_b - \epsilon_c) = r \cos 2\alpha,$$

$$\epsilon_b - c = \frac{1}{3}(-\epsilon_a + 2\epsilon_b - \epsilon_c) = r\left(-\frac{1}{2}\cos 2\alpha - \frac{\sqrt{3}}{2}\sin 2\alpha\right).$$

Dividing these two eliminates r and leaves us with an equation in which 2α is the only unknown. After some algebra we find:

$$\tan 2\alpha = \sqrt{3}\,\frac{\epsilon_c - \epsilon_b}{2\epsilon_a - \epsilon_b - \epsilon_c}, \tag{38b}$$

so that we can calculate 2α from the test data. With the center of the circle and with 2α and ϵ_a known, the point A in Fig. 176 can be constructed and with it the circle and the points B and C at 120 and 240 deg distance. The distances ϵ_b and ϵ_c so found then must agree with the test results, and this constitutes a check on the correctness of our calculations.

After the principal *strains* ϵ_1 and ϵ_2 are so determined, the *stresses* can be calculated from Eq. (18) (page 75).

From Eq. (18) it is clear that the stress in a given direction is equal to the strain times E *only* if the stress at right angles to it is zero, *i.e.*, only in the case of "one-dimensional stress," such as occurs in pure tension or bending. Suppose then we ask whether or not it is possible for a two-dimensional case to make a *strain* gage by which we can measure directly the *stress* in a given direction. We find that it *is* possible, but not in too simple a way. The gage we construct is illustrated in Fig. 177. The resistance variation of the long leg is proportional to $l\epsilon_x$, and that of the short leg is proportional to $\mu l\epsilon_y$, so that the gage reading is $l(\epsilon_x + \mu\epsilon_y)$. Substituting Eq. (18) into this, we find for the reading of the gage:

$$\frac{l}{E}[s_x - \mu s_y + \mu(s_y - \mu s_x)] = l\,\frac{1 - \mu^2}{E}\,s_x,$$

which is independent of the stress in the direction of the short leg of the gage. This is always true, whether s_x and s_y are principal stresses or not.

Resistance-electrical-strain gages are now widely used in many industries, particularly in aircraft industries. In an experimental airplane, hundreds of such gages are attached all over the wing and the rest of the structure, which is then subjected to definite test loads.

Another very useful device of recent development is brittle lacquer coating. The test piece in its unstressed state is painted with a special clear varnish, which dries up brittle. Upon loading the piece, the varnish will crack when the local strain exceeds a

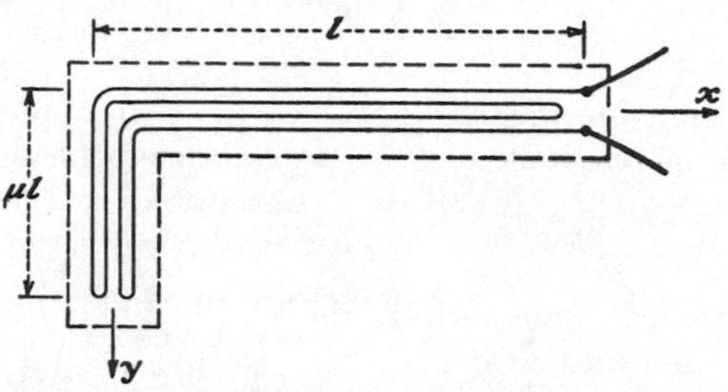

FIG. 177. An L-shaped strain gage in which the short leg has μ times the length of the long one, when μ is Poisson's ratio of the test piece. The reading of this gage is proportional to the stress along the long leg and is independent of the stress along the short leg. Thus the gage is a *stress* gage, and not a strain gage.

certain limit. Thus, by gradually loading the piece and by carefully watching for cracks in the varnish coating, we can determine the location of maximum strain on the piece and the direction of such strain. The cracks, however, are difficult to see, and the method is not so well perfected as that of the electrical-resistance gages. It serves as a useful preliminary to an elaborate strain-gage test by guiding us in placing the gages in suitable locations and directions.

Problems 329 *to* 337.

41. Fatigue. It is common experience that a piece of steel can be broken more easily by alternating or reversing stresses than by a constant stress, a fact that comes in handy when we want to break something and when no pliers or other tools are around. Many structures and machine elements in daily use are subjected to alternating stress so that the fact that such parts will eventually break under an alternating load two to four times smaller than the steady load of ultimate failure is of very great practical importance. Under alternating stress the material is said to *fatigue*, and the fatigue-strength properties are determined experimentally by the simple machine of Fig. 178. A dozen or more identical specimens of the material in question are made, and one of them is inserted in the machine, which is rotated by an electric motor at considerable speed. The load W is divided into equal halves so that the specimen is loaded in "pure bending." The upper fiber of the beam is in compression, and the lower fiber is in tension, but since the beam rotates, each individual fiber passes through a

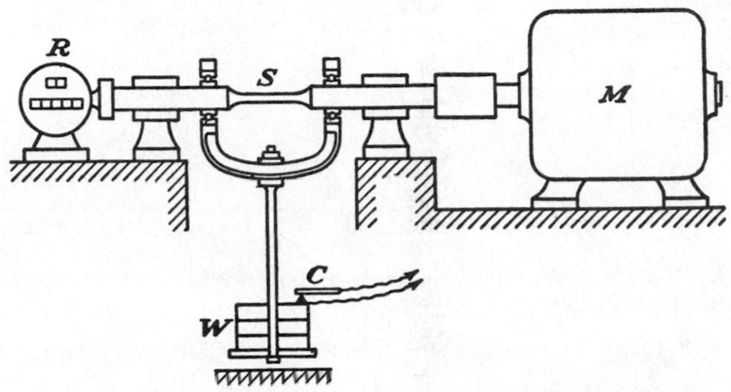

Fig. 178. Revolving-beam fatigue-testing machine, driven by an electric motor M. The test specimen S is loaded in pure bending by the weight W. The revolution counter R registers the number of cycles. The electric contactor C stops the motor when the specimen breaks.

complete reversal of stress for each revolution of the shaft. Thus if the motor runs at 1,750 rpm, as usual, the specimen can be subjected to a million cycles in about 10 hr. A revolution counter is attached to the beam, and upon failure the load W drops down $\frac{1}{4}$ in. and breaks a contact, shutting off the electric motor. When the experimenter arrives in his laboratory in the morning, he finds the broken test pieces with the counters indicating the number of cycles before failure. For a test on a given material, usually a dozen or so identical standard test pieces are made and set running, with different loads W, until failure occurs or until a million cycles accumulate, whichever happens first. A plot of the stress against the number of cycles gives a diagram like that of Fig. 179. It has been found by long experience that, if a piece can outlive a million stress reversals, it can stand these reversals indefinitely. This alternating stress is called the "fatigue strength" or the "fatigue limit" of the material, and it appears as a horizontal asymptote in Fig. 179. Although there is no definite relation between "ultimate strength," "yield point," and "fatigue limit," so that it is necessary to determine all three separately by test, it is found that for most ductile materials the yield point is ap-

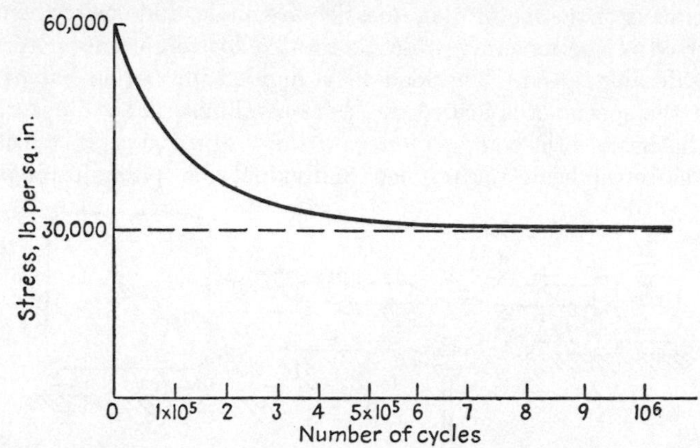

Fig. 179. Fatigue curve of mild steel with an ultimate strength of 60,000 lb/sq in. and a fatigue strength of 30,000 lb/sq in.

proximately equal to the fatigue strength, each being about half the ultimate strength.

Figure 180 shows the cross section of a typical fatigue break, from which we can more or less read the history of the failure. It starts with a small crack at the surface of the piece, at or near the point of maximum stress, of course. The sharp bottom of that crack then immediately acts as a focus of large stress concentration (Fig. 166), and the crack develops further. While this is going on, the two opposite faces of the crack rub over each other, making these faces even smoother than when they were formed first. The crack develops to a point where the remaining cross section becomes so weak that it can be broken by a single application of the load itself, and then it breaks off all at once. This last part of the failure is not smooth but quite rough, as shown by the figure. The semicircular curves about the start, or "focus," of the fatigue

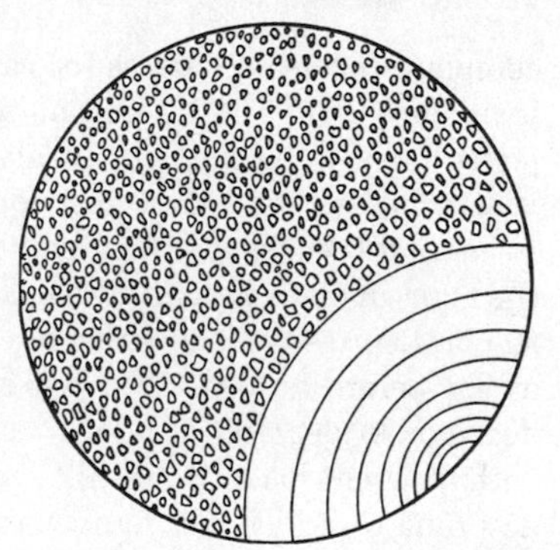

Fig. 180. Fatigue fracture of a shaft, showing beach marks around a focus in a smooth portion, and a rough portion where the final break occurred suddenly.

break, often called "beach marks," are very characteristic and are a sure indication of whether or not a given failure is caused by fatigue or by a single application of the load. The fact that a fatigue crack progresses owing to the stress-concentration effect at its bottom is related to the fact that *the plane of the fatigue failure is perpendicular to the direction of maximum stress.* Thus the direction of the fatigue failure is perpendicular to the direction of maximum stress. Hence the direction of the fatigue failure is another clue by which it can be distinguished from a break by a single

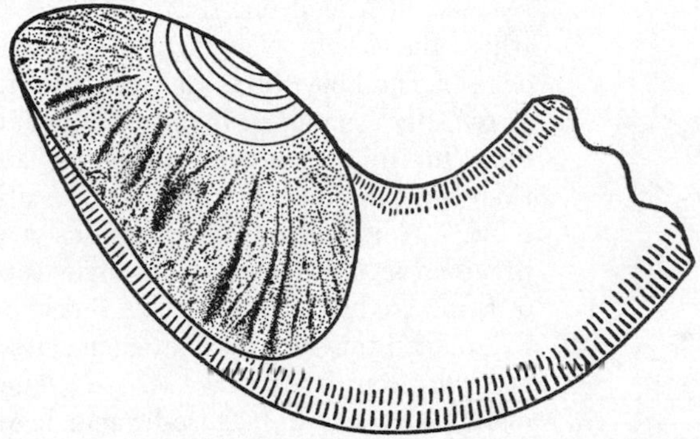

Fig. 181. The focus of the fatigue failure in a coil spring is on the inside of the coil because there the torsion stress and the direct shear stress are additive.

application of a load in a ductile material. *A single application failure* starts with slipping of layers over each other by shear, and the ultimate failure is a shear failure, the direction of the plane of failure therefore being that of the maximum shear stress, *at 45 deg with respect to the maximum principal stress.* Thus a coil spring that has failed in fatigue always shows a 45-deg break, with the focus of the break at the inside of the coil (for the reason explained on the top of page 28). All of this applies only to *ductile* materials.

In a *brittle* material, failure occurs, not by shear, but rather by direct separation due to the maximum tensile principal stress. A convenient way to show this in a classroom is with a piece of schoolteacher's chalk (Fig. 182). For such a material the direction of the break for single or repeated load applications is the same. It is advisable that the reader carefully study these relations, because this will enable him to diagnose the cause of a failure, which is the first important step in deciding on how to redesign the piece in order to avoid repetition of the trouble.

It is principally in connection with fatigue that stress concentration is so extremely important. Figure 168 shows that at the edge of a small hole the stress is about three times as large as in the full piece itself, and this is true even for a very small hole. Now when we put a flat steel bar in a tensile-testing machine and compare its (single-application) ultimate load with that of a similar bar with a small hole in it (Fig. 168), we find that the two loads are almost the same, which seemingly contradicts our notion of stress concentration. What happens is that the yield point at the edge of the hole is indeed reached at a load one-third of that where the whole bar would yield. At that low load the edge of the hole starts to yield, and the local stress there cannot go higher. With increasing load, finally the entire bar yields, and from there on the influence of a small hole is negligible. Thus stress concentrations are of no practical significance for the single application of loads to ductile materials. But suppose we had placed the same two bars, one with and the other without hole, in a tensile-fatigue-testing machine; then the failure-fatigue load of the perforated bar would really be about one-third of that of the non-holed one.

FIG. 182. Failure of a stick of school-teacher's chalk when twisted off. In bending, the piece will break off perpendicular to the length direction.

At the small alternating failure-fatigue load of the holed piece the small regions near the edges of the hole are being subjected to a push-pull yield stress, which keeps the material continuously working, developing a small local crack. Once this happens the original stress concentration is increased, and the piece ultimately breaks altogether. Therefore one of the most important principles of practical design is the scrupulous avoidance of sharp corners or other "stress raisers" if there is any suspicion of alternating stress. In this connection we can now appreciate why there is often such a large discrepancy between the ultimate strength of a material and the stress at failure. If the applied load is a shear or torsional load, the actually measured or calculated stress at failure often is about $\frac{1}{8}$ of the ultimate strength. The factor 8 can be broken up into three factors 2, the first for the ratio of fatigue strength to ultimate strength, the second for the ratio of shear stress to tensile stress (Fig. 57, page 66), and the third for unavoidable stress concentration.

Figure 180 gives us a means to decide that a failure after it has occurred is a fatigue failure, but it is of greater importance to find that out experimentally before the failure has taken place. An "incipient fatigue failure" consists of a crack in the piece, which on the outside surface is seen only as a very fine line of short length. Unless the surface is mirror polished, it is almost impossible to detect such cracks even under bright light with a good magnifying glass. For locating such cracks two processes are in existence, known by their trade names as "Magnaflux" and "Zyglo." The Magnaflux process works only with steel or similar magnetic materials. The piece is brought between the jaws of a powerful electromagnet and is magnetized. Then very fine iron powder is applied to the surface, either dry or wet, suspended in oil. The magnetic field near a crack or other sharp corner is accentuated by it so that the iron particles "go for it" and show up the crack quite clearly.

The Zyglo process does not depend on magnetism and hence can be applied to aluminum and brass as well as to steel. It depends on capillary action of the crack. The piece in question is painted with a thin oil or kerosene in which a finely powdered fluorescent substance is suspended. After painting, it is carefully dried off by wiping with a rag. The only place where the oil can still remain after this operation is in the cracks. The piece is then viewed under ultraviolet light, which is no "light" at all, and leaves the whole piece dark, except the fluorescent material in the crack, which appears illuminated. Any one of these two procedures is often applied if fatigue cracks are suspected.

Another troublesome phenomenon is corrosion fatigue, which occurs when a machine element is subjected to alternating stress in corrosive surroundings, even mildly corrosive ones, such as sea water. As soon as the slightest crack is formed, the corrosive liquid penetrates it and works on its sharp bottom edges. The combination of the corrosive action and alternating stress at the bottom of the crack is vicious and reduces the safe stress very materially. This is a large subject in itself, on which active research is being conducted in several places.

Figure 179, defining the "fatigue strength" of a certain material, is based on a total reversal of stress (Fig. 183*a*), where the maximum tensile and compressive stresses are equal. In most practical cases the tensile and compressive stresses are different in magnitude, and in that case the total stress can be looked upon as the superposition of a steady, or constant, component s_{steady} and an alternating component s_{alt}. If s_{steady} is larger than s_{alt}, the total combined stress never reverses but only fluctuates on the same side of zero; when s_{alt} is the larger of the two, the stress does reverse but it is greater on one side of zero than on the other. There is no logical way whereby the failure stress in the case of Fig. 183*b* can be calculated from the fatigue limit of Fig. 183*a* and from the steady ultimate stress; the only way to find out is by experiments. Many such experiments have been conducted on various ductile

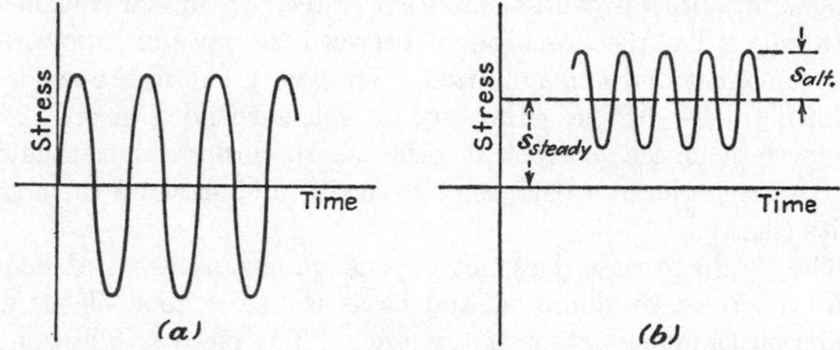

FIG. 183. (*a*) Complete stress reversals; (*b*) the superposition of a steady stress and an alternating stress.

materials, and the result of such tests for one particular material has been plotted in the diagram of Fig. 184 as curve 1. For different ductile materials different curves exist, but they all are somewhat above the straight line 2, connecting the ultimate steady stress with the fatigue limit of completely alternating stress. The curve 2, therefore, is usually regarded as a conservative interpretation of the test results for all ductile materials. Thus, for a combination of stress above curve 2, the piece will fail, and below curve 2 it will not fail. However, this is not quite satisfactory. For example, a point in the diagram just below curve 2 and close to the vertical axis represents a case that will not fail, but it yields all over the place and hence is not acceptable in most engineering constructions. For that purpose the curve 3 connecting the yield point with the fatigue limit is more useful. This was first suggested by Soderberg, and the line 3 is usually called the "Soderberg criterion." For actual design these stresses, which are just barely satisfactory, must be divided by a factor of safety, as shown in curve 4. The equation of that curve is:

$$\frac{s_{steady}}{\text{Yield Stress}} + \frac{s_{alt}}{\text{Fatigue Limit}} = \frac{1}{\text{Factor of Safety}}. \qquad (39)$$

In a good design this equation is satisfied at the most dangerous point of the construction.

Before leaving Fig. 184 it is interesting to note that for the case that $OF = OY$ (which is about right in practice) the meaning of the straight line FY, curve 3, is that the highest peaks of Figs. 183*a*

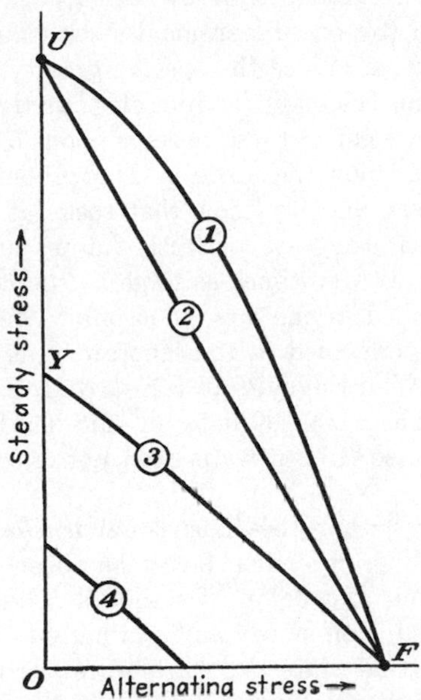

Fig. 184. Diagram in which the coordinates are the steady-and alternating-stress components at a point of a machine element or of a test piece. The letters U, F, and Y mean the ultimate, fatigue, and yield stresses; curve 1 is the experimental-failure curve for one particular material; curve 2 is a conservative interpretation of curve 1, good for all ductile materials; curve 3 is an acceptable curve where the yielding or deformation is not too large; curve 4 is curve 3 multiplied by a factor of safety and represents the "working stress."

and *b* are at the same total stress. The reader should prove this for himself.

Problems 338 *to* 345.

42. Strength Theories; Conclusion. On pages 73 to 77 we discussed several "theories of failure" or "theories of strength," among which the maximum-shear theory is the most important one because it agrees quite well with the first appearance of yield in a ductile material. We saw that the most pronounced characteristic of that theory is that no yield will occur for three-dimensional hydrostatic compression or tension, no matter how high the stresses become. Yield appears only as a consequence of shear stress, which exists only when the difference between the principal stresses has an appreciable value. A practical example of this is seen in rocks at great depth in the earth. Such rocks have a crushing strength (by one-dimensional short-column crushing) of about 3,000 lb/sq in. Since the specific gravity of rock is about 2.5, the reader can calculate for himself that the pressure inside the earth by the weight of these rocks is about 5,500 lb/sq in. for each mile of depth below the surface. This pressure is hydrostatic (three-dimensional), and we know that rocks at great depths do not yield or crush; they are perfectly happy under hydrostatic compressive stresses fifty times as large as the stress that would crush them if applied in one dimension only. Tests at high pressures have been performed in the laboratory as well; the Nobel Prize winner P. W. Bridgman at Harvard University has made pressures as large as 1,000,000 lb/sq in. and has found that weak materials such as ice (frozen water) did not crack up under that stress.

An experiment with three-dimensional *tension* is much more difficult than with compression, but it has been performed once, by the Russian scientist Joffe. He took a solid glass sphere (a marble) and cooled it off slowly and gradually to the temperature of liquid air. In that state the marble presumably was without stress. Then the marble was taken out of the liquid-air bottle, exposed to room temperature, and carefully watched. The outside layers of the marble would warm up while the center was still cold. The thermal expansion of the outer layers was prevented by the cold inside, thus putting the center of the sphere in hydrostatic tension. Although this tension could not be measured directly, it could be calculated by the theories of heat conduction and we have sufficient faith that such calculations are not too far from the truth. The calculation showed hydrostatic tension at the center of the sphere far greater than the ordinary, one-dimensional tensile strength of glass, but the center of the marble remained clear and did not crack or tear.

These experiments and observations convince us fairly well that the maximum-shear theory of the first appearance of yielding is a reasonably accurate description of fact. The main practical conclusion of that theory is that, in pure shear, yielding will start at a shear stress half as large as the tensile stress required for yielding in a tensile-test piece (Fig. 57, page 66). Many accurate experiments have been carried out, and it has been found that this ratio of shear stress to tensile stress for yielding actually is somewhat larger than 0.50; it is more nearly 0.57. In order to explain this, another strength theory has come to the fore during the last twenty-five years: the *theory of maximum distortion energy.* If a small element $dx\ dy\ dz$ (Fig. 185) is subjected to three principal stresses s_1, s_2, s_3, it contains a certain amount of strain energy. For the special case that these stresses were all alike, there would be hydrostatic stress, and any 90-deg angle in the element would remain 90 deg after stressing: there would be no shear anywhere. In case the three principal stresses s_1, s_2, s_3 have different values, there will be some shear. Now we can think of the stress $(s_1 + s_2 + s_3)/3$ as the average principal stress, and if such a stress should apply hydrostatically, the element would contain less strain energy than when s_1, s_2, and s_3 were acting on it. The total energy stored in it can be thought of as consisting of the sum of the hydrostatic stress energy and the shear or distortion energy. The new theory now claims that the first yielding occurs when this distortion energy reaches a critical value. Obviously the distortion energy is zero for a hydrostatic stress condition of any value so that the criterion is substantially the same as that of maximum shear. It differs only in detail.

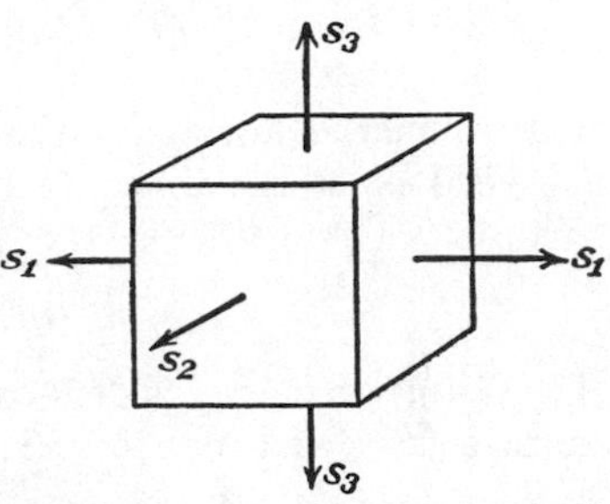

FIG. 185. An element $dx\ dy\ dz$ subjected to three different principal stresses.

The analytical expression for the new theory is considerably more complicated than that for the maximum shear theory. First we calculate the energy in the element of Fig. 185 for three different stresses. The stress s_1 corresponds to a force $s_1\ dy\ dz$. By Eq. (18*a*) (page 75), the elongation in its direction is $[s_1 - \mu(s_2 + s_3)]\ dx/E$, so that the work done by the stress s_1 is:

$$\frac{1}{2}\, s_1[s_1 - \mu(s_2 + s_3)]\,\frac{dx\ dy\ dz}{E},$$

or

$$[s_1^2 - \mu(s_1 s_2 + s_1 s_3)]\,\frac{d\,\text{vol}}{2E}.$$

The stresses s_2 and s_3 do amounts of work that can be found from the above expression by interchanging the subscripts 1, 2, and 3. Adding the three amounts we find for the strain energy in the element:

$$dU_{\text{total}} = [s_1^2 + s_2^2 + s_3^2 - 2\mu(s_1 s_2 + s_1 s_3 + s_2 s_3)] \frac{d \text{ vol}}{2E}. \qquad (a)$$

This is the actual strain energy. For the case that all the stresses are equal to s, it reduces to:

$$dU = 3s^2(1 - 2\mu) \frac{d \text{ vol}}{2E},$$

and if that stress s is the "average principal stress" $s = (s_1 + s_2 + s_3)/3$, the energy is:

$$dU_{\text{hydrostatic}} = \frac{1 - 2\mu}{6E} d \text{ vol } (s_1 + s_2 + s_3)^2. \qquad (b)$$

The distortion energy is defined as the total energy less the hydrostatic energy, and the reader is asked to carry out the subtraction and find:

$$dU_{\text{distortion}} = \frac{1 + \mu}{3E} (s_1^2 + s_2^2 + s_3^2 - s_1 s_2 - s_1 s_3 - s_2 s_3) \, d \text{ vol}. \qquad (40)$$

This expression becomes zero for the special case of three equal stresses, as it should (see Problem 346).

Let us now apply Eq. (40) to the case of simple tension with stress s and also to the case of pure shear with stress s_s. For the first case we have $s_1 = s$, $s_2 = s_3 = 0$, so that the energy is:

$$dU_{\text{distortion}} = \frac{1 + \mu}{3E} s^2 \, d \text{ vol}.$$

For pure shear s_s the principal stresses are (see Fig. 60, page 68) $s_1 = s_s$, $s_2 = -s_s$, $s_3 = 0$, so that the distortion energy is:

$$dU_{\text{distortion}} = \frac{1 + \mu}{E} s_s^2 \, d \text{ vol}.$$

By the new theory, yield will occur when the distortion energy reaches a certain value. Equating these energies for the tensile and shear tests, we find:

$$\left(\frac{s_s}{s}\right)^2 = \frac{1}{3} \quad \text{or} \quad \frac{s_s}{s} = \frac{1}{\sqrt{3}} = 0.57,$$

which agrees well with experimental evidence.

For the important simplified case of plane stress, where $s_3 = 0$, Eq. (40) simplifies to:

$$dU_{\text{distortion}} = \frac{1 + \mu}{3E} (s_1^2 + s_2^2 - s_1 s_2) \, d \text{ vol} \tag{40a}$$

For the still simpler case of the usual tensile stress, where yield occurs at a stress $s_1 = s_y$; and $s_2 = s_3 = 0$, we have:

$$dU_{\text{distortion}} = \frac{1 + \mu}{3E} s_y^2$$

Comparing this with Eq. (40a), we see that for a two-dimensional-stress case (s_1, s_2) yielding will start according to the distortion-energy theory when

$$s_y^2 = s_1^2 + s_2^2 - s_1 s_2.$$

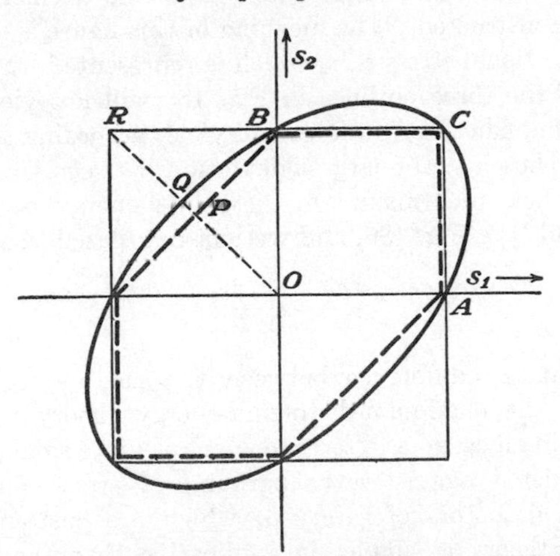

FIG. 186. Various strength theories for a two-dimensional stress combination s_1, s_2. The theory of maximum stress is represented by the square figure in thin outline; the maximum shear theory is shown by the dashed figure; the distortion energy theory is shown by the ellipse in heavy outline.

If this equation is plotted in a s_1-s_2 diagram, as in Fig. 186, where s_y is considered constant, and s_1 and s_2 variable, we find an ellipse. In the diagram the distances OA as well as OB are equal to the simple yield stress s_y. The maximum stress theory plotted on this same diagram is represented by a square with O as center and s_y for the length of half its side because it states that if the larger of the two stresses s_1 and s_2 reaches the value s_y, yielding will take place.

When plotting the maximum shear theory on this diagram we must distinguish two cases, depending on whether the stresses s_1, s_2 have the same or opposite signs. If they have the same sign, then Mohr's circle is as the solid one of Fig. 66, page 72, and the maximum shear stress is half the larger of s_1 or s_2. Thus yielding will occur if either s_1 or s_2 reaches the value s_y, and the criterion is the same as the maximum stress theory. On the other hand, if s_1 and s_2 are of opposite sign, Fig. 66 is modified in that the fully drawn circle is the largest of the three and the maximum shear stress is $\frac{1}{2}[s_1 + (-s_2)]$, which then should be $\frac{1}{2}s_y$ when yield starts. This gives the equation:

$$s_1 - s_2 = s_y = \text{constant},$$

a 45-deg straight line in Fig. 186. Thus the dashed outline in Fig. 186 is constructed. The meaning of this figure is that a state of two-dimensional stress s_1, s_2 which is represented by a point inside any of the three outlines in Fig. 186 will not yield, while a point s_1, s_2 outside these outlines will yield, according to whatever criterion is chosen. The largest deviation between the criteria of maximum shear and maximum distortion energy occurs at the points P and Q in Fig. 186, and we just calculated above that

$$\frac{OQ}{OP} = \frac{0.57s_y}{0.50s_y}$$

We see that the difference between the maximum-shear-stress theory and the maximum-distortion-energy theory is less than 15 per cent in all cases, a difference which is quite small, especially when considered against a background of stress-concentration factors 3 and factors of safety 4, which are customary. The shear-stress theory is simpler in application than the distortion-energy theory, and gives adequate answers, so that it still is being widely used as a criterion for the appearance of first yielding.

Problems 346 *to* 350.

PROBLEMS

1. A propeller shaft in the largest and most powerful ships transmits about 50,000 hp. Assume that the propeller transforms this power into a forward push on the ship with an efficiency of 70 per cent and that the ship's speed then is 30 knots (1 knot is 6,080 ft/hr). The shaft is 300 ft long from the propeller to the thrust bearing in the engine room, where the thrust is transmitted to the ship's hull structure. The diameter of the (solid, circular) shaft is 24 in. Calculate the elastic contraction of the shaft at full power.

2. In the short period of history in which streetcars and concrete road pavements existed simultaneously, the rails were embedded in concrete. Both rails and concrete being integral with the ground, there is no temperature expansion in either. Suppose the rails are without stress at the average temperature of 50° F, what is the stress in them at the extreme temperatures of (*a*) 100° F; (*b*) $-$ 20° F? The thermal-expansion coefficient of steel is 6.7×10^{-6} in./in./° F.

3. In the late 1930's, when the range of aircraft was limited, the technical press contained serious proposals of large floating landing fields to be located in mid-ocean and to be anchored at considerable depth of water. Suppose these anchors were held by cables of constant cross section. The top of the cable would have to carry the entire weight of the cable (less buoyancy) down to the ocean bottom. The weight of the steel is 0.28 lb/cu in., and that of sea water is 64 lb/cu ft. Assuming constant cross section, how long a cable could be let down into the deep ocean before its top end would yield? The yield point for soft steel is 30,000 lb/sq in., and for the best piano wire it is 180,000 lb/sq in.

4. Two rods of the same material, of equal length l, and of cross section A and $2A$, respectively, are mounted between two rigid (*i.e.*, nondeformable) cross frames. The frames are pulled by a pair of forces P, located at distance x from the thin bar. Derive an expression for the "deflection" between the two forces P, that is, for the elastic increase in distance between them. Check this general formula for the two special cases $x = 0$ and $x = a$ before you look at the answer in the back of the book.

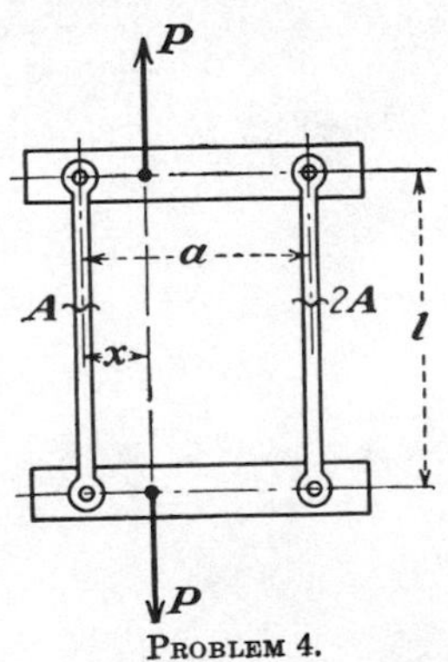

PROBLEM 4.

5. An aluminum bar of circular cross section ($E = 10 \times 10^6$ lb/sq in.) consists of a 2-ft piece of 2 in. diameter and an 18-in. piece of 1 in. diameter, in series. Calculate the elongation as a result of a force of 5,000 lb

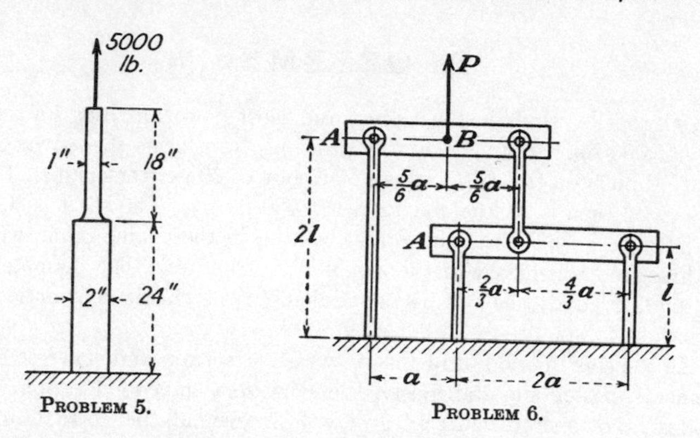

PROBLEM 5. PROBLEM 6.

6. Four bars of the same cross section and of the same material with lengths l and $2l$ are arranged as shown. The cross pieces A are rigid. Calculate the stiffness k at the point B, which is the "load per unit deflection" at point B.

7. The figure shows one form of strain gage used in the laboratory for measuring the strain in test pieces. At point A a small diamond-shaped piece with two knife-edges is pressed against the test piece. The other knife-edge of the diamond rests against a light rod R, of which the other

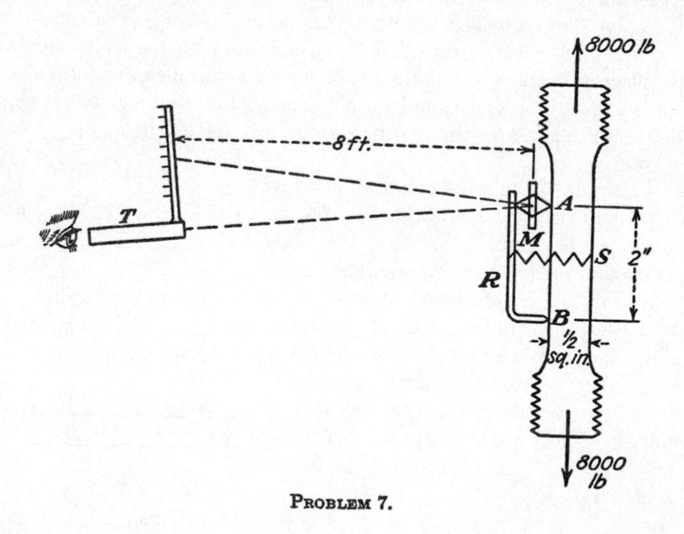

PROBLEM 7.

end with a knife-edge rests against the test piece at B. The assembly is held in place by a spring S. When the gage length AB elongates, the rod R retains its length and the relative motion between A and B turns the diamond through a small angle. The diamond carries a small mirror, which turns with it. The mirror is viewed through a telescope 8 ft from it, on which is attached a graduated scale, which is seen in the mirror. When the apparatus is set up, a reading of the scale is made through the telescope T via the mirror M while the test piece is without stress. Then the test piece is loaded, and the reading repeated. The difference in scale readings is a highly magnified reading of the elongation of AB. One advantage of this construction is that, if the entire test piece AB shifts up or downward without strain, no tilt of the mirror M results and no shift on the scale is observed. Thus the shift in scale reading is caused only by a change in the length AB.

a. Derive the magnification factor of the strain gage from the dimensions shown.

b. Between loads 0 and 8,000 lb on the test piece a shift in scale reading of 1.65 in. is observed. Deduce from it the modulus of elasticity of the test piece.

8. The deepest oil well drilled so far is about 18,000 ft deep. The drilling is done by means of a shaft, which goes down all the way from the surface to the drill tool below. Independent of the work this shaft does, there are large stresses in it due to its own weight, even when the shaft does not rotate and is at rest.

a. Assuming a cross section of 1 sq in. uniformly along the length of the steel drill rod, calculate the stress in the rod at the top due to its own weight, neglecting the weight of the drill tool below and of the various couplings along the rod attaching its various sections to each other.

b. Repeat this calculation for a rod of which the bottom 6,000 ft has a 1-sq-in. cross section, the middle 6,000 ft a 1.5-sq-in. section, and the top 6,000 ft a 2-sq-in. section. Where does the maximum stress occur?

9. A bar of length l, constant cross section A, modulus E, and weight per cubic inch γ is suspended from its top and is hanging downward under the influence of gravity.

a. Plot the stress distribution against the lengthwise location, and calculate the maximum stress.

b. The elastic deformation of the bar due to gravity causes each point x (except the top one) to move downward with respect to its location in the unstressed state. Plot the distribution of this elastic displacement against the lengthwise location, and calculate the maximum value of this displacement.

c. Calculate numerically the maximum sag of the drill rod of Prob. 8*a*.

d. The same for the drill rod of Prob. 8*b*.

10. The previous problems suggest this one. What shape should a freely hanging cable have in order to be of constant stress? More precisely, how does the cross-sectional area A have to vary with the depth x in order that the tensile stress remain constant all along the length?

a. Solve this problem by considering a piece dx of the bar, where the cross section is A, and by writing the equation of vertical equilibrium of this piece dx as a free body. Then integrate the equilibrium equation, and express the area A as a function of x, using such other letters as may be necessary. Don't forget the integration constant!

b. When a steel rod has to be designed for 10,000 lb/sq in. stress and its diameter at considerable depth is 1 in., what should the diameter be 5,920 ft higher up?

c. What is the elastic elongation of the 5,920-ft rod of question *b*?

11. A uniform bar of length $2l$ is rotated in a horizontal plane about its mid-point as center with an angular velocity ω. This case must be reduced to a problem in statics by the application of d'Alembert's principle, the "forces" acting being "centrifugal" forces.

a. Plot the stress distribution and the elastic displacement against the radial distance from the center, and derive formulae for the maximum stress and the maximum displacement.

b. Let the rod be made of steel of 2 ft total length, spinning at 3,000 rpm. Calculate the maximum stress and the total elongation.

12. A 1-ton wall hoist consists of a horizontal beam, supported by a tension rod at 30 deg as shown. The rod is to be designed with a factor of safety 5 with respect to its yield point of 30,000 lb/sq in. Determine its necessary cross section.

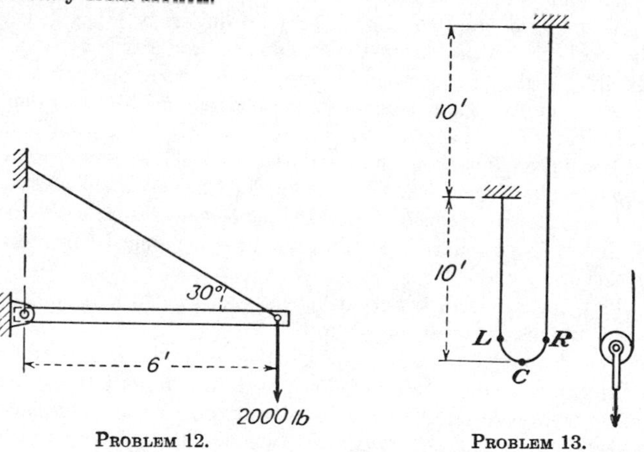

PROBLEM 12. PROBLEM 13.

13. A flexible manila rope of 30 ft length is attached with its ends to two points 10 ft apart in height and close together horizontally. Consider three points on the rope: the point C, exactly 10 and 20 ft from the ends, and the points L and R, each 2 in. to the left and right of point C. First we attach a small weight, say 10 lb, at C so that both sides of the rope are straight. Then we hang 500 lb at C.

a. Calculate the stresses and deflection of C caused by these 500 lb.

b. Do the same for the 500 lb being attached at L instead of at C.

c. The same for R instead of C.

d. The original 10 lb is removed from C and a 10-lb pulley (of small dimensions and without friction) is hung inside the rope. Now 500 lb is suspended from the pulley axle. Calculate the stresses and deflection.

For all cases, the cross section of the rope is ½ sq in., and its "modulus" is 150,000 lb/sq in.

14. Two equal steel wires of length $l = 1$ ft, modulus $E = 30 \times 10^6$ lb/sq in., and cross section $A = 0.001$ sq in. are attached to each other at B and to solid, immovable walls at A and C. A force $P = 50$ lb is imposed at B along the wires. A "wire" is a structural element that can take only tension; when the attempt is made to put a compressive force on it, the wire bends out sidewise and the compressive load remains zero.

a. For $P = 0$ the wires have no stresses in them. Find the deflection of point B for piano wire with a yield point of 150,000 lb/sq in.

b. Again with no internal stresses at zero load find the deflection at B for soft iron wire with a yield point of 30,000 lb/sq in.

c. At zero load the piano wires ($YP = 150{,}000$) have an "initial tensile stress" of 60,000 lb/sq in. in them. How far will the point B deflect under the influence of the 50-lb load P?

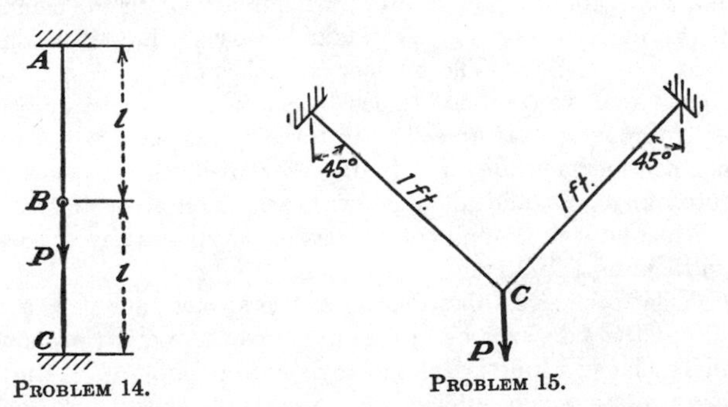

PROBLEM 14. PROBLEM 15.

15. Two piano wires ($A = 0.001$ sq in.; $E = 30 \times 10^6$ lb/sq in.; $YP = 150{,}000$ lb/sq in.) of 1-ft length each are arranged at 45 deg with respect to the vertical. What is the deflection of the center point C under the influence of a force $P = 100$ lb?

16. Three piano wires of the same dimensions as in Prob. 15 are arranged 120 deg apart, without internal initial stress.

a. A force of 100 lb is applied at C in the direction along one of the wires, say vertically downward. Calculate the downward deflection and also the horizontal deflection, if any.

b. Let the 100-lb force at C be acting in a direction perpendicular to one of the wires, say horizontally. Calculate the horizontal and vertical components of the deflection of C.

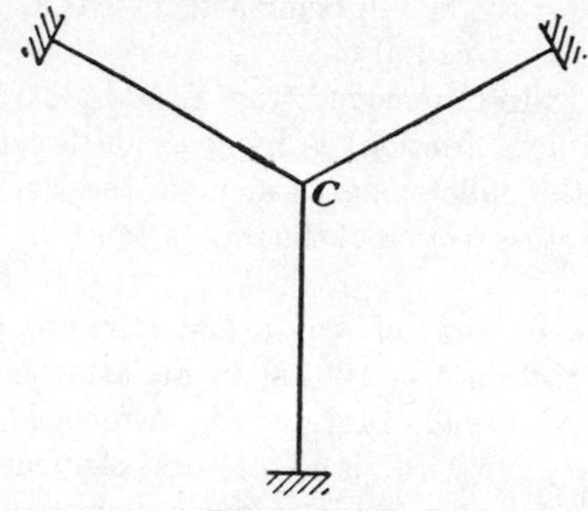

PROBLEM 16.

17. In Prob. 16 let the wires have an initial ("locked-up") tensile stress in them of 100,000 lb/sq in. Find the deflection of the center under the influence of

a. A vertical force of 100 lb.

b. A horizontal force of 100 lb.

c. A force of 100 lb at angle α with respect to the vertical.

18. The previous problem is almost that of a bicycle wheel. The spokes of such a wheel are not quite "wires," but almost so, for the compressive force that a bicycle-wheel spoke can take is very small, and it is justified to neglect it altogether. The number of spokes of the wheel is so large that a good analysis can be obtained by considering that number to be infinite. Moreover we assume that all these spokes lie in one plane.

Hence consider a wheel of 2 ft diameter with many spokes equally distributed angularly and all together having a cross section A of 0.1 sq in. When no load is applied to the axle, the spokes have a locked-up stress in them of s_0 lb/sq in.

a. Calculate the stress distribution in the spokes caused by a load $P = 100$ lb at the axle, expressing the answer as the stress (in pounds per square inch) being a function of the angle α from the top. Assume that the locked-up stress s_0 is sufficient to keep all the spokes stiff, including the bottom one.

b. What is the necessary s_0?

c. What is the deflection?

d. What is the weight of all the spokes combined?

19. This problem is preliminary to Prob. 21 about a bicycle wheel. Consider a stiff rod AA, 4 in. long. The ends of it are connected with four wires to two solidly anchored points B, all of it lying in the plane of the paper. The bar AA is now subjected to a couple F, F as shown. If there are no locked-up tensions in the wires, this moment is taken by tensions in two wires only, the other two going slack. When there is sufficient initial tension in the wires, all four wires help carrying the moment.

a. Derive formulae for the stresses in the wires and for the deflections at points A, assuming sufficient locked-up tension.

b. Find numerical answers for stress and deflection for the case of steel wires, each of 0.001 sq in. cross section, with an initial tensile force of 30 lb. The forces F are 20 lb each.

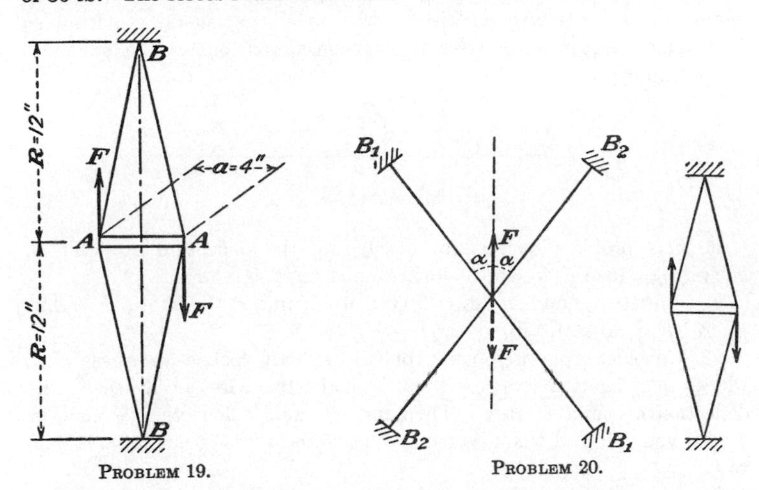

PROBLEM 19. PROBLEM 20.

20. In this problem we come a little closer to the bicycle wheel by considering two sets of spokes, each like the one of Prob. 19, and angle 2α apart in the plane of the wheel. The moment FF on the axle is in a plane halfway between the planes of the two sets of spokes. From symmetry we conclude that the deformation consists of a small angle of turn of the axle in the plane FF.

a. Derive formulae for the stresses and deformations.

b. Substitute the same numbers as in the previous problem.

21. Consider a bicycle wheel with an infinite number of spokes uniformly distributed in two shallow cones, of dimensions a, R, as sketched with Probs. 19 and 20. The combined cross section of all the spokes (in both cones) is $A_0 = 0.1$ sq in. The initial locked-up tensile stress in the spokes is 3,000 lb/sq in. The wheel rim is assumed not to deform and is held in place by proper means while the (rigid) axle is subjected to a moment of 600 in.-lb.

a. Derive a formula for the stress distribution in the spokes in terms of an angle α along the periphery.

b. Substitute the above numbers, and find the maximum stress in the construction.

22. Two wires or rods of length l, cross section A, and modulus E are hinged together at their center point C. There is a "large" initial (= locked-up) tensile stress in the wires. The meaning of the word "large" will now be explained. When there is a locked-up stress in the rods, the length AC is greater than the stressless length by the "locked-up elongation." When a load P is applied, as shown, the point C moves, thereby elongating AC by an amount called the "elongation due to loading." The locked-up stress is said to be "large" when the locked-up elongation is large in comparison with the elongation due to load, so that as a consequence the stress in AC is considered constant, independent of the load P.

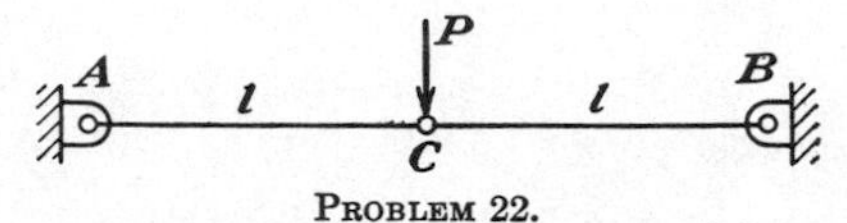

PROBLEM 22.

a. This problem consists in calculating the deflection of point C, caused by a load P, when the locked-up stress s_0 is "large."

b. Substitute numbers: $s_0 = 20{,}000$ lb/sq in.; $P = 100$ lb; $A = 0.25$ sq in.; $E = 30 \times 10^6$ lb/sq in.; $l = 1$ ft.

23. Consider the previous problem without locked-up stress, *i.e.*, whatever elongation or stress there is in the two wires is caused by the deflection of point C only. Then for a "small" deflection δ (that is, $\delta \ll l$) you will find that the elongation of the wires is not proportional to δ.

Set up the problem by assuming a small δ, calculate the elongations and stresses in the wires, then calculate the load P. By making such simplifications as are feasible for small δ, derive a formula relating the load P to the deflection δ. HINT: $\sqrt{1+\epsilon} = 1 + \frac{1}{2}\epsilon + \cdots$.

24. A problem related to the previous one is that of the toggle joint. Here we do not have wires, but two stiff rods, hinged at both ends, on immovable foundations at A and B, and enclosing a small angle between

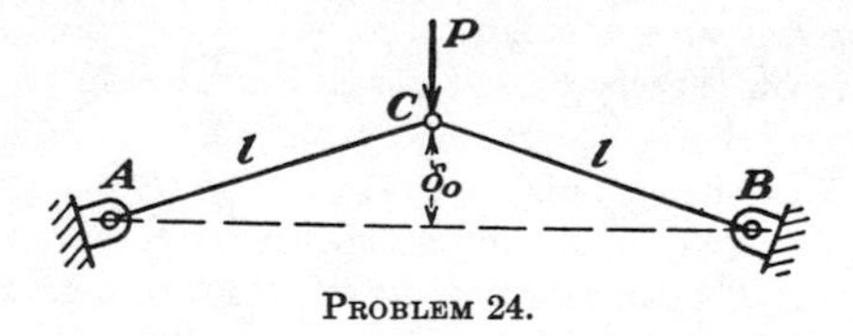

PROBLEM 24.

them in the stressless state. In that state the elevation of point C above the line AB is δ_0. When the load P is applied, the deflection at P is δ and the elevation of C above AB becomes $\delta_0 - \delta$. It is obvious that P cannot be proportional to δ, because when δ becomes equal to δ_0 the bars are in line at high compression, but $P = 0$.

a. Calculate P as a function of δ, assuming δ as well as δ_0 to be "small" with respect to l. Retain terms δ/l, $(\delta/l)^2$, and $(\delta/l)^3$.

b. If the yield point of the steel bars is 30,000 lb/sq in. and the length $l = 1$ ft, what is the maximum value of δ_0 for which the joint can be flattened out without yielding?

25. This problem is a preliminary to the automobile-tire problem (Prob. 26). Consider two flat steel plates, 6 by 6 in. square, joined together by a square canvas bag 1 in. high all along its 24-in. periphery. The bag is supposed to be rubberized so that it can hold air pressure. The upper plate carries a load of 750 lb (roughly one-fourth the weight of an automobile). Without air pressure the bag is collapsed, and the upper steel plate rests directly on the lower one.

a. Calculate the air pressure just necessary to get the weight air-borne.

b. Assuming the canvas sides to be straight without bulging out, what is the stress in the canvas for an air pressure of 32 lb/sq in. above atmospheric?

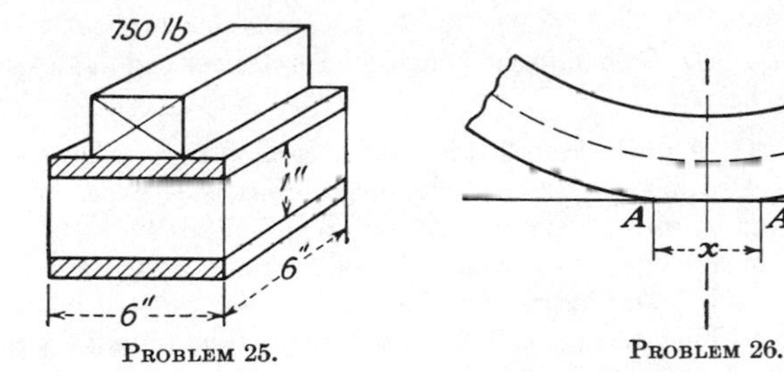

PROBLEM 25. PROBLEM 26.

26. An automobile tire is 6 in. wide at its widest point (in a direction perpendicular to the plane of the wheel). It carries 32 lb/sq in. pressure (above atmospheric), and the wheel load is 750 lb. When it rests on the ground, the tire is flattened over a length of x in. Assume that the side wall of the tire has no stiffness and that the section AA of x in. circumferential length has no stress, whereas the rest of the tire side wall carries such stress as is caused by the air pressure.

a. Calculate the length x.

b. Calculate the (radial) tension in the side wall of the tire outside the region AA.

27. A steel steam pipe of 24 in. inside diameter and 1 in. wall thickness connects two steam turbines (high pressure and low pressure) that are 12 ft apart. Such a connection is never made straight, for a reason that will be obvious after this problem has been solved. The assembly is supposed to be without stress at 60° F. Now steam of 500° F is passed through the pipe, which carries insulation on its outside, so that the steel assumes the steam temperature.

Make three unreasonable assumptions: (1) The designer has made the pipe straight. (2) Both turbines are rigid (do not give). (3) The pipe does not yield or buckle at any stress.

a. Calculate the force exerted on the turbines.

b. Calculate that force, assuming a yield stress of 20,000 lb/sq in. in the hot pipe, but retaining assumptions 1 and 2.

28. In the previous problem the turbines are no longer considered rigid, but at the points of attachment of the crossover pipe they have a "stiffness" of $k = 10^7$ lb/in. The two turbines are supposed to be identical, and again the yield point of the pipe material is 20,000 lb/sq in. Find the force on the turbines.

29. A steel bolt of 1.5 sq in. cross section has a bronze bushing shrunk over it of 1 sq in. cross section. The assembly is 40 in. long. The moduli of the two materials are 30 and 12×10^6 lb/sq in., respectively. Calculate the force required to elongate the bar by 0.020 in., and find the stresses in both materials under that load.

30. The bolt of the previous problem is heated 200° F. The coefficient of expansion of steel is 6.67×10^{-6} in./in./° F, and for bronze it is 10.0×10^{-6} in./in./° F. Calculate the (longitudinal) stresses and the total elongation of the bolt.

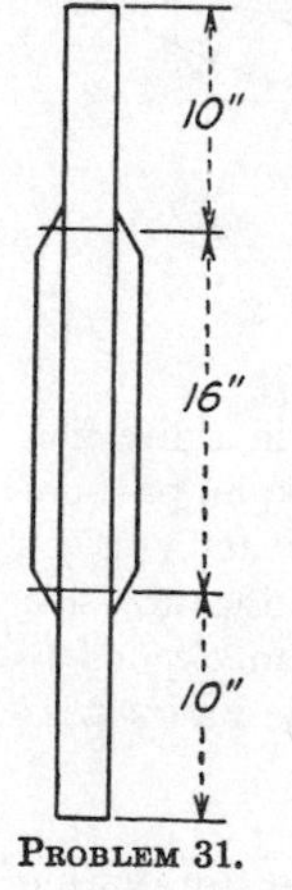

PROBLEM 31.

31. A steel rod of 3 ft length has a circular cross section of 1 sq in. area. On it is shrunk a bronze sleeve, also of 1 sq in. cross-sectional area and of 16 in. length.

a. Find the elongation of the entire rod of 36 in. length, caused by end forces of 10,000 lb.

b. Find the elongation of the entire rod caused by a temperature rise of 50° F.

c. Find the elongation caused by 5,000 lb pull and a 50° F temperature rise simultaneously.

d. If the temperature rise is 100° F and the elongation is completely prevented, what is the compressive force resulting from it? Use the following constants:

For steel, $E = 30 \times 10^6$ lb/sq in.;
$\alpha = 6.7 \times 10^{-6}$ in./in./° F.
For bronze, $E = 12 \times 10^6$ lb/sq in.;
$\alpha = 10.0 \times 10^{-6}$ in./in./° F.

32. Two steel plates are butt-welded together along an edge, which is a common operation in ship construction. Idealizing the situation we assume at the completion of the welding that a thin strip in the plates is hot, while the large plates themselves are cold. It is further assumed that there are no internal stresses at that moment. Now the plate is allowed to cool down to uniform temperature, which will cause a tensile stress in the strip and a negligibly small compressive stress in the large plates to balance the strip tension statically.

Calculate the initial temperature difference between the strip and the plates required to cause a stress of 30,000 lb/sq in. (the yield stress) in the weld. Take $\alpha = 6.67 \times 10^{-6}$ in./in./° F.

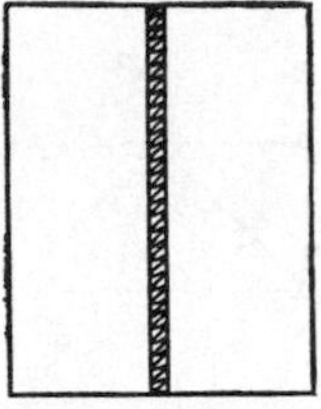

PROBLEM 32.

33. If the hoist of Prob. 12 has a steel tension rod of 1 sq in. cross section and a steel horizontal beam of 2 sq in. cross section, determine the vertical deflection of the end under a load of 2,000 lb. Assume the supporting wall to be immovable.

34. The truss shown in the figure consists of six bars all of equal length $l = 3$ ft and equal cross section 1 sq in. All bars are at 60 deg with respect to each other in the unstressed state. Assuming no motion of the two supports A and B and a load of 2,000 lb at the end point C, find

a. The stresses in all bars.

b. The deflection of the end C.

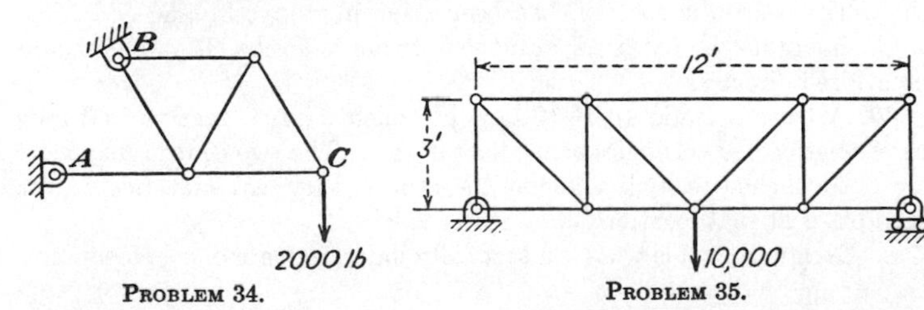

PROBLEM 34. PROBLEM 35.

35. A truss is made up of 15 bars, all of the same cross section 1 sq in., and all being horizontal, vertical, or at 45 deg in the unstressed state.

a. Determine all stresses under a 10,000-lb central load.

b. Find the vertical deflection at the center.

36. The truss of Prob. 35 is to be designed for a constant stress of 5,000 lb/sq in. in all bars.

a. Find the necessary cross sections of the bars.

b. If the truss were built that way, find what the central deflection would be.

c. What would happen if in addition to the 10,000 lb vertical central load we should impose a horizontal force of 100 lb on the truss anywhere?

37. A truss consists of 12 horizontal or vertical bars of length a, 6 diagonals of length $a\sqrt{2}$, and 1 bar of length $2a$ as shown. It is supported at A and B and loaded by a force P in the center, as shown. Owing to this load the legs A and B will be spread apart, *i.e.*, the point B will move somewhat to the right. Determine this horizontal spread, assuming all bars to have equal cross section A and to be made of the same material E.

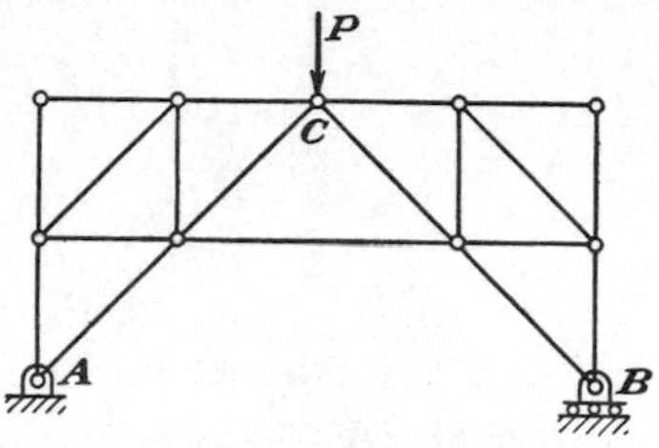

PROBLEM 37.

38. Omit the load P of Prob. 37, and replace it by a load Q, acting at B and directed horizontally to the right. This load Q also will spread the points A and B apart elastically. Determine by how much.

39. Omit the roller support at B of Prob. 37, and replace it by a hinge support as at A. This makes the truss statically indeterminate; *i.e.*, it can be made to take equal and opposite horizontal reactions at A and B. Load this truss by a central load P at C. Determine the value of the horizontal reactions at A and B as well as all the bar stresses. Most of the work toward the solution has been done in Probs. 37 and 38. This work should not be repeated, and the answers to Probs. 37 and 38 should be utilized directly.

40. A truss is made up of 10 bars of which 6 have length a and four have length $a\sqrt{2}$, all angles being 45 or 90 deg. The two central diagonals are not attached to each other at the center: they can slide freely over each other at that location.

a. Explain that this truss is a statically indeterminate one. How many bars are redundant?

b. The truss is loaded symmetrically with two loads P as shown. Hence the bar forces must also be symmetrical. Calculate all bar forces, using these properties of symmetry and the properties of statics only (without calculating any deformations).

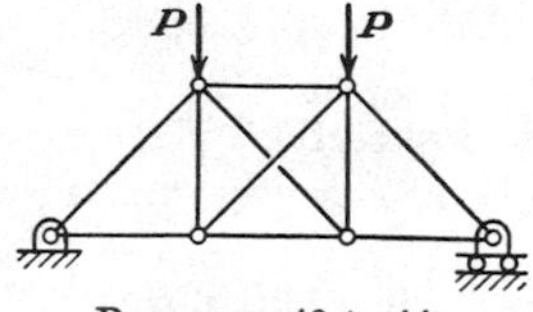

PROBLEMS 40 to 44.

41. Find the deflection under each of the loads, using the answers of Prob. 40 and assuming all bar cross sections to be equal to A.

42. The truss of Prob. 40 is now loaded antisymmetrically, *i.e.*, with a downward load P at the right-hand upper corner and with an upward pulling force P at the left-hand upper corner.

Again find all bar forces by using statics and symmetry only, without calculating deflections.

43. Find the deflections under the loads in the previous problem.

44. Superpose the loadings of Prob. 40 and 42 on each other, which gives a single load $2P$ on the right-hand upper corner only. Find all the bar forces, and find the deflection under the load.

45. A flat cantilever beam, carrying an end load of 100 lb, is attached to a flat wall by two bolts of circular cross section. Assume that the vertical shear load of 200 lb is equally divided between the two bolts. Assume further that the bending moment is carried by the bolts in the form of a pair of equal and opposite horizontal forces. Assume in the third place that the shear forces are uniformly distributed across the cross sections of the bolts. Design the bolts for an average shear stress of 2,000 lb/sq in. What is their diameter?

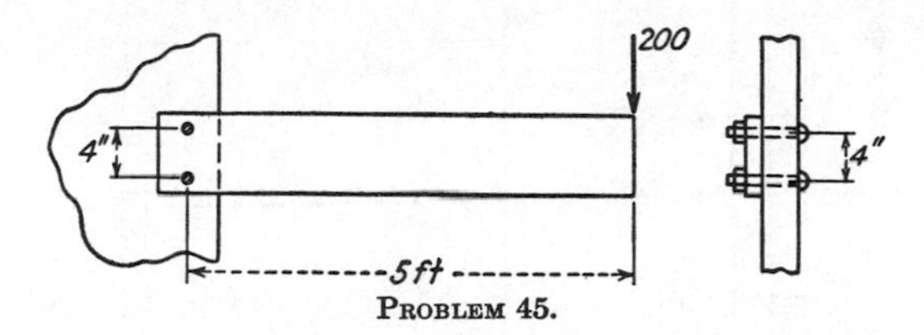

PROBLEM 45.

46. A flange-and-bolt coupling between two shafts carries 12 bolts at a pitch diameter of 18 in. The shafts carry a torque of 50,000 ft-lb.

Assuming that this torque is transmitted by shear across the 12 bolts and that these bolts are to be designed for a shear stress s_s = 3,000 lb/sq in., what is the bolt diameter, rounded off to the nearest ⅛ in.? For this size recheck the stress, and state what it is.

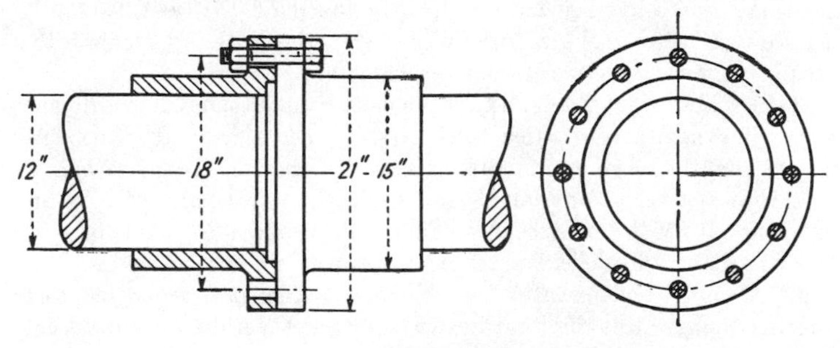

PROBLEM 46.

47. Referring to the previous problem, another way of designing the bolts is by assuming that the torque is transmitted by friction between the flanges, which are pressed together by tension in the bolts, the shear stress in the bolts being assumed as having no part in the torque transmission. Let the bolts be pulled up so that their tensile stress is 15,000 lb/sq in. For simplicity assume that the friction force acts at the bolt pitch diameter of 18 in. (This is accomplished in practice by relieving the contact between the flanges in the central part, as shown in the sketch.) If the coefficient of friction between the flanges is 25 per cent, what bolt diameter is required?

48. Two steel boiler plates are joined to each other by a riveted "double-plate butt joint," as shown. The thickness of the plates is t, the diameter of the rivets $d = 2t$, and the center-to-center spacing of the rivets is p, the "pitch," of a value to be determined.

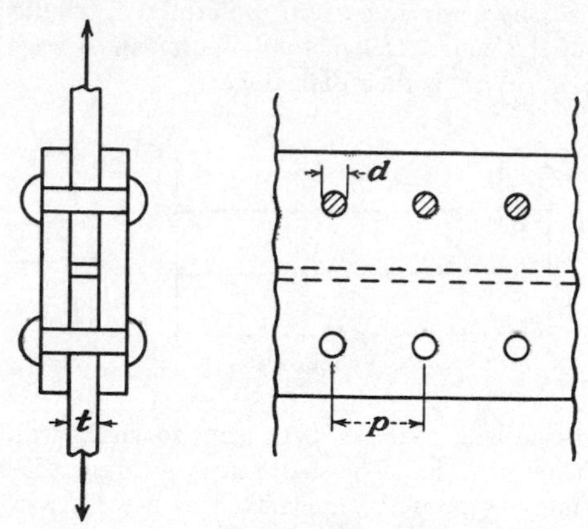

PROBLEM 48.

Assume that the tensile pull of the plates is transmitted by shear across the rivet shanks and that the average shear stress in the rivets is made the same as the average tensile stress in the plate (between two rivet holes of the plate). Neglect friction between the plates. Calculate the necessary pitch p in terms of the rivet diameter d.

49. A rubber sheet of 8 in. height, somewhat more than 8 in. width, and ½ in. thickness carries three steel bars on each side of it, which are screwed tightly together in pairs, as shown. The center pair of bars is pulled down with a force of 100 lb, while the outer pairs of bars are pulled up with 50 lb. The shear modulus of the rubber is $G = 500$ lb/sq in. Find the downward deflection of the center with respect to the outsides.

50. A solid circular shaft of length l and radius r_0 is subjected to a torque. Imagine this shaft cut in two along a flat meridian plane passing through the diameter. By the property of Fig. 11a (page 15), there will be shear stresses in the flat section, as shown in the sketch, and these stresses will have a moment about axis AA. Prove that the moment about AA of the shear stresses acting on the top and bottom half circles is equal and opposite to the moment about AA of the shear stresses in the flat section.

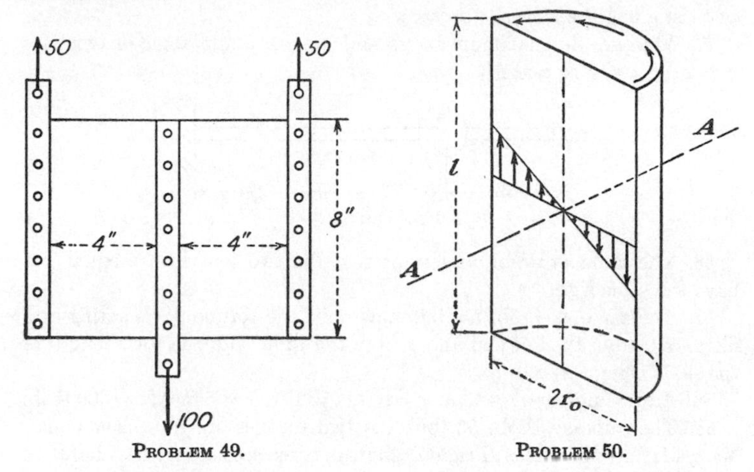

PROBLEM 49. PROBLEM 50.

51. If in a transmission problem the horsepower (hp), the rpm, and the maximum allowable torsion stress s_s are given, the shaft diameter d follows. Derive the formula for d as a function of these three variables.

What is the (automobile-engine shaft) diameter required to transmit 100 hp at 2,000 rpm with a stress $s_s = 5{,}000$ lb/sq in.?

52. Derive a formula for the angle of twist of a solid circular shaft in terms of the horsepower (hp), the rpm, the material G, and the shaft dimensions l and d.

53. In drilling a deep oil well the drill shaft, consisting of sections of "drill rod," is powered from the top, while the load or counter torque is taken off at the bottom. The deepest hole drilled so far is 3½ miles deep. Assuming a solid drill rod of circular cross section and 1 in. diameter, designed to a working shear stress of 5,000 lb/sq in., turning at 700 rpm, calculate

a. The wind-up, *i.e.*, the twist angle between top and bottom.

b. The horsepower of the driving engine.

54. A small inboard motorboat is powered with an automobile engine running at 1,500 rpm and developing 60 hp at that speed. The propeller shaft is to be made of bronze ($G = 6 \times 10^6$ lb/sq in.) and is to be designed with a shear stress of 2,000 lb/sq in. Determine its necessary diameter, rounding it off upward to the nearest ⅛ in. Also determine the angular wind-up at that diameter when the shaft length is 10 ft.

55. A steel shaft of solid circular cross section r and total length $3l$ is built in solidly at both ends. At the point distant l from the left end a torque M_t is imposed on the shaft, while at the other one-third-length point, distant l from the right, an equal and opposite torque M_t acts.

a. Calculate and plot the maximum-stress distribution and the angle of twist along the length.

b. What are the maximum stress and the maximum angle of twist for $r = 2$ in., $l = 2$ ft, and $M_t = 300$ ft-lb?

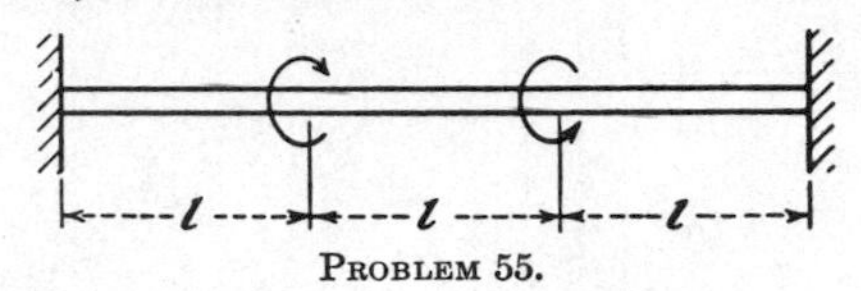

PROBLEM 55.

56. The same as Prob. 55, except that the two torques are equal and have the same sign.

57. The same as Prob. 55, but now only one torque M_t is acting at a distance a from the left end and b from the right end, the total length of the shaft being $l = a + b$.

Substitute numbers: $r = 2$ in., $l = 6$ ft, $a = 1$ ft, $b = 5$ ft, $M_t = 300$ ft-lb.

58. The same as Prob. 55, but the two torques are no longer equal, being M_l and M_r (left and right), and the three sections of the shaft are no longer of equal length, being a (left), b (center), and c (right).

Substitute numbers: $a = 1$ ft, $b = 2$ ft, $c = 3$ ft, $M_l = 200$ ft-lb, $M_r = 300$ ft-lb.

59. A strong mechanic in tightening a bolt is pulling with a force of 100 lb at the end of a 1-ft wrench.

a. What must be the diameter of the bolt (in the root of the thread) so that the bolt just starts to yield? Assume that the shear force is distributed uniformly over the cross section and that the yield point in shear of mild steel is 15,000 lb/sq in.

b. If the torque to break the bolt off is twice as great as that necessary to just start local yielding, what diameter bolt can be broken by a wrench torque of 100 ft-lb?

60. Calculate the necessary diameter of a shaft between a Diesel engine and an electric generator, transmitting 1,000 hp at 720 rpm, if the working shear stress is to be 4,000 lb/sq in.

61. The drive shaft of a conventional automobile between the transmission gearbox and the rear differential usually has the form of a thin-walled circular tube. Design this tube for an engine rated 100 hp at 3,500 rpm. The transmission-gear ratio for first gear is 3.5; hence the drive shaft runs at 1,000 rpm when the engine is at 3,500 rpm. The working stress is to be $s_s = 12,000$ lb/sq in., permissible in a good alloy steel.

Make the outside diameter = 2 in., and calculate the wall thickness.

62. Two bronze shafts ($G = 6 \times 10^6$) are coupled by a sleeve coupling of steel ($G = 12 \times 10^6$). Determine the ratio of the diameters D to d for which the shafts and coupling sleeve have the same factor of safety in torsion, when the yield point of steel is twice that of bronze.

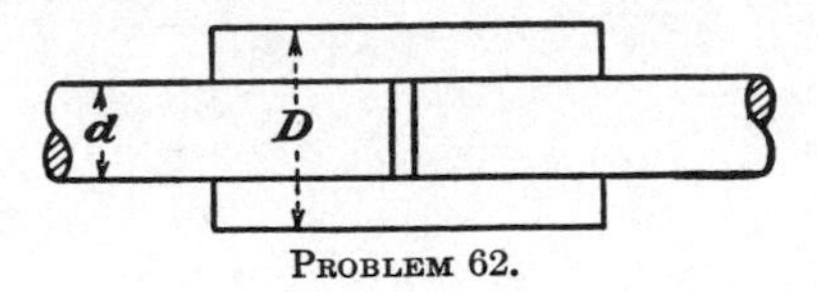

PROBLEM 62.

63. The same as the previous problem, but now the shafts are of steel and the sleeve is made of bronze.

64. A shaft is to be cut up into four concentric elements, the inner one of which is solid and the outer three are tubes snugly fitting over each other. If the outer diameter of the largest tube is D and the four elements are to have the same "torsional stiffness," *i.e.*, to twist through equal angles for equal torques, find the three inner diameters.

65. Calculate the stresses and the angle of twist at the free end of a solid shaft of 6 ft length, having diameters of 1 and 2 in. as shown, subjected to a torque of 100 ft-lb.

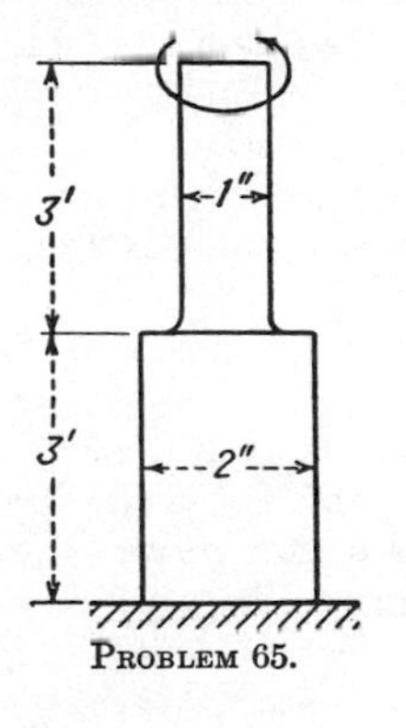

PROBLEM 65.

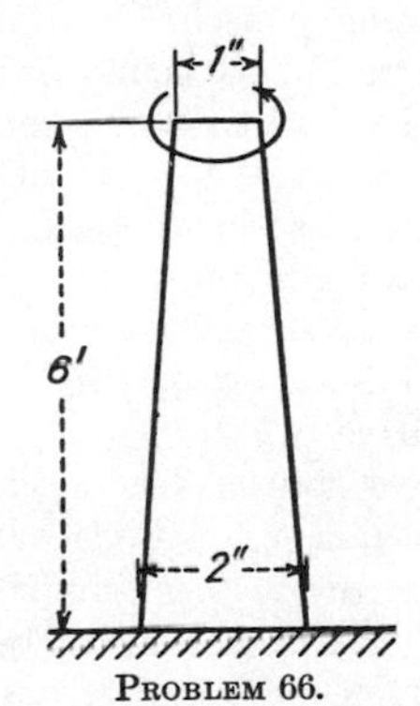

PROBLEM 66.

66. *a.* Derive a formula for the angle of twist of a shaft of gradually varying diameter, conical in shape, as shown. The diameters at the two ends are d_{max} and d_{min}, the length is l, and the torque is M_t.

b. Substitute numbers: $d_{max} = 2$ in., $d_{min} = 1$ in., $l = 6$ ft, $M_t = 100$ ft-lb.

67. *a.* Derive a formula for the stress in a compound shaft, subjected to a torque M_t and made up of a conical steel core of end diameters d_1, d_2 and length l, on which is shrunk a bronze sleeve of outside diameter d_3.

b. Calculate the maximum stress numerically in a shaft where d_1

$= 1$ in., $d_2 = 2$ in., $d_3 = 2\frac{1}{2}$ in., $l = 3$ ft, $M_t = 100$ ft-lb, $G_{steel} = 12 \times 10^6$ lb/sq in., $G_{bronze} = 6 \times 10^6$ lb/sq in.

PROBLEM 67.

68. An aluminum bar of a square cross section 1 by 1 in. and 30 in. length is subjected to a torque of 50 ft-lb. Calculate the approximate angle of twist. The shear modulus of aluminum is about $G = 4 \times 10^6$ lb/sq in.

69. Calculate the polar moment of inertia of a cross section having the shape of an equilateral triangle of side length a, about its center of gravity. HINT: $r^2 = x^2 + y^2$.

70. Using the result of Prob. 69 calculate the approximate angle of twist of a wooden bar of triangular cross section $a = 1$ in. and 30 in. length, subjected to a torque of 2 ft-lb. The shear modulus of the wood is 400,000 lb/sq in.

71. Design a steel coil spring with 10 coils to carry a load of 40 lb, at a shear stress of 100,000 lb/sq in. What is the deflection at full load?

a. Using a wire diameter of 0.092 in. (No. 13 wire).

b. Using a wire diameter of 0.105 in. (No. 12 wire).

72. A steel spring of $\frac{3}{16}$-in.-diameter wire and a coil diameter of 2 in. is to be designed with a maximum shear stress of 100,000 lb/sq in.

a. What load can this spring support?

b. If this spring must have a stiffness $k = 60$ lb/in., how many turns must it have?

73. Two springs, one inside the other, not touching each other, have the same over-all length and the same number of turns. The coil diameter of the outside spring is twice that of the inner one. The outer spring is of steel wire ($G = 12 \times 10^6$) and the inner one of bronze ($G = 6 \times 10^6$). When they are being compressed, they obviously have the same deflection.

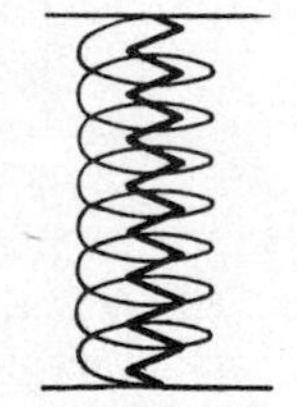

PROBLEM 73.

a. If the two springs are to be designed each to take half of the total compressive force, what is the ratio of their wire diameters?

b. In that case, what is the ratio of their stresses?

74. A rigid horizontal bar is supported by two springs, distance $(a + b)$ apart. The springs are made of the same wire and have the same number of turns and differ from each other only in that the coil diameter D of the right spring is 50 per cent larger than the coil diameter of the left spring ($D_r = \frac{3}{2}D_l$). Where must the load P be placed in order to deflect the horizontal bar downward parallel to itself?

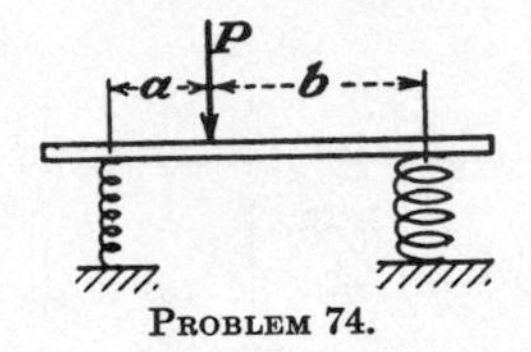

PROBLEM 74.

75. A stiff square plate is supported at its four corners by four springs, which are identical in every respect except that springs 1 and 2 have a wire diameter d which is 18.6 per cent larger than that of springs 3 and 4 ($d_{1,2} = 1.186d_{3,4}$). The plate is loaded by a vertical load P in one corner, just above one of the stiff springs, No. 1.

Determine the ratio y of the downward (or upward) deflections at the four springs.

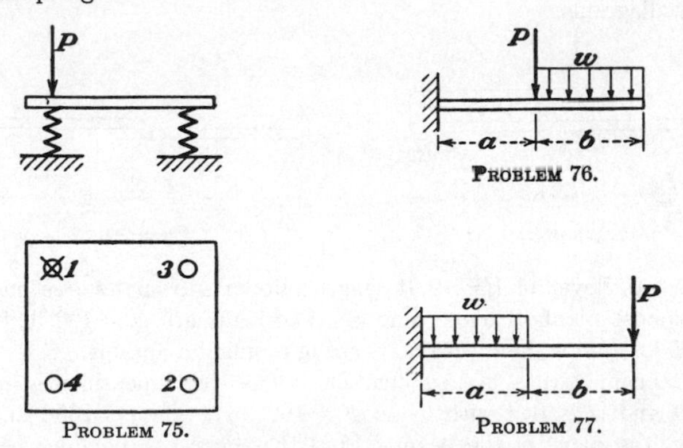

PROBLEM 75.

PROBLEM 76.

PROBLEM 77.

76. A cantilever beam of dimensions $a = 2$ ft, $b = 3$ ft is loaded with $w = 200$ lb/ft and $P = 500$ lb. Calculate and plot the shear-force and bending-moment diagrams.

77. Plot the shear and bending-moment diagrams (indicating the scales clearly on the sketch) for $a = 3$ ft, $b = 2$ ft, $w = 100$ lb/ft, and $P = 100$ lb.

78. A beam on two supports of total length 8 ft is loaded by two equal concentrated forces $P = 500$ lb. Plot the shear and bending-moment diagram.

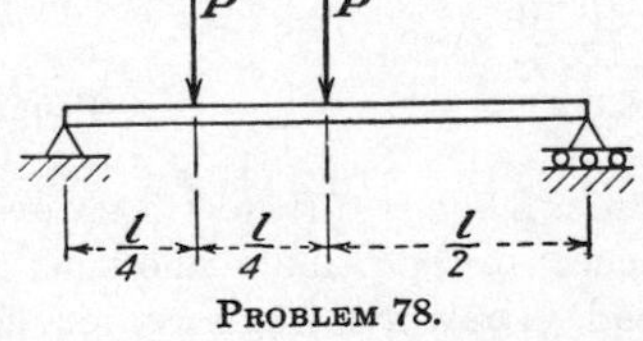

PROBLEM 78.

79. Let $l = 10$ ft, $P = 1{,}000$ lb, and $w = 200$ lb/ft. Find the magnitude and location of the maximum bending moment.

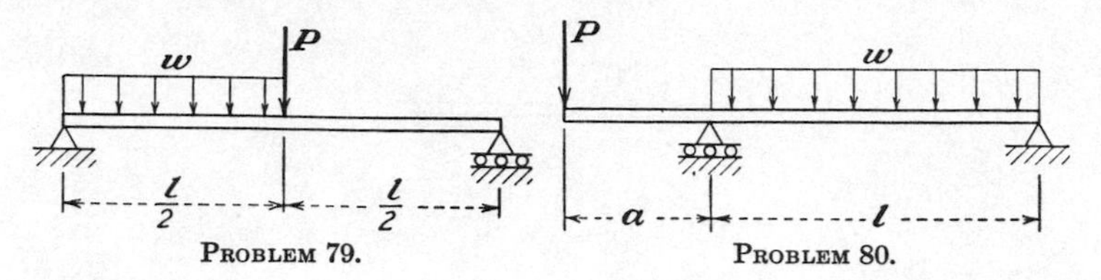

PROBLEM 79. PROBLEM 80.

80. A beam on two supports of span $l = 10$ ft and overhang $a = 3$ ft is loaded with $w = 100$ lb/ft and $P = 500$ lb. Plot the shear and bending-moment diagrams.

81. A beam of span $l/2 = 8$ ft and with an overhang of equal length is loaded with $w = 300$ lb/ft. What load P can it carry so that the right support just lifts off? For this condition, plot the shear and bending-moment diagrams.

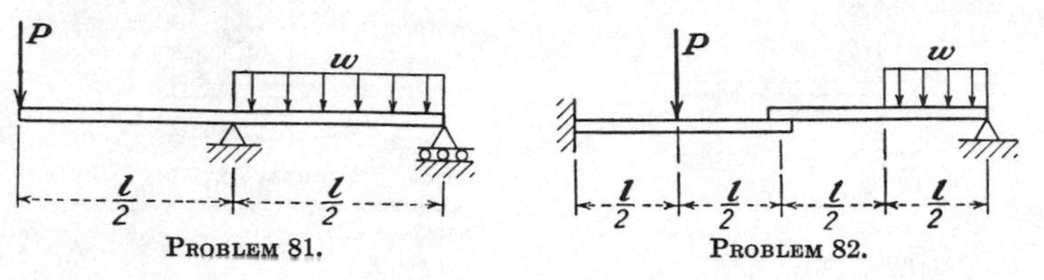

PROBLEM 81. PROBLEM 82.

82. A cantilever of $l = 10$ ft length supports from its free end the tip of a loose plank of equal length. The loads are $w = 100$ lb/ft and $P = 500$ lb. Draw the shear and bending-moment diagrams.

83. A beam with a 90-deg bend in it has the dimensions $a = 6$ ft, $b = 5$ ft and is loaded with loads $w = 100$ lb/ft and $P = 300$ lb, both acting in the plane of the beam. Find the shear and bending-moment diagrams.

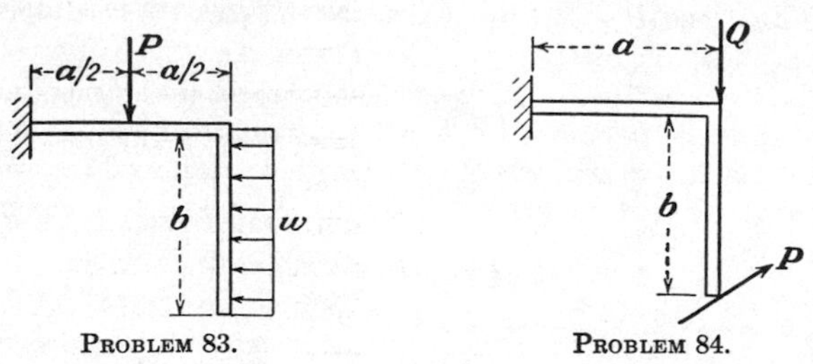

PROBLEM 83. PROBLEM 84.

84. The beam of Prob. 83 ($a = 6$ ft, $b = 5$ ft) is subjected to a load $P = 100$ lb, perpendicular to the plane of the bend and another load Q in the plane of the bend. Draw five diagrams: for the shear force in the plane of the bend, for the shear force in the plane perpendicular to it, for the bending moment in those two planes, and for the twisting moment.

85. A quarter-circle cantilever of radius $R = 4$ ft carries an end load $P = 500$ lb in its own plane. Plot three diagrams: for the shear force, for the tensile force, and for the bending moment in the plane of the beam. What is the bending moment in the plane perpendicular to that of the beam?

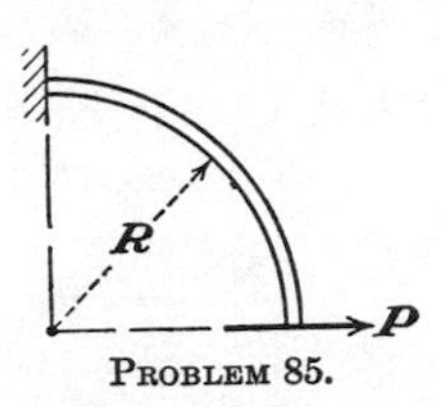

PROBLEM 85.

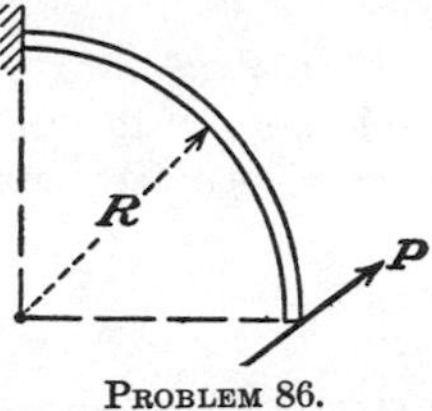

PROBLEM 86.

86. The beam of the previous problem ($R = 4$ ft) is now loaded by $P = 100$ lb in a plane perpendicular to that of the quarter circle. Draw six diagrams: for the shear force in the plane of the circle and in the plane perpendicular to it, for the tensile force, for the bending moments in the two planes, and finally for the twisting moment. Some of these quantities are zero throughout; which ones?

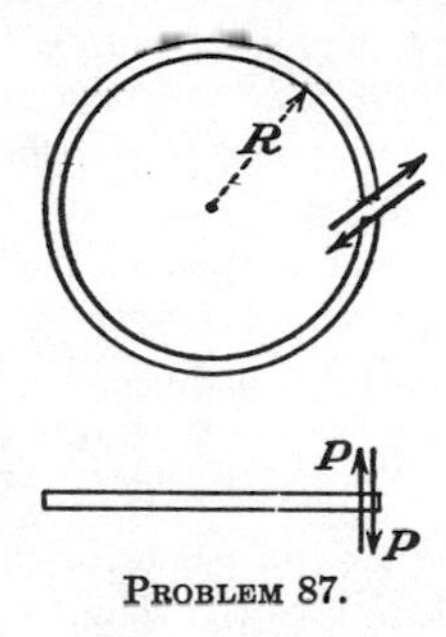

PROBLEM 87.

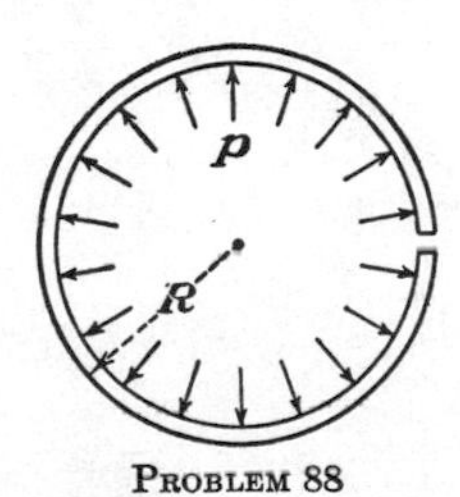

PROBLEM 88

87. A beam in the shape of a full circle of radius R is acted upon by a pair of equal and opposite forces P, push-pulling the ends into and out of the plane of the circle. Plot diagrams for whatever forces and moments occur in the circle, all along its length.

88. The slit circular beam of Prob. 87 is now loaded by a uniform radial pressure p all along its inside periphery and lying in the plane of the circle. Plot all diagrams.

89. From a table for structural sections (for example Marks' Handbook) we take the following figures for a 12-in. Bethlehem "girder" I beam:

Over-all height 12 in.	Thickness of web 0.38 in.
Width of flange 10 in.	Area of section 16.35 sq in.

Idealize the section as consisting of three rectangles, thus neglecting fillets and rounded corners. From the above calculate the thickness of the flanges and then the two principal moments of inertia. See how close your answers come to the actual values listed in the tables.

90. The same as Prob. 89, now applying to a light 12-in. American standard channel section:

Height 12 in. Flange width 2.94 in.
Area 6.03 sq in. Thickness of web 0.28 in.

Find the location of the center of gravity and the moments of inertia for two perpendicular axes through G.

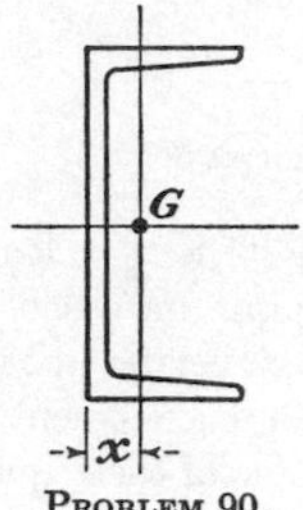

PROBLEM 90.

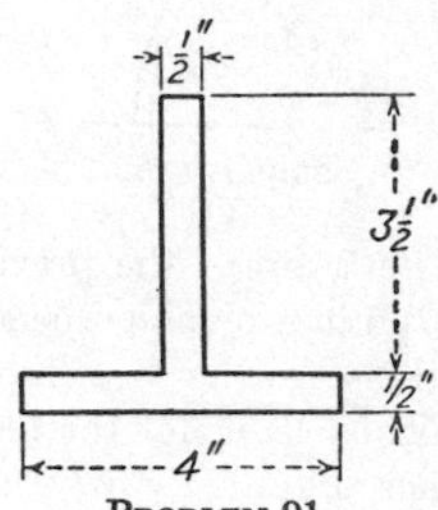

PROBLEM 91.

91. A cantilever beam of 6 ft length is made of steel, with a T-shaped cross section as shown. If this beam is to be designed for a maximum bending stress of 10,000 lb/sq in., tension or compression, what is the permissible end load?

92. A steel channel section is used to cover a span of 10 ft. The properties of this channel are found in the table as follows:

$h = 6$ in.; area $= 5.54$ sq in.; center-of-gravity distance $x = 0.55$ in.; width of flange $= 2.28$ in.; moment of inertia $I_{AA} = 0.53$ in.^{4}; $I_{BB} = 19.5$ in.^{4}

The channel is to be designed to a working load of 8,000 lb/sq in., and there is a uniform load w over the entire length of the span. Calculate the permissible load w for

a. The channel standing upright, 6 in. high.

b. The channel lying flat with the 6-in. flat face horizontal.

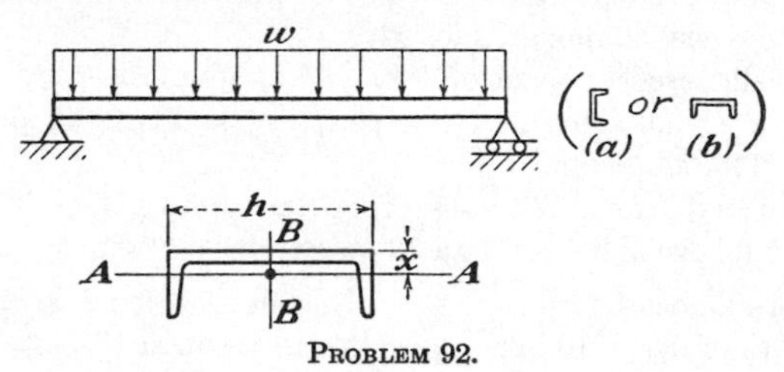

PROBLEM 92.

93. In the previous problem the channel is replaced by a beam consisting of two of such channels riveted together to form a solid beam. Again answer questions *a* and *b* for the beam standing upright 6 in. high or lying horizontally 4.56 in. high.

Read page 49 of the text if necessary.

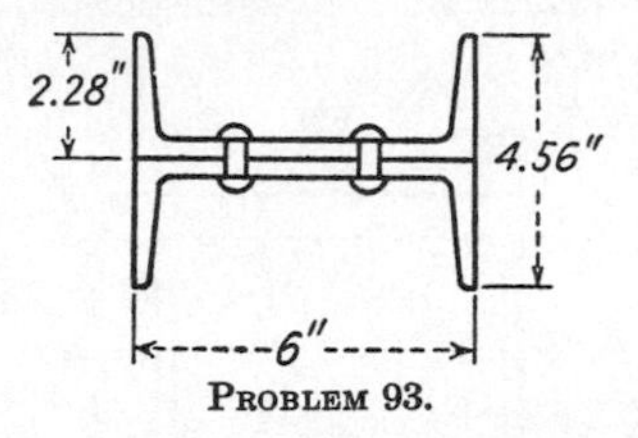

PROBLEM 93.

94. A common type of steel-wire fence is attached to steel pipes having an outside diameter of 2½ in. and an inside diameter of 1¾ in. These pipes are set into holes dug in the ground, which are then poured full of concrete, so that the pipes are solidly built in. Assuming that this foundation holds, what horizontal force applied at the top of the pipe, 6 ft above the ground, will cause the beginning of yield at a stress of 30,000 lb/sq in.?

95. In simple house construction the floors are laid on "joists," being timber beams spanning the width of the room and simply supported at both ends. Suppose one such joist carries the weight of three persons (450 lb) in its center. The joist is made of hemlock, usually 2 in. wide and of suitable height, 4 in., 6 in., etc., up to 12 in. The allowable stress in hemlock is 1,000 lb/sq in. Calculate the allowable room widths (spans) for the various timber sizes, 2 by 4, 2 by 6, etc., up to 2 by 12.

96. The floor loading in a light machine shop is 80 lb/sq ft. Design the floor joists (see Prob. 95) for such a shop of 20 ft width, in which the joists are made of hemlock of a permissible stress of 1,000 lb/sq in., spaced 12 in. apart center to center.

97. If the floor of the previous problem is to be designed with steel I-beam joists to a stress of 15,000 lb/sq in., also spaced 1 ft apart, calculate the necessary section modulus Z, and select a suitable section from a handbook table.

98. A steel beam (or wire) of circular cross section, of diameter d and length l, is simply supported at both ends, and is loaded by its own weight only.

a. What is the relation between the length and diameter when the beam is so long that it just starts yielding under the influence of the bending stress? The yield stress in steel is 30,000 lb/sq in.

b. In particular, what is the length for $d = 1$ in.?

99. Generalize the previous problem to a beam of any cross section. The properties of that section are described by the constants A (area),

I (moment of inertia about the bending axis in question), and y_{max} (distance from the center of gravity to the yielding fiber). Introduce the radius of gyration k, defined as $k^2 = I/A$, and

a. Express the dangerous length l of the beam as a function of k, y_{max}, s_{yield}, and γ, the weight per cubic inch.

b. Substitute numbers for an 8-in. Carnegie steel I beam with $A = 7.06$ sq in. and $I = 84.3$ in.^{4}

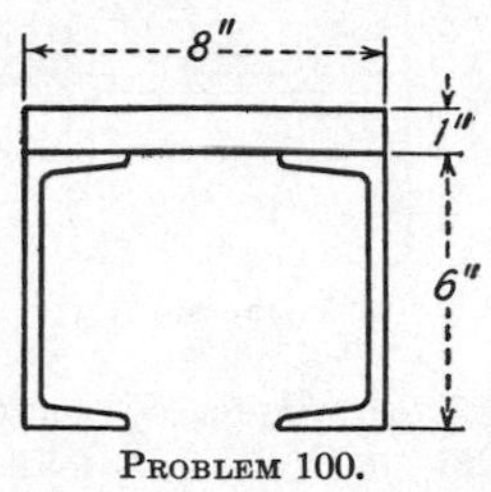

PROBLEM 100.

100. A beam is made up of two 6-in. steel channel sections and a 1- by 8-in. flat steel plate. For the channels we read off the table:

$h = 6$ in.; area $= 2.39$ sq in.; flange width $= 1.92$ in.;
thickness of web $= 0.20$ in.; weight per foot $= 8.2$ lb;
$I = 13.0$ in.^{4} about a horizontal axis through the center of gravity.

Determine the center of gravity G of the combination, its moment of inertia, and its section modulus Z about a horizontal axis through G.

101. Referring to page 46 of the text, show by integration of the shear-stress distribution over a circular cross section that the answer comes out equal to the shear force S, as it should.

102. Refer to Prob. 95. If it is desirable to have the floor low with respect to the sill beam A, the carpenter will cut a rectangular piece out of the joist B and lay it on the sill with a thin overhanging lip (Fig. *a*), instead of laying the joist right on top of the sill (Fig. *b*). The construction *a* looks frightening.

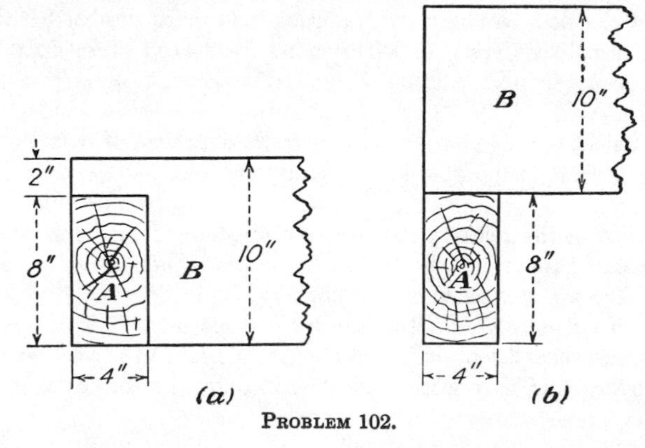

PROBLEM 102.

Consider a 2- by 10-in. joist of 16 ft span, loaded by a single concentrated load at mid-span. If the permissible tensile stress in bending is 1,200 lb/sq in. and the permissible shear stress is 200 lb/sq in., for what height of overhanging lip will the lip detail be equally dangerous with the main beam?

103. A long cantilever beam of solid circular cross section is loaded with a load P at its free end. In case a this load P passes through the center of the circular cross section, while in case b the load is tangent to the periphery of the cross section. By what ratio does the maximum shear stress of case b exceed that of case a?

104. A wooden beam of rectangular cross section bh and total length l is simply supported at both ends and is loaded in the center by a load P. For no reason at all somebody makes two saw cuts across the section, at the neutral line of bending, diminishing the width there from b to the much smaller value c. The saw cuts have no width to speak of, and being at the neutral line they do not change the bending stresses in the least. However, they do increase the local shear stresses. If the permissible stress in shear is one-fifth of the permissible stress in tension, find the critical width c at which the shear stress becomes just as dangerous as the push-pull bending stress.

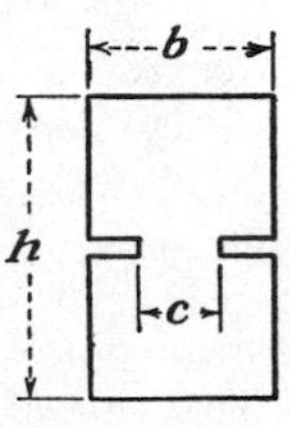

PROBLEM 104.

105. Calculate the maximum shear stress in an 8-in. Bethlehem girder I beam of 12 ft span, simply supported at both ends and loaded with a uniform load w of such intensity that the maximum push-pull bending stress in the beam is 20,000 lb/sq in.

The properties of the beam are as follows (taken from Marks' Handbook):

$h = 8$ in.; $A = 9.69$ sq in.; $I = 116.1$ in.^{4};
thickness of web $= 0.29$ in.; width of flange $= 8.00$ in.

For calculating $\int y\, dA$ assume the section to consist of three rectangles.

106. A beam having the cross section shown in Prob. 100 transmits a shear force S of 5,000 lb. Utilizing the answers to Prob. 100, calculate the shear *force* (not stress) transmitted by the rivets between the top plate and one channel, per inch length of the beam.

107. Consider a beam of T-shaped cross section, as illustrated with Prob. 91. Let this beam have a shear force $S = 7{,}500$ lb. Calculate and plot the horizontal as well as the vertical components of shear stress across this section. Where does the maximum shear stress appear, and how many times larger is it than the average shear stress $S/A = 2{,}000$ lb/sq in.?

108. Suppose you were working for the Bethlehem Steel Company and you were asked to review critically the web thicknesses of the company's line of I beams. Those webs could be made very thin if the beams were to be loaded in pure bending only, but they must have a certain thickness if shear forces are to carried.

As an example consider their 8-in. "girder," for which the specifications are given in Prob. 105. From this we calculate for the thickness of the flanges 0.48 in. Let the beam be of length l, simply supported at the ends with a load P in the center; keep the flanges 8 by 0.48 in., but let the web thickness be the variable t.

a. Design the web thickness t as a function of l, so that the maximum bending stress is twice the maximum shear stress (the reason for this will be explained on page 66). Keep the calculation simple by observing that the influence of a variable web thickness on I as well as on $\int y\, dA$ is small, and neglect that small influence by assuming these two quantities constant.

b. From the general formula just derived draw a graph and find to which beam length l the actual Bethlehem design corresponds.

c. What web thickness would be sufficient if the company could be sure that nobody would ever use a beam shorter than 12 ft?

109. For the 8-in. Bethlehem girder of the previous problem the table in Marks' Handbook states that the maximum allowable shear force to be carried is 27,800 lb. For what shear stress has this beam been designed?

110. For the girder beam of Probs. 108 and 109, loaded with the maximum permissible shear force of 27,800 lb, calculate the distribution of the (horizontal) shear stress in the flanges.

111. A beam of the cross section shown in Prob. 100 (and with the constants as given in the answer to that problem) of total length 20 ft, simply supported at both ends, is uniformly loaded by a load w of such intensity that the maximum bending stress is 15,000 lb/sq in.

The rivets joining the angles to the top plate have a diameter of $\frac{5}{8}$ in. and can safely carry a shear force of 2,000 lb each. Calculate the necessary spacing (center-to-center distance) of the rivets.

112. Refer to Prob. 93. The rivets shown are of $\frac{5}{8}$ in. diameter and can safely transmit 2,000 lb shear force each. There are two rows of such rivets, as shown in the sketch. Calculate their necessary center-to-center spacing for the loading w shown in the answer to Prob. 93. Answer the question for both alternatives *a* (I cross section) and *b* (H cross section).

113. A beam is built up of two 8-in. Bethlehem I girders riveted together as shown. The constants of each individual beam are stated in Prob. 105.

a. Calculate the moments of inertia about a horizontal and vertical axis through G of the combination.

b. Find the maximum permissible uniform loading w that this beam can carry on a 30-ft simply supported span with a maximum bending stress of 15,000 lb/sq in.

PROBLEM 113.

114. The rivets holding together the beam of the previous problem have a 1-in. diameter and can carry a shear force of 8,000 lb each. With the loading of Prob. 113 and with the beam in the stiff position only (16 in. vertical and 8 in. horizontal), calculate the necessary rivet spacing.

115. A miscellaneous lot of electric cables has to span a river by being suspended from a bridge. For neatness all cables are enclosed in a steel cylinder of 9 in. inside diameter and ½ in. wall thickness, as shown. The cables have considerable weight but do not contribute to the strength. The weight loading thus is $w = 120$ lb/ft. Assume that the pipe spans a distance l by being simply supported at its ends.

a. What is the maximum possible span when the bending stress is to be held to 15,000 lb/sq in.?

b. If the rivets between the two half tubes each can transmit 200 lb shear force, what must be their spacing?

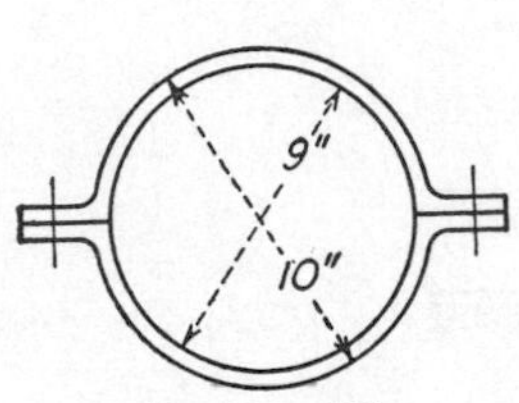

Problem 115.

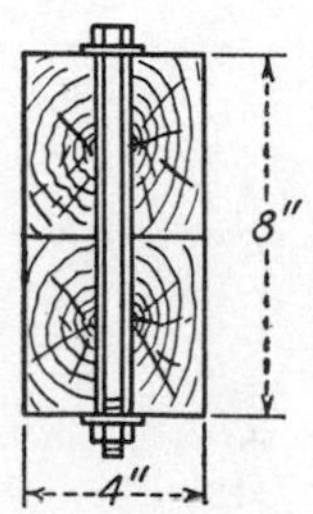

Problem 116.

116. A wooden girder is made up of two four-by-fours, joined together by a number of steel bolts. This girder is covering a 16-ft span, being simply supported at both ends, and carrying a concentrated load P at mid-span.

a. What load P is possible for a bending stress not exceeding 1,000 lb/sq in.?

b. Calculate the necessary bolt diameter if the bolts are spaced 1 ft apart and are capable of transmitting an average shear stress of 10,000 lb/sq in.

c. Calculate the same bolt diameters, based on the assumption that the shear between the two timbers is transmitted by friction only, that the bolts are pulled up to a tension of 24,000 lb/sq in., and that the friction coefficient between the timbers is 0.25.

117. Replace the beam of the previous problem by one built up of three two-by-fours, as shown. Answer the same questions *a*, *b*, and *c*.

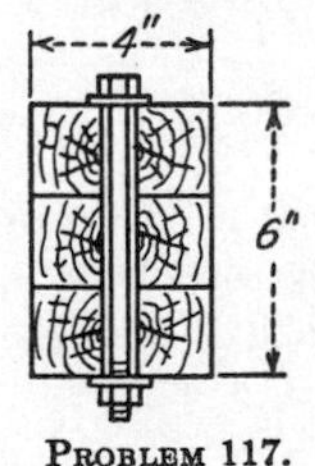

Problem 117.

118. A steel girder is made up of four angles and a plate of 8 by ½ in. cross section. The properties of each angle (American standard 2 by 2 by ¼) are taken from Marks' Handbook as follows: Area = 0.94 sq in.; $I_{AA} = I_{BB} = 0.35$ in.^{4}; $x = 0.59$ in.; $I_{\min} = 0.14$ in.^{4} (at 45 deg). Find the maximum and minimum (vertical and horizontal) moments of inertia of the combined girder.

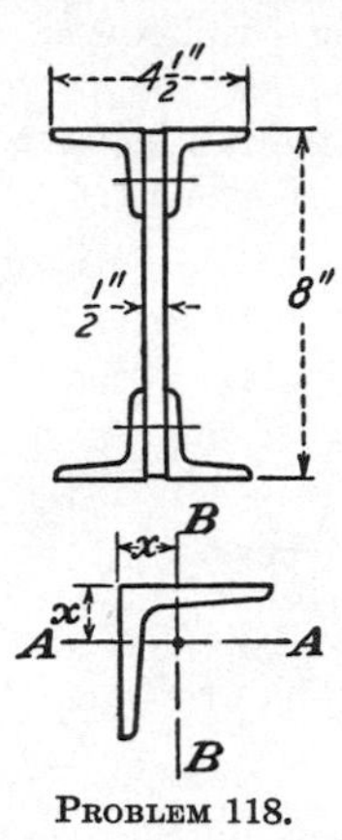

PROBLEM 118.

119. If the girder of the previous problem is to be designed for a maximum shear stress of 10,000 lb/sq in. (where does it occur in the section?) and if the rivets can transmit a shear force of 1,000 lb each, what must be their spacing?

120. Instead of using a simple 2 by 8 timber for a beam, a not too smart engineer designs one made up of a 2 by 6 and a 2 by 4 held together by lag bolts screwed into the 2 by 4 as shown.

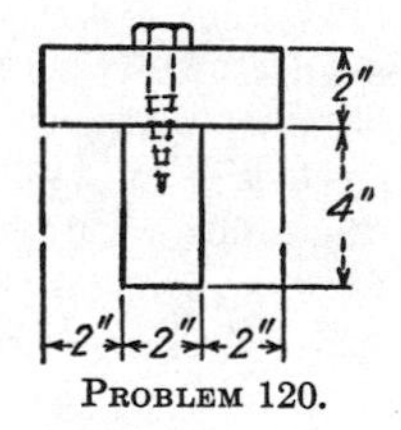

PROBLEM 120.

a. Find the center of gravity G and the moment of inertia about a horizontal axis through G.

b. Find the shear force that can be taken by the section if the shear stress in the wood does not exceed 200 lb/sq in.

c. If each lag bolt can stand a safe shear force of 1,000 lb, find their center-to-center spacing.

121. A beam of rectangular cross section $b = 3$ in. and $h = 6$ in. is subjected simultaneously to a central compressive force of 18,000 lb and a bending moment of 3,000 ft-lb, acting in the "stiff plane." Find the location of the neutral line and the maximum tensile and compressive stresses in the beam.

122. A column consists of a Bethlehem 16-in. "girder" I beam, for which Marks' Handbook gives the following constants:

height = 16 in.; width of flange = 11.5 in.;
area = 23.82 sq in.; thickness of web = 0.42 in.;
moment of inertia with neutral axis $\perp$ web at center = 1,131 in.^{4};
moment of inertia with neutral axis along web at center = 164.6 in.^{4}

Calculate and sketch the "core" for this section (page 57).

123. A concrete dam of height h and constant thickness d all along its height holds back a reservoir of water on one side. The water reaches all the way to the top. If the thickness d is too small, that water pressure, in bending the cantilever dam, will cause tension in the concrete at A. Calculate the necessary thickness d of the dam to avoid this, assuming that the specific gravity of concrete is $2\frac{1}{2}$ times that of water.

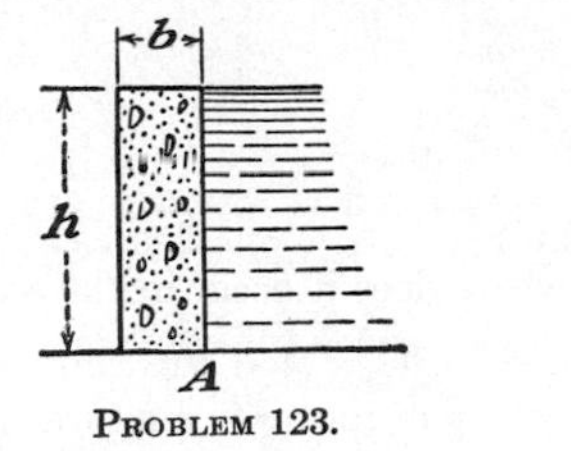

PROBLEM 123.

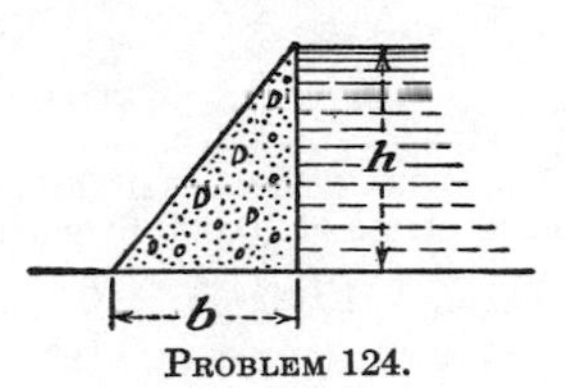

PROBLEM 124.

124. A dam can be made with less concrete than that of Prob. 123 by thinning it off toward the top. Carry this process to the limit by assuming a dam of triangular cross section. Find the necessary base width b in terms of the height h in order to ensure compressive stress along the entire base. As before, let the weight of concrete be $2\frac{1}{2}$ times that of water.

125. A perfectly constructed brick factory chimney of constant outside diameter d and height h will carry its weight uniformly distributed over its foundation. Let the wind pressure during a storm exert a force on this chimney equal to p times the projected area. Here p is an average wind pressure, which we assume to be constant with height. When the chimney is too high, the wind bending moment may cause tension in the brickwork at the base. Investigate whether there is such a limiting height h; and if so, express it in terms of the variables d, p, and w_1, where the latter quantity is the weight of the chimney per foot of height.

126. Many persons have looked with astonishment at the leaning tower of Pisa or at least at a picture of it. What is the maximum angle of inclination α of a tower of cylindrical shape of dimensions d and h made of material unable to have tensile stress? Assume that the weight of the tower is uniformly distributed along its height.

(The tower of Pisa and all other buildings constructed before the discovery of Portland cement about 1830 are piles of stones unable to take any tension. From a photograph of the tower we find $h/d = 3.4$ and $\alpha = 0.06$ approximately.)

127. A column has for its cross section a solid regular hexagon. Find the "core" of this section. First reason without calculating, until you have the entire shape but for one dimension; then calculate that dimension.

128. Let, in the figure, $s_1 = +10{,}000$ lb/sq in., $s_2 = +10{,}000$ lb/sq in., and $s_s = 0$. What is the stress on the section $\alpha = 45$ deg?

129. Let $s_1 = +10{,}000$, $s_2 = -10{,}000$, $s_s = 0$. Find the stress on a section $\alpha = 60$ deg.

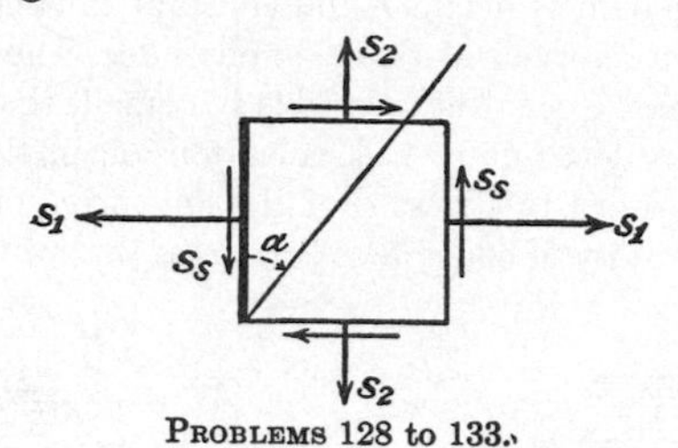

PROBLEMS 128 to 133.

130. Let $s_1 = 5{,}000$, $s_2 = 10{,}000$, and $s_s = 4{,}000$. Find the stress on a section $\alpha = 60$ deg.

131. Let $s_1 = 10{,}000$, $s_2 = 0$, and $s_s = 5{,}000$. Find the value of the maximum principal stress and the direction α of the section on which that stress acts.

132. Let $s_1 = +10{,}000$, $s_2 = -5{,}000$, and $s_s = -4{,}000$. Find the maximum principal stress and the angle α of the face on which it acts.

133. For $s_1 = +4{,}000$, $s_2 = -6{,}000$, and $s_s = +8{,}000$ find the compressive principal stress and the angle α of its face of action.

134. The stress at a point is caused by two actions, one of which produces a pure shear $s_s = 5{,}000$ lb/sq in. on a set of faces shown in the left

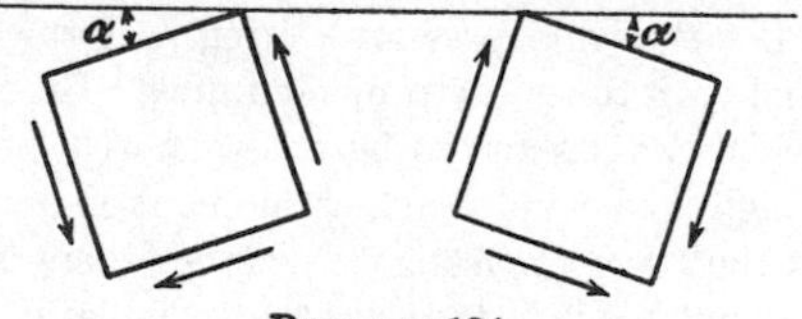

PROBLEM 134.

figure with $\alpha = 30$ deg, and the other of which is responsible for the stresses at the right, again 5,000 lb/sq in. with $\alpha = 30$ deg. In order to "add" these two states of stress at the same point it is necessary to reduce the two states to the same angular directions, say $\alpha = 0$ deg. Find the stresses on $\alpha = 0$ and 90 deg for both cases, and then add them directly. For the combined state of stress, what are the principal directions and the principal stresses belonging to them?

135. The state of stress at a point is caused by three separate actions, each of which produces a pure, unidirectional tension of 10,000 lb/sq in. individually, but in three directions, differing by 60 deg among each other. Superpose these three stresses on each other by the procedure hinted at in the previous problem, and find the total state of stress at that point.

PROBLEM 135.

136. A state of stress at a point is the sum of the stresses described by Probs. 131 and 132. Find the principal stresses and their direction.

137. Consider a rectangular area of dimensions b and h. Write the expression for the area moments of inertia about the horizontal and vertical axes and for the product of inertia $\int xy\, dA$ about those axes. From them deduce the moment of inertia about the 30-deg line AA by means of a Mohr-circle diagram.

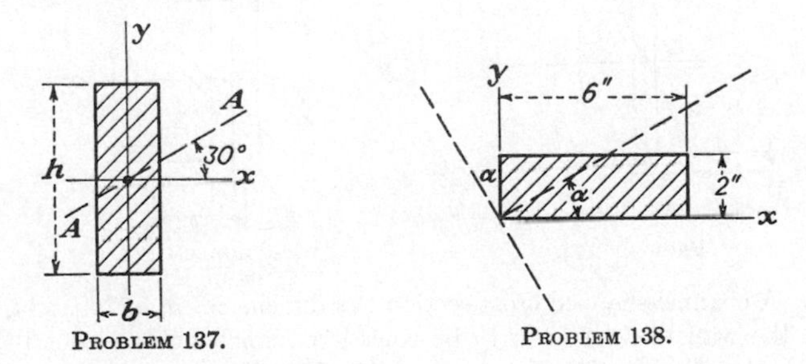

PROBLEM 137. PROBLEM 138.

138. Calculate the moments of inertia and the product of inertia of a rectangular area of 2 by 6 in. about one of its corners, as shown. Find the direction α of its principal axes of inertia and the values of the maximum and minimum moments of inertia.

139. Generalize Prob. 138 by letting the dimensions be a and b instead of 2 and 6 in.

140. Calculate the moments and product of inertia of a cross section looking like an angle of 1 in. width, as shown, about a pair of axes through the corner O. From that find the principal directions through the corner O and the corresponding values of the principal moments of inertia.

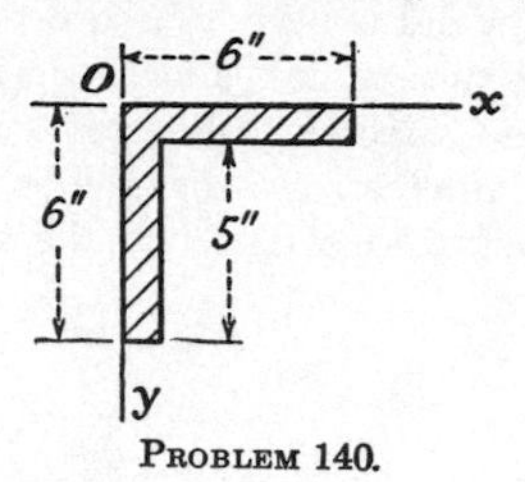

PROBLEM 140.

141. Find the location of the center of gravity G of the cross section of Prob. 140, and repeat the calculations of that problem, this time for a set of axes passing through G instead of through O. Use the answers to Prob. 140.

142. A cross section of Z shape has dimensions a, a, $2b$ as shown and a thickness t, which can be considered "small" with respect to a and b. Calculate the moments and product of inertia for the horizontal and vertical axes through the center of gravity O. From them find the direction α of the principal axes of inertia and the values of the principal moments of inertia.

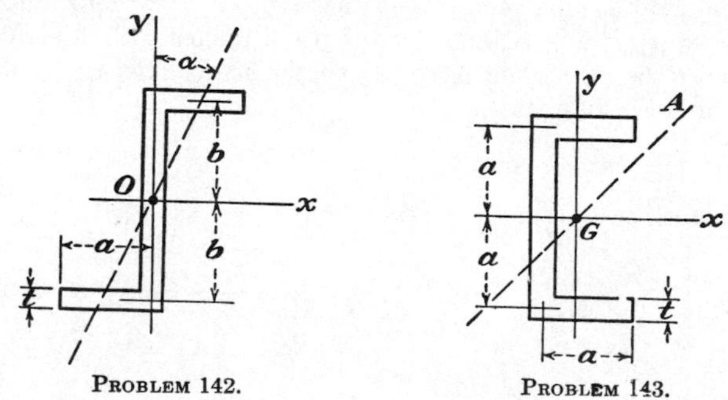

PROBLEM 142. PROBLEM 143.

143. A channel-shapped cross section has dimensions a, a, $2a$, and t, in which the wall thickness t is to be considered small with respect to a. Find the location of the center of gravity G and the inertia properties of the horizontal and vertical axes through G. From them deduce the value of the moment of inertia about the 45-deg axis AA.

144. Using the answers of Prob. 143, find the shape and size of the "core" of that cross section.

145. In a circular shaft of radius r a keyway is cut, of which the cross-sectional area A is supposed to be small with respect to the shaft cross section πr^2. Calculate the principal moments of inertia about axes through the center of gravity, and also I about a 45-deg axis through that point.

146. An element on the periphery of a 24-in.-diameter propeller shaft for a very large ship is subjected simultaneously to a torsional stress of 4,000 lb/sq in. caused by the torque, a bending stress of 1,000 lb/sq in. caused by its own weight midway between two bearings spaced far apart, and a compression stress of 500 lb/sq in. caused by the propeller thrust. Find the maximum principal stress and also the maximum shear stress at that element.

147. A 6-in.-diameter shaft is subjected simultaneously to a torque $M_t = 10{,}000$ ft-lb and to a bending moment $M_b = 8{,}000$ ft-lb. Find the maximum principal stress and the maximum shear stress in the shaft.

148. A 5-in.-diameter shaft is subjected simultaneously to a torque of 5,000 ft-lb and a compressive force of 50,000 lb. Find the maximum shear stress.

149. A beam of rectangular cross section 2 by 6 in. is subjected to a bending moment $M_b = 12{,}000$ ft.-lb and to a shear force $S = 12{,}000$ lb. Find the maximum shear stress at the three points 1, 2, and 3 shown in the figure.

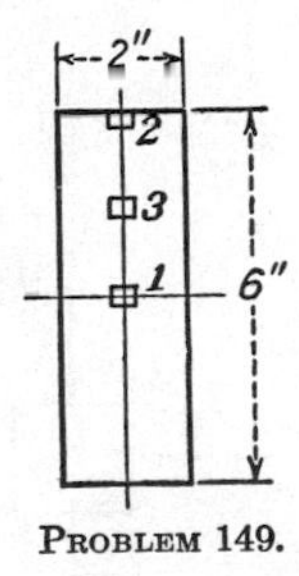

PROBLEM 149.

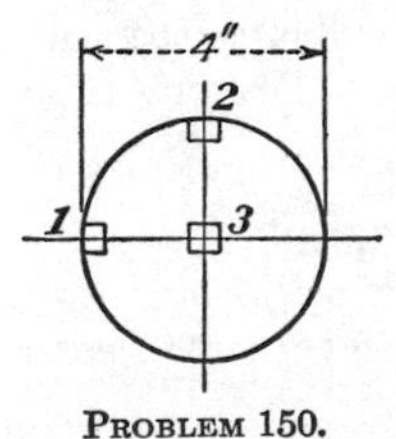

PROBLEM 150.

150. A 4-in.-diameter circular shaft is subjected to a shear force of 12,000 lb and a torque $M_t = 5{,}000$ ft-lb. The bending moment is zero at the cross section considered. Determine the maximum shear stress and the maximum principal stress at the points marked 1, 2, and 3 in the sketch. The shear force has the direction 2-3.

151. An 8-in. Bethlehem I girder has the constants described in Prob. 105. It is subjected to a bending moment of 6,000 ft-lb and a shear force of 1,000 lb. Determine the maximum shear stress at three different points,

a. The top corner of the flange.

b. At the horizontal center line, *i.e.*, at the neutral line.

c. At a point of the web 3 in. above the neutral line.

152. An American standard 10-in. channel has the following characteristics:

weight = 30 lb/ft; area = 8.80 sq in.; flange width = 3.03 in.;
web thickness = 0.67 in.; center-of-gravity distance $x = 0.65$ in.;
I about horizontal axis through center of gravity = 3.95 in.[4]

A 2-ft length of this channel is arranged as a cantilever beam and loaded with a 1,000-lb end load, bending it in the limber plane, as shown.

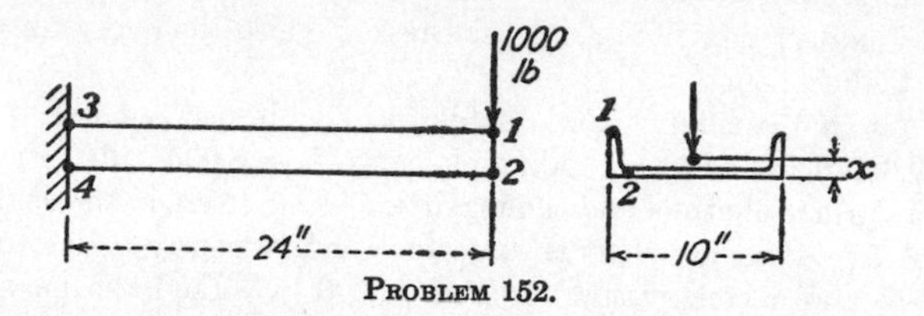

PROBLEM 152.

Find the maximum shear stress in the points 1, 2, 3, and 4, shown in the sketch. Points 1 and 3 are on the upper end of the U arm of the section. Points 2 and 4 are in the horizontal flange, close to the elbow corner.

153. Three elements are subjected to the states of stress a, b, and c. Which one of the three will yield first by

1. The maximum-stress theory?
2. The maximum-strain theory?
3. The maximum-shear theory?

The value of Poisson's ratio $\mu = 0.25$.

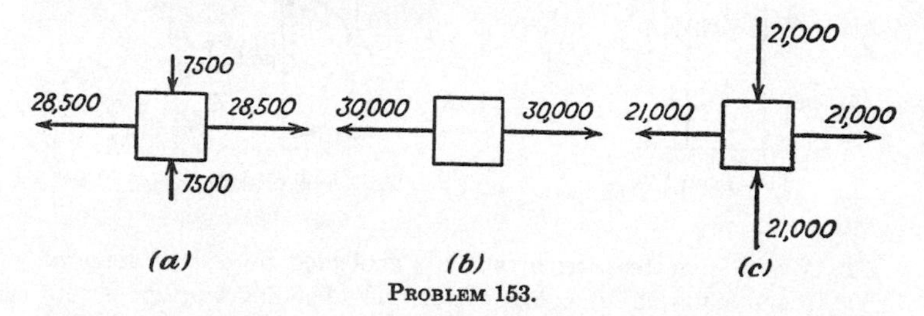

PROBLEM 153.

154. Calculate the necessary diameter of a steel shaft of 20 ft length that must transmit 50 hp at 600 rpm. The shaft is unsupported along its length and rests in bearings (simply supported) at its extremities only. It experiences bending caused by its own weight in addition to the torsion.

The material is cold-rolled steel with a tensile yield point of 40,000 lb/sq in. and hence with a shear yield point of 20,000 lb/sq in. The design is to be based on the maximum-shear theory with a factor of safety 2 with respect to yielding. (This problem leads to a cubic equation, which should be solved numerically by trial and error.)

155. A ring of 16 in. diameter and square cross section is split at one point and loaded with a pair of shear forces of 1,000 lb. Find the location of the maximum bending moment, and calculate the necessary dimension of the square cross section, based on the maximum-shear theory with a working stress of $s_s = 8{,}000$ lb/sq in.

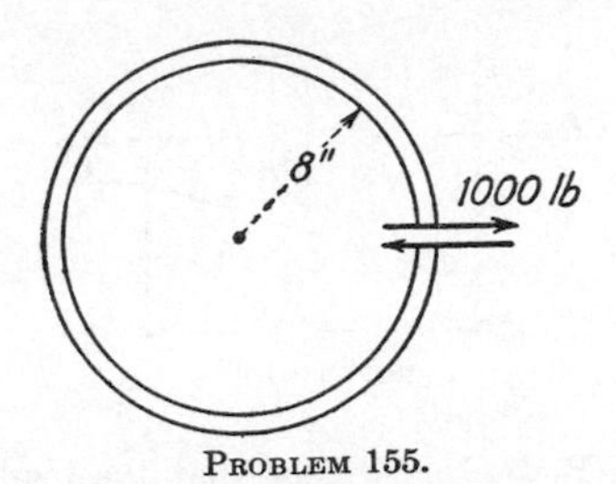

PROBLEM 155.

156. A steel shaft of circular cross section is subjected simultaneously to a torque of $M_t = 10{,}000$ ft-lb and a bending moment of $M_b = 10{,}000$ ft-lb. Design the necessary diameter of this shaft on three assumptions

a. Maximum-shear theory: $s_s = 5{,}000$ lb/sq in.
b. Maximum-stress theory: $s = 10{,}000$ lb/sq in.
c. Maximum-strain theory: $\epsilon = \frac{1}{3{,}000}$.

The constants of the steel are $E = 30 \times 10^6$ and $G = 12 \times 10^6$, as usual.

157. A wooden floor beam of 2 in. width and 20 ft span, freely supported at both ends, is loaded with a uniform load of 200 lb/ft, all along its length. Calculate the necessary height of this beam, based on

a. The maximum-shear theory with $s_s = 250$ lb/sq in.
b. The maximum-stress theory with $s = 500$ lb/sq in.
c. The maximum-strain theory with $\epsilon = \frac{1}{2{,}000}$.

The constants of the material are $E = 1 \times 10^6$ lb/sq in. and $\mu = 0.25$.

158. A uniform beam EI on two supports is loaded with a triangularly distributed load with an intensity growing from zero at the left to a maximum value of w_0 lb/in. at the right. Find the equation of the deflection curve and from it the location and magnitude of the maximum deflection.

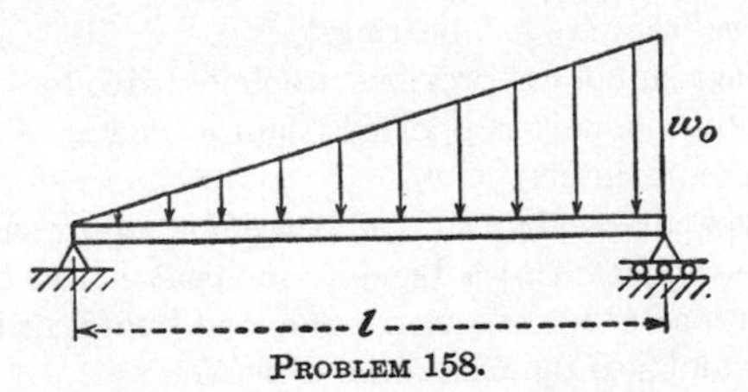

PROBLEM 158.

159. A cantilever beam has a constant width b, but a height which varies linearly from h_e at the end to h_b at the base. Derive the equation of the deflection curve, and find the end deflection under the influence of a concentrated end load P.

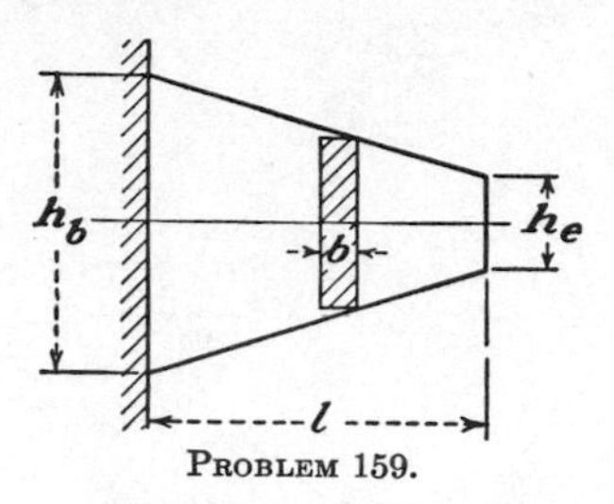

PROBLEM 159.

160. The same as Prob. 159, except that the loading consists of an end moment M_b only.

161. Find the end deflection of a cantilever of length l, constant width b, and a height h which grows linearly from zero at the free end to h_o at the built-in end, loaded by a uniformly distributed loading w_o.

162. A cantilever beam of length l and bending stiffness EI is loaded with a uniform load of intensity w along half its length only. Set up the differential equations for the two separate halves, and integrate them, each with the proper number of integration constants. Then calculate these integration constants from the conditions at the two ends and at the middle. From this find the deflection at the free end.

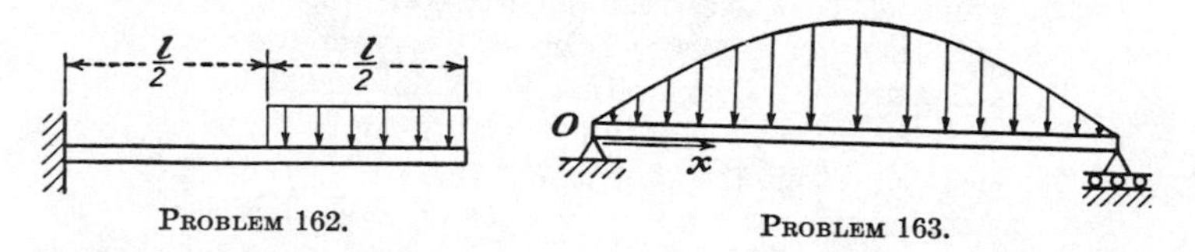

PROBLEM 162. PROBLEM 163.

163. A beam of constant cross section EI and length l, simply supported at its ends is loaded by a distributed loading varying like half a sine wave, with a maximum intensity w_0 in the center. Calculate the reactions at the ends, and derive the expression for the bending moment as a function of x, the distance from the left bearing.

164. From the result of the previous problem derive the equation of the deflection curve of a simply supported beam loaded by a half-sine-wave loading of maximum intensity w_0.

165. A beam of cross section EI and length l is simply supported at the two ends. It is subjected to a bending moment M_0 at the end $x = l$. Derive the equation of the deflection curve, and from it deduce the slopes at the two ends and also the maximum deflection.

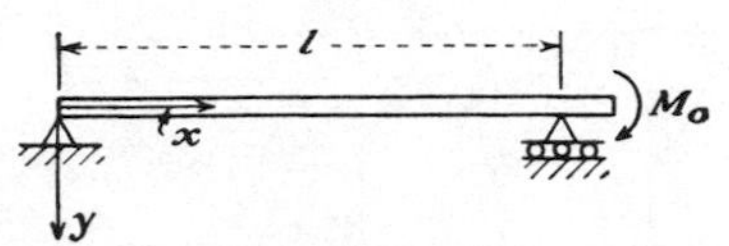

PROBLEM 165.

166. A beam EI is loaded by end moments M_0 and $2M_0$ as shown. Determine the location and magnitude of the maximum deflection, starting from the answer to Prob. 165.

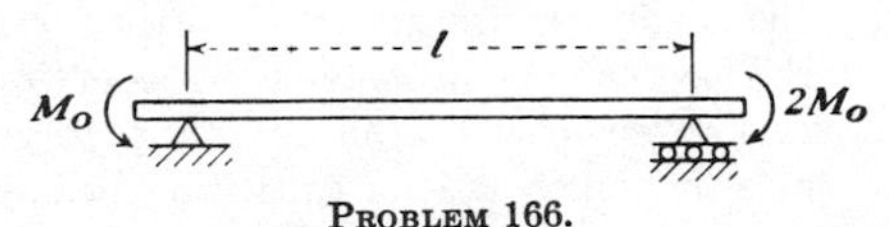

PROBLEM 166.

167. A beam EI, l is loaded with two moments M_0 of equal magnitude and sense of rotation. Sketch the shape of the deflection curve before calculating, and then determine by calculation the location and magnitude of the maximum deflection.

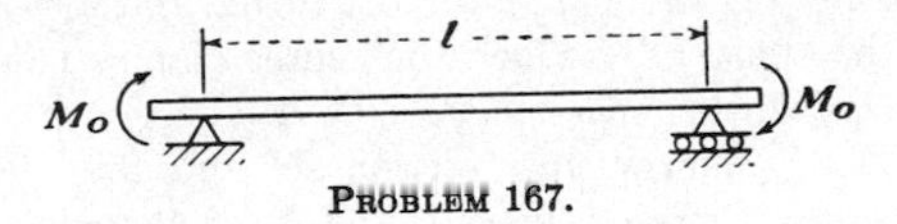

PROBLEM 167.

168. A solid steel rotor of the dimensions shown in the sketch is loaded with a central load of 10,000 lb. Calculate the deflection in the center.

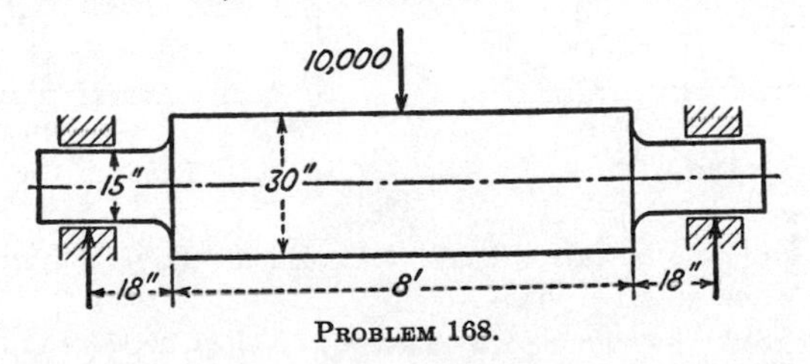

PROBLEM 168.

169. The same as Prob. 168, in which the central load of 10,000 lb is replaced by 10,000 lb distributed uniformly all along the central portion of 8 ft length.

170. A roll in a rolling mill experiences a uniform pressure p from the metal sheet that is being rolled down. The reaction is taken by two bearings, which are considered as concentrated forces in the sketch. The dimensions are a, l, r, and $2r$ (see figure on next page).

a. Derive an expression for the maximum bending stress.

b. Calculate the deflection of the roll in the center O and at the ends A, assuming that the bearings B are immovable.

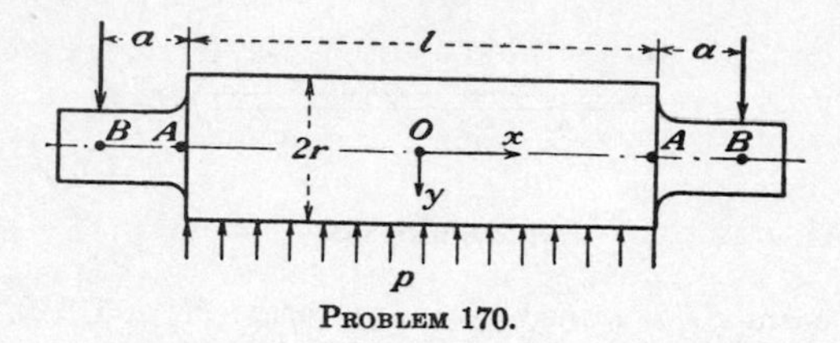

PROBLEM 170.

171. Refer to the previous problem. Owing to the deflection of the roll under the load p the conditions are not quite as they were assumed in Prob. 170. If the roll has the same diameter along its entire length, either the metal sheet being rolled must be thicker in the center than on the sides (with uniform p) or the sheet is of equal thickness and p is larger near the bearings than in the center of the roll. This undesirable condition can be avoided by making the roll diameter slightly larger in the center than near the ends, offsetting the deflection curve, so that, when the roll center line is deflected owing to uniform load p, the bottom edge of the roll becomes straight.

For $l = 5$ ft, $a = 1$ ft, $r = 9$ in., $p = 3{,}000$ lb/in., and $E = 30 \times 10^6$ lb/sq in., calculate the amount by which the center diameter of the roll at O must exceed the diameter near the ends, at A.

172. A beam of uniform cross section EI and total length $3\ l/2$ is loaded with two concentrated forces P and X. Calculate the value of X that gives zero deflection at the end X.

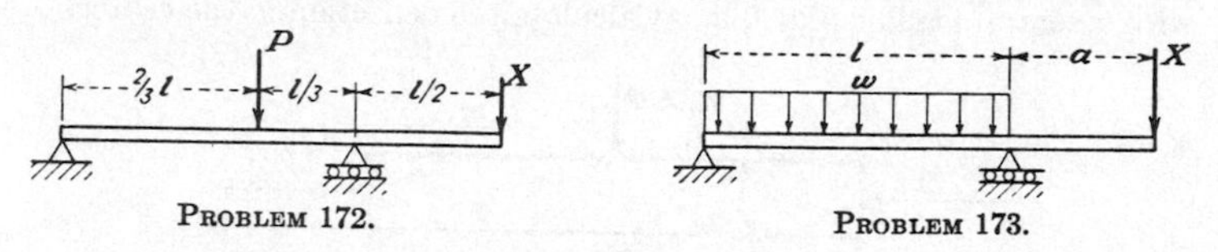

PROBLEM 172. PROBLEM 173.

173. A uniform beam EI of length $l + a$ is loaded by a uniform loading w and by a concentrated force X.

a. Calculate the value of X that gives zero slope at the end X.

b. Calculate the value of X that gives zero deflection at X.

174. A uniform beam of stiffness EI and total length $4l/3$ is supported and loaded as shown.

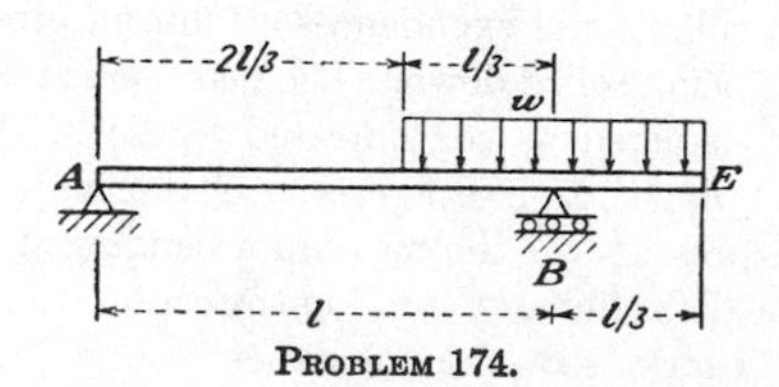

PROBLEM 174.

a. Calculate the deflection of the end E, assuming the supports A and B to remain in place.

b. Calculate the location and magnitude of the maximum deflection between A and B.

175. A uniform beam of stiffness EI and total length $5l/4$ is supported and loaded as shown. Calculate the end deflection at E.

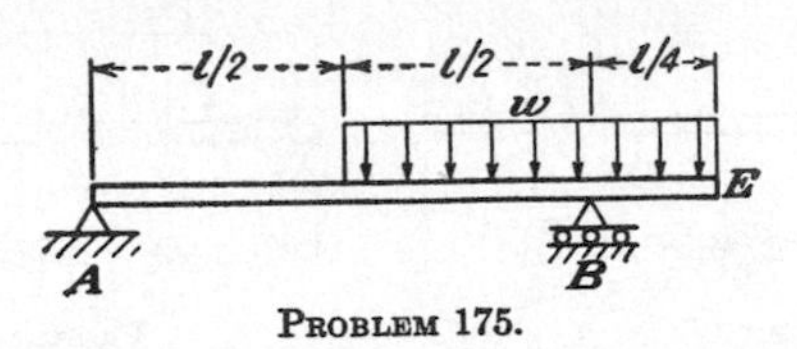

PROBLEM 175.

176. The design of wooden floor beams of simple houses (see Prob. 95) on the basis of stress is not always satisfactory. A beam may be sufficiently strong but may deflect considerably when a man walks on it at mid-span, which gives a feeling of flimsiness and hence is unsatisfactory. Make a list of the necessary height of cross section of 2-in.-wide beams covering various spans (simply supported at both ends) if the center deflection is to be kept at $\frac{1}{16}$ in. under a central load of 200 lb. The material constant is $E = 10^6$ lb/sq in.

177. A beam EI of total length $2a + 2b$ is symmetrically supported on two supports and symmetrically loaded by two overhung loads P. Set up the equation of the deflection curve from O to A and also the one from A to B. From it determine the maximum upward deflection at O and the maximum downward deflection at B.

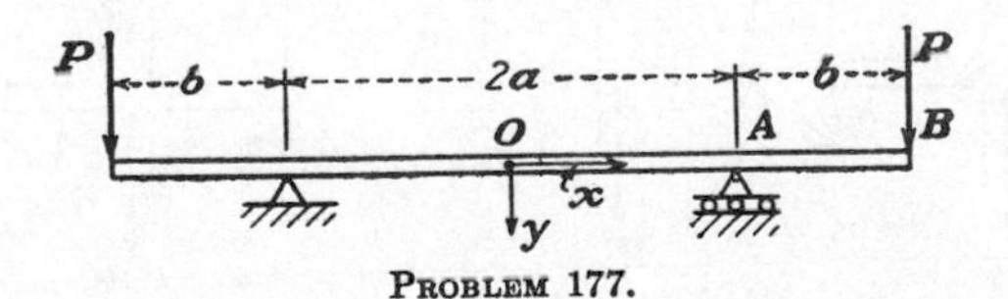

PROBLEM 177.

178. A beam of length l and cross section EI is subjected to a bending moment $M_0 = 2Pa$ in its center. Sketch the deflection curve, and calculate the location and magnitude of the maximum deflection.

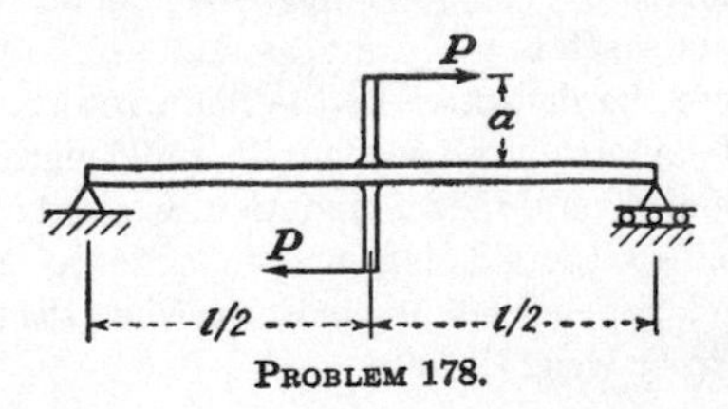

PROBLEM 178.

179. A beam EI of length $l + a$ on two supports is loaded as shown.

a. Derive expressions for the deflections under P and at the extremity of the overhang.

b. Substitute numbers: $l = 20$ ft; $a = 8$ ft; $w = 500$ lb/ft; $P = 3{,}000$ lb; the beam is the 8-in. Bethlehem girder of Prob. 105.

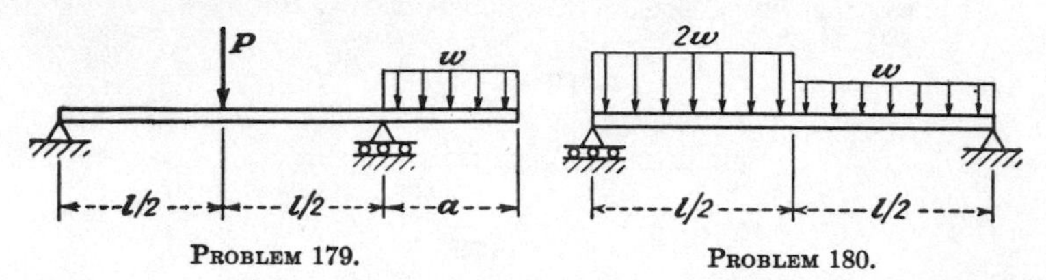

PROBLEM 179. PROBLEM 180.

180. A beam EI, l is loaded as shown. Find the location and magnitude of the maximum deflection.

181. Two rotors of weights W_1 and W_2 on shafts EI_1, l_1 and EI_2, l_2, respectively, are supported in bearings. The bearings II and III are close together. The rotors have to be coupled together by a sleeve coupling between II and III. If all four bearings are at equal height, the elastic deflections will cause an angular deviation between the two shaft stubs at the coupling, which is undersirable.

If the bearings II, III, and IV are kept at the same elevation, calculate by how much the bearing I must be raised in order to line up the shafts properly at the coupling.

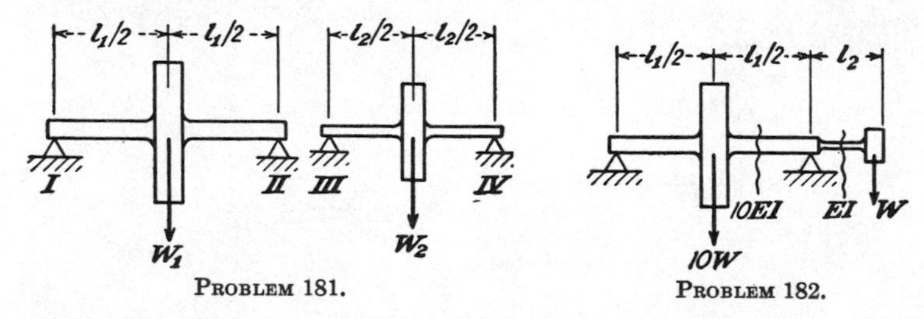

PROBLEM 181. PROBLEM 182.

182. A direct-current generator rotor on its shaft is mounted in two bearings, distance l_1 apart. On an overhang of length l_2 it carries an exciter rotor. The main generator weighs ten times as much as the exciter W. Also its shaft is ten times as stiff as the overhung exciter shaft EI. Calculate the deflection at the main rotor.

183. A beam of uniform cross section EI and length $a + b$ is simply supported at one end A on a rigid foundation, while the other support B is a spring of stiffness k (pounds per inch deflection). Find the stiffness k for which the end C goes neither up nor down when the load P is applied. Assume the beam to be weightless.

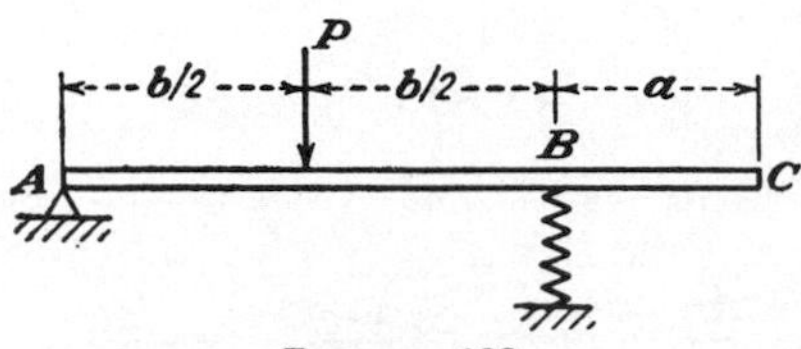

PROBLEM 183.

184. *a.* Answer the question of the previous problem when the concentrated load P is zero and is replaced by the weight w of the beam itself, uniformly distributed over the length $a + b$.

b. For which length ratio a/b does the end C stay where it is when the spring at B is stiff ($k = \infty$)?

c. Answer question *a* numerically for $a = 2$ ft, $b = 12$ ft; for a 6-in. standard I beam. (Marks' Handbook says: "Weight per foot 12.5 lb; $I = 21.8$ in.^{4}")

185. The sketch shows a simplified version of the nests of leaf springs used in locomotives. Here there are only two such springs, laid loosely on one another. The loading, being symmetrical, reduces the problem to two cantilevers, built in together at the center. The individual springs are equal in cross section EI each. Assume that the reaction between the two springs consists of concentrated forces between the two at A and at the center, while there is no contact force between them in the region from the center to A.

a. Calculate the end deflection in terms of P, l_1, l_2, and EI.

b. Substitute numbers: $l_1 = 18$ in.; $l_2 = 12$ in.; thickness $t = 3/8$ in.; width $b = 3$ in.; $P = 300$ lb.

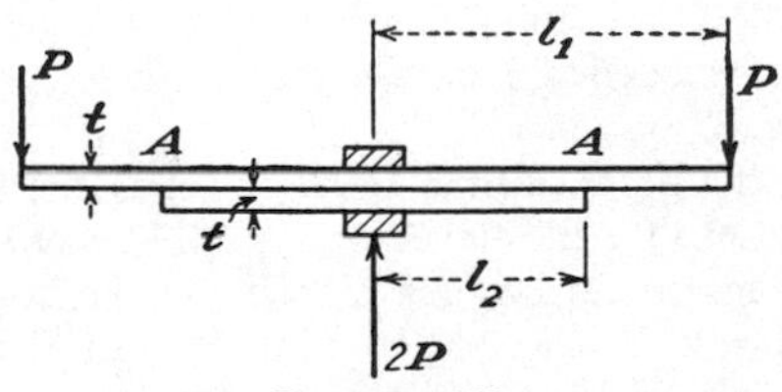

PROBLEM 185.

186. A slightly curved leaf spring of (large) radius of curvature R, total length l, and cross-sectional dimensions t and b lies on a rigid, flat foundation. It is loaded with two equal forces P at the ends. These forces P will have to acquire a definite value before the curvature of the spring at C has been reduced to zero (see figure on next page).

a. Calculate this value of P. Then in the next stage, for larger P, the spring will lie flat on the foundation over a central portion of length $2x$.

b. Determine the relation between P and x.

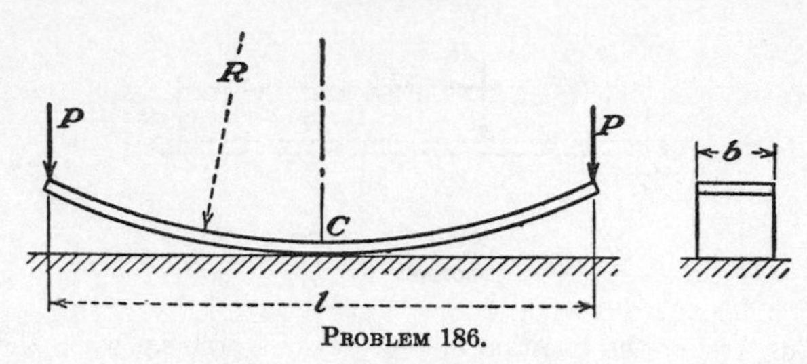

PROBLEM 186.

187. A shaft of length l, bending stiffness EI, and torsional stiffness GI_p is simply supported on two bearings A and B. In its center it carries an overhang of the same bending stiffness EI and of length a, with a load P at its end perpendicular to the plane of l and a. It is kept in equilibrium by a suitable torque at B (no torque at A) and by suitable reaction forces at A and B. There is no rotation of the shaft at B. Calculate the angle of rotation at A and the deflection under the load P.

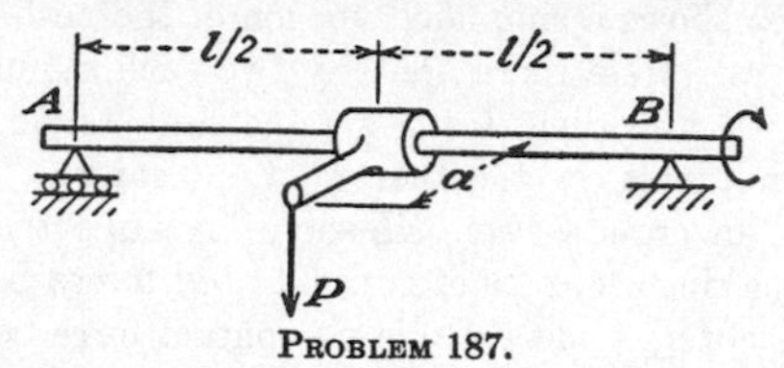

PROBLEM 187.

188. For the cantilever beam of Fig. 84 (page 92), calculate

a. The bending moment at the built-in end as a function of a.

b. The bending moment under the load P as a function of a.

Plot both of these against a. The diagrams so obtained are the "influence diagrams" of the bending moments.

189. Calculate the location x and the magnitude y_{max} of the deflection for the case of Fig. 87 (page 95).

190. Complete the problem of Fig. 90, begun in the text, by finding the general expressions for the end reactions in terms of P, a, and b.

Then, find the maximum bending moment in the beam for the particular case that $l = 12$ ft, $a = 9$ ft, $b = 3$ ft, $P = 1{,}000$ lb.

191. A cantilever beam of stiffness EI and total length l is loaded with distributed loads of intensities w and $2w$, as shown. There is an additional end support at such height that it is just touching the bar when the loads w and $2w$ are absent. Calculate the end reaction under those loads, and sketch the bending-moment diagram.

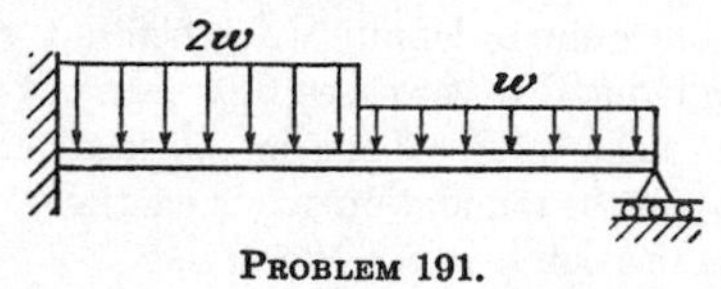

PROBLEM 191.

192. A beam EI on two end supports is subjected to two equal loads P, as shown. It has an additional support in the center in the form of a spring of stiffness k (pounds per inch). Calculate the three support reactions.

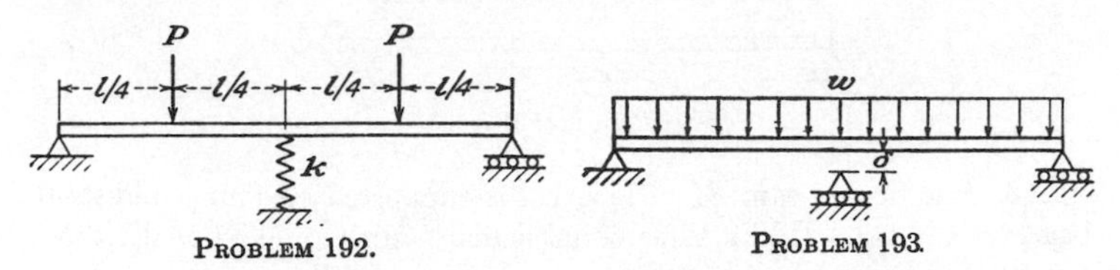

PROBLEM 192.　　PROBLEM 193.

193. A uniform beam EI of length l is laid on three supports, of which the center one is at a small distance δ below the straight line connecting the two end supports. The load on the beam is a uniform w. Calculate the value of δ for which the bending moment at the center of the beam is just zero.

194. If in the previous problem the center support is pushed slightly higher, the center of the beam acquires a bending moment opposite in sign to that of most of the rest of the beam. The most efficient use of the beam material is made when the center support is so adjusted that the bending moment just above it is equal and opposite to the maximum bending moment in the two side spans. Calculate the necessary value of δ to accomplish this and the three support reactions that go with it.

195. A straight steel shaft of 1 in. diameter, of total length 8 ft, runs in three bearings, two at the ends and one in the center of the shaft.

a. Calculate the reaction of the center bearing and the maximum bending moment in the shaft if the center bearing is offset ¼ in. sidewise from the straight line connecting the end bearings. Assume the shaft to be weightless.

This problem is identical with that of an initially curved shaft, ¼ in. off at the center, laid in three bearings which are lined up properly. At least, the two problems are identical as long as the shafts do not rotate.

b. Discuss the difference between the two cases for a rotating shaft in the behavior of the force at the bearing as well as in the behavior of the stress in one particular fiber of the shaft.

196. A uniform beam EI on three supports of equal height is loaded by two forces P in the middle of the two spans. Calculate the three bearing reactions.

PROBLEM 196.

197. A beam EI on three supports (which can take upward as well as downward reactions) is subjected to a bending moment M_0 at one end. Determine the three bearing reactions.

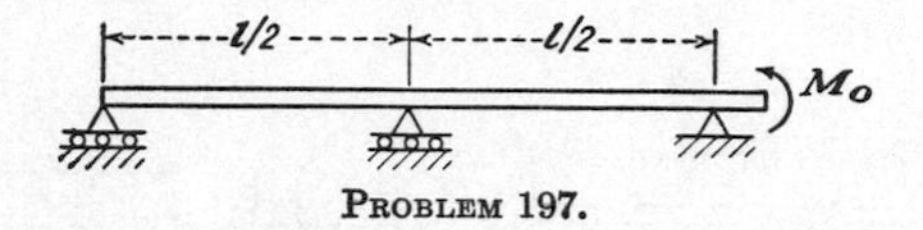

PROBLEM 197.

198. A uniform beam EI of length l is supported on four equidistant bearings and is loaded by three equal loads P in the middle of the three spans. Determine the four bearing reactions. Use the symmetry of the figure to full advantage before you begin calculating.

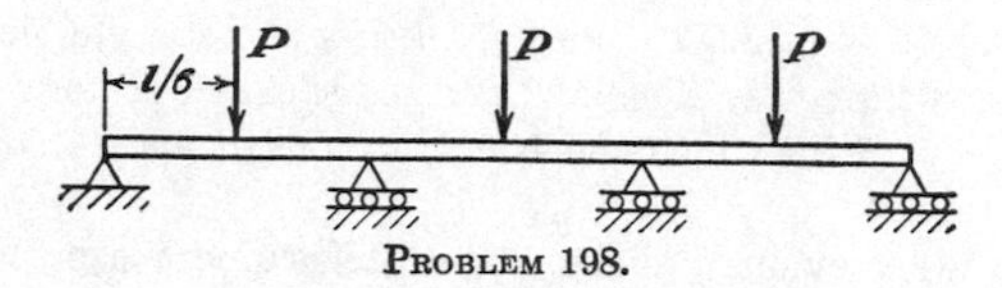

PROBLEM 198.

199. Continue the analysis of Fig. 95 (page 104) by calculating the deflection and slope at the point E (where the value of I jumps), both by the area moment and by the Myosotis method.

200. An I beam of height h, flange width h, and thickness $t = h/10$ is laid on two supports and is loaded by a concentrated force in the center. Calculate the length ratio l/h for which the central deflection caused by shear becomes 20 per cent of the central deflection due to bending.

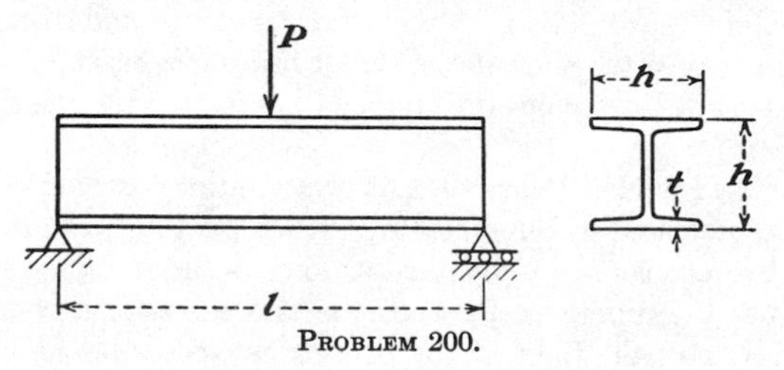

PROBLEM 200.

201. Answer the same question as in Prob. 200, in which the central concentrated force is replaced by a loading w uniformly distributed along the span l.

202. A beam is made up of a 4- by 6-in. wooden core ($E = 1 \times 10^6$ lb/sq in.) and two 6 by ⅜ steel plates bolted to it. Calculate the equivalent section modulus Z in cubic inches and the equivalent bending stiffness EI in pound inches squared about a horizontal neutral line.

203. A 6 by 8 wooden cross section ($E = 1 \times 10^6$ lb/sq in.) is "reinforced" by two thin steel plates ⅛ in. thick. If the maximum allowable

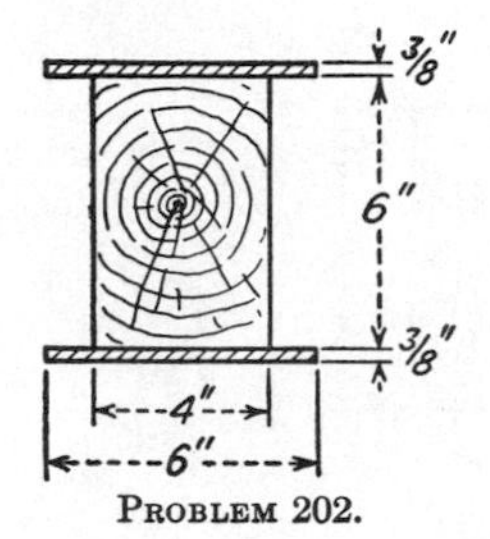

PROBLEM 202.

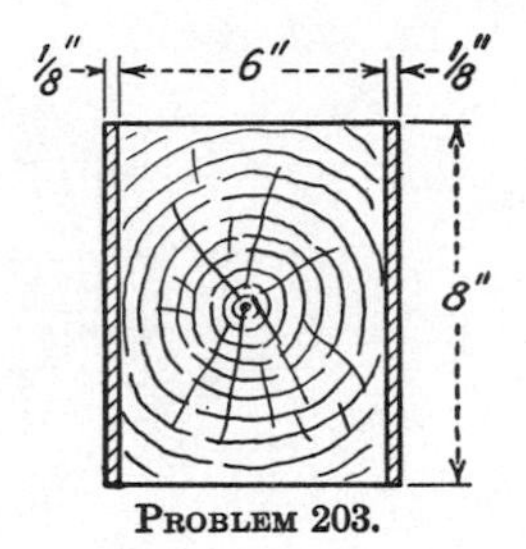

PROBLEM 203.

stress in the wood is 1,000 lb/sq in. and that in steel is 10,000 lb/sq in., calculate the allowable bending moment (about a horizontal neutral line) of

a. The composite beam as shown.

b. The wooden core by itself without steel plates.

204. A steel bolt of 2 sq in. cross section is inside a copper tube, also of 2 sq in. cross section. There are nuts on the bolt ends, as shown, and the length of the copper tube is 10 in.

a. Starting from the unstressed state with zero clearance, one of the nuts is screwed up through a distance of 0.01 in. along the steel bar. Calculate the stresses in the assembly.

b. Starting from the previous state, an extra pulling force of 5,000 lb is exerted on the bolt. Find the stresses.

c. Starting from the unstressed state with zero clearance, the system is heated 50° F. Find the stresses.

The material constants are, for steel:

$E_s = 30 \times 10^6$ lb/sq in.; $\alpha_s = 6.7 \times 10^{-6}$ in./in./° F.;

PROBLEM 204.

and for copper:

$$E_c = 10 \times 10^6 \text{ lb/sq in.}; \alpha_c = 10.0 \times 10^{-6} \text{ in./in./° F.}$$

205. A tailshaft for a ship has a solid steel core of 20 in. diameter, on which a 2-in. bronze sleeve is shrunk, so that the total outside diameter is 24 in. The shaft is 24 ft long. Calculate its torsional constant k, expressed in pounds-inches per radian.

206. A column in a building has the cross section of a 16-in. Bethlehem girder cast in concrete. The girder has the properties described in Prob. 122, and the concrete is a rectangle covering all steel parts with a 1-in.-thick layer, as shown in the figure on the next page. The modulus of concrete is $E_c = 2.5 \times 10^6$ lb/sq in. Calculate

a. The equivalent "compressive stiffness" AE.

b. The equivalent "bending stiffness" EI in both principal planes.

c. The shape and dimensions of the core of the cross section (see page 57).

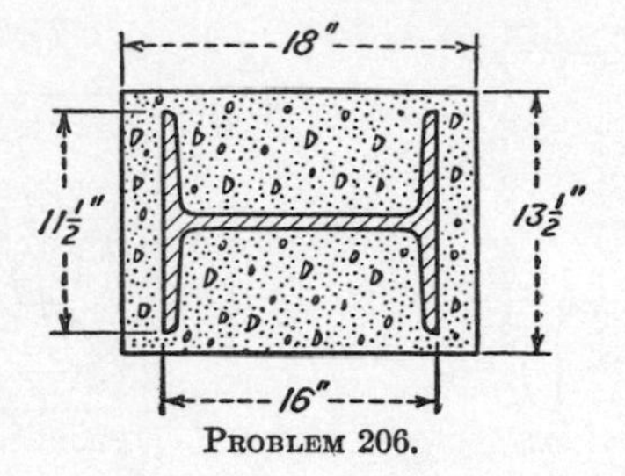

PROBLEM 206.

207. A "bimetallic" strip for use as a thermostat consists of a thin layer of steel soldered to a layer of bronze over its entire surface. The thickness of the steel as well as of the bronze is 0.030 in.; the width of the spring is ½ in. At room temperature the assembly is straight. It is heated 50° F (which tends to make it a slightly curved beam) and is kept straight by the application of equal and opposite bending moments at both ends. $\alpha_b = 10.0 \times 10^{-6}$; $\alpha_s = 6.7 \times 10^{-6}$ in./in./° F; $E_b = 15 \times 10^6$ lb/sq in.

Calculate these bending moments, the stresses in the two materials, and the location of the neutral line of bending.

208. A cantilever beam of length l, loaded with a force P at its free end, has the cross-sectional shape of an angle with one leg of length a, the other of length $2a$. Find

a. The angle between the principal axes through the center of gravity G and the vertical and horizontal directions.

b. The bending-stress distribution at the built-in end.

c. The angle between the neutral line of bending and horizontal axis.

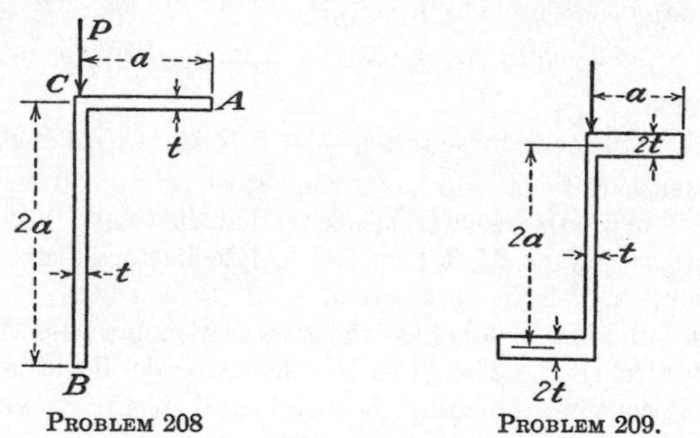

PROBLEM 208 PROBLEM 209.

209. A cantilever beam of Z section, as illustrated, is subjected to a force along its vertical web. As a result it will deflect downward and sidewise.

a. If the picture shows a view on the free end of the cantilever, will it deflect to the right or to the left?

b. Determine the angle between the horizontal line and the neutral line of bending.

210. A cantilever is made up of a 10-in. American standard channel section, having the following properties:

weight per foot = 35 lb; area = 10.27 sq in.; width of flange = 3.18 in.; $I_{11} = 115.2$ in.^{4}; $I_{22} = 4.63$ in.^{4}; $x = 0.69$ in.

The cantilever is loaded with a 45-deg force. Determine the angle between the axis 1-1 and the neutral line of bending.

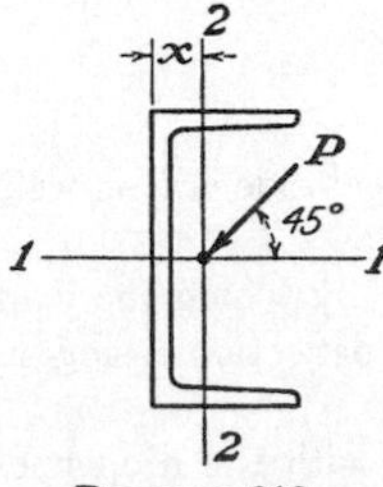

PROBLEM 210.

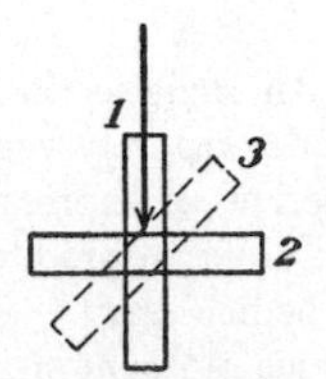

PROBLEM 211.

211. A cantilever beam has a rectangular cross section $b \times h$, where $b < h$. In the unstressed state, it is twisted through 90 deg along its length; at the built-in end, the large dimension h is vertical (1 in the sketch); at the free end, h is horizontal (2 in the sketch); and halfway between, it is at 45 deg (the dotted figure 3 in the sketch). The load is an end load P, acting on section 2. Derive a formula for the end deflections by a process of integration involving the steps below:

a. Express the geometrical angle of inclination as a function of the length x along the beam ($0 < x < l$).

b. At a piece dx resolve the bending moment into its two principal components, and find the two end deflections caused by a deformation of dx alone, the rest of the bar being considered rigid (see page 28 or page 102).

c. Resolve these end deflections into horizontal and vertical components.

d. Integrate.

212. An I beam is simply supported at both ends and subjected to a central concentrated load. The beam is accidentally set nonvertical by a small error angle φ. This error causes the bending stresses in the beam to be greater than without the error (see figure on next page).

a. Derive a formula for the quantity $(ds/d\varphi)/s$, that is, the percentage increase in stress per radian error. Use such letters as may be necessary.

b. Substitute numbers for a 12-in. American standard I beam of 55 lb/ft;

$I_{max} = 319.3$ in.^{4}; $I_{min} = 17.3$ in.^{4}; width of flange = 5.6 in.

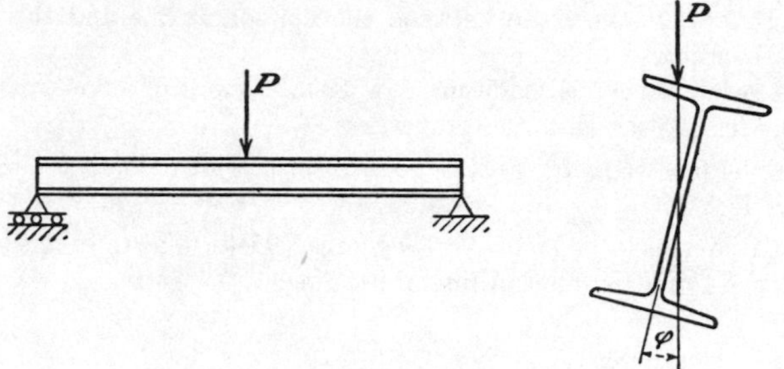

PROBLEM 212.

213. An angle-section cantilever will deflect sidewise as well as downward under a purely vertical load.

a. Derive an expression for the angle α by which the load must be offset in order to produce a purely vertical deflection, using such letters as may be necessary.

b. Calculate α numerically for the angle with the properties listed in Prob. 118.

c. A cantilever is made up of two such angles riveted together as shown, loaded by a central vertical load. State any conclusions you may reach from the foregoing on the subject of the tensile forces in the rivets holding the two angles together.

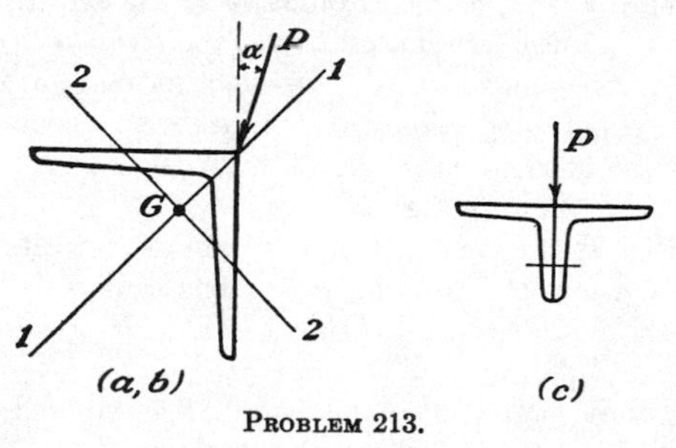

PROBLEM 213.

214. Determine the location of the center of shear of a channel section of uniform (small) wall thickness t, web dimension b, and flange height a.

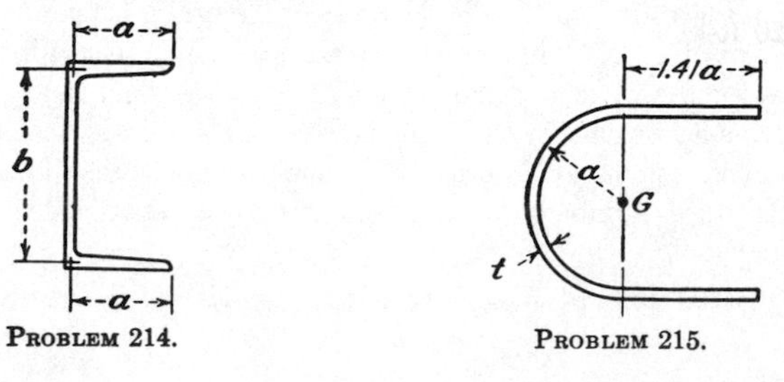

PROBLEM 214. PROBLEM 215.

215. A thin-walled cross section of uniform wall thickness t consists of a half circle of radius a and two straight pieces of length $a\sqrt{2}$.

a. Verify that the center of gravity of the cross section coincides with the center of the circle.

b. Find the center of shear of this cross section.

216. Find the location of the center of shear of a thin-walled cross section having the shape of a rectangle of wall thickness t, split on one side. Let t be small with respect to the other dimensions a and b.

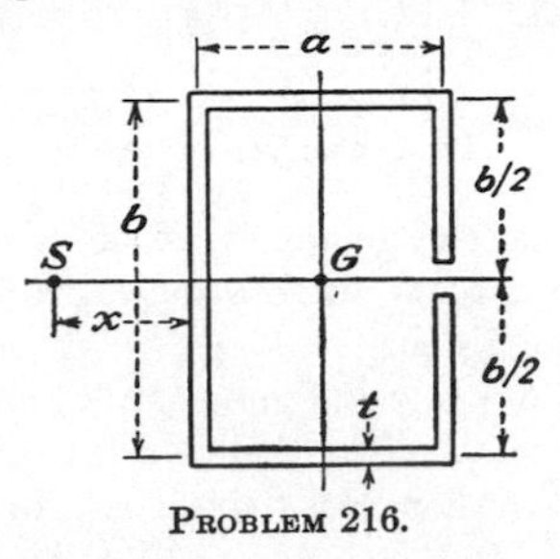

PROBLEM 216.

217. Referring to Fig. 113 (page 126), and assuming a thickness t of the cross section small with respect to the leg length a, calculate the points where the neutral line of bending intersects the legs, and, for a given bending moment M, calculate the stress in the corner as well as the stress in the two ends of the legs of the angle.

218. Referring to Figure 114*d*, assuming a "thin" angle ($t \ll a$) and a given shear force S, calculate the peak values of the a and b parabolas and the distance CN of the neutral point of shear stress.

219. Derive the set of equations (a_1), (b_1), and (c_1) of page 131 for a T-shaped reinforced-concrete beam (Fig. 116).

220. Design a reinforced-concrete beam of rectangular cross section, of width b = 10 in., in which the permissible stress in steel is s_s = 18,000 lb/sq in.; in concrete, s_c = 1,200 lb/sq in. The beam has to support a bending moment of M = 25,000 ft-lb, and the ratio of the moduli of elasticity is $E_s/E_c = 10$.

221. A reinforced-concrete beam of rectangular cross section has a width b = 17 in. and an "effective" height h = 31 in. It has A_s = 6.93 sq in. of steel in it, the ratio $E_s/E_c = 10$, and it carries a bending moment M = 320 ft-kips (= 3,200,000 ft-lb). Calculate the stresses in steel and concrete.

222. Derive a set of simplified formulae for a reinforced-concrete beam of T-shaped section (Fig. 116, page 130), in which the compressive stress in the concrete is supposed to exist only in the horizontal flange (*i.e.*, where the compression in the vertical web is neglected). The linear strain distribution diagram of Fig. 116 still remains as sketched; the "neutral line" is still somewhere in the web and not at the point where web and flange meet.

223. Apply the simplified formulae of Prob. 222 to the case shown in the sketch. The steel cross section is $A_s = 2.40$ sq in., the moment carried is $M = 72{,}000$ ft-lb, and $E_s/E_c = 10$.

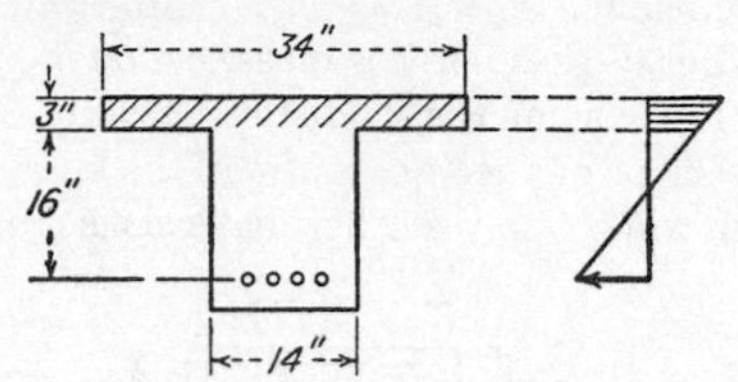

PROBLEMS 222 and 223.

224. Calculate the stresses in the beam of Prob. 223 on the basis of the equations of page 131, which *do* include compressive stresses in the upper part of the vertical web.

225. Refer to Fig. 120 (page 135), and assume a cross section in which the flange width a equals the web height a and in which the thickness t (equal in the web and the flange) is to be considered "small" with respect to a.

a. Calculate the location of the neutral line of bending for completely developed plasticity.

b. Calculate the ratio of the bending moment in the completely plastic state to the maximum possible elastic bending moment.

226. Three long steel bars, each 1 sq in. in cross section, are brazed together along their entire lengths. All bars have the same $E = 30 \times 10^6$ lb/in.; one of them is of ordinary steel with a yield point of 30,000 lb/sq in.; the other two are of alloy steel with a yield point of 60,000 lb/sq in. They are being pulled by a gradually increasing force, until complete plasticity is reached; then the force is gradually backed off. Draw diagrams of the distribution of the total force between the three bars for the various stages of the development, and in particular calculate the residual stress left in the structure after the load is back to zero.

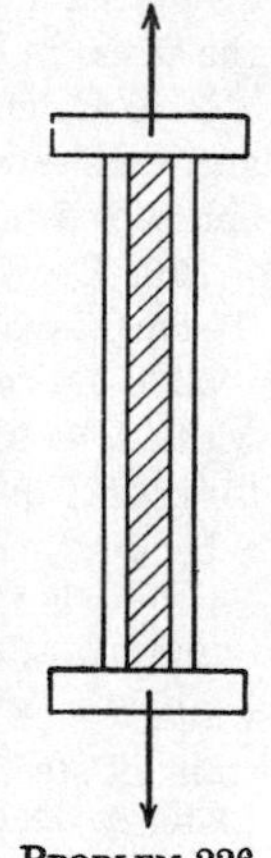

PROBLEM 226.

227. Develop diagrams like Figs. 118 and 119 and formulae like those on page 133 for the case of torsion in a circular shaft of uniform material. In particular, calculate the ratio of the torque for total plasticity and for just impending plasticity at the outer fibers.

228. Repeat Prob. 227 for a hollow shaft of outer radius r_o and inner bore radius r_i.

229. A long shaft consists of a central core of steel and a bronze concentric sleeve around it. There is sufficient pressure between them to prevent relative slip, but the shrink stresses are to be considered negligible, or at least without influence on what happens afterward. The properties are as follows:

$r_{steel} = 2$ in.; $G_{steel} = 12 \times 10^6$ lb/sq in.; $s_{s\ yield} = 15{,}000$ lb/sq in.;
$r_{o\ bronze} = 3$ in.; $r_{i\ bronze} = 2$ in.; $G_{bronze} = 6 \times 10^6$ lb/sq in.;
$s_{s\ yield\ bronze} = 5{,}000$ lb/sq in.

The shaft is subjected to a gradually increasing torque, until complete plasticity occurs; then the torque is backed off. Calculate the residual stresses left in the setup.

230. Calculate the ratio of the bending moment for complete plasticity to the maximum elastic bending moment that can be carried by an 8-in. Bethlehem I girder having the properties described in Prob. 105.

231. The cylindrical drum of a steam boiler is subjected to an internal steam pressure of 600 lb/sq in. (above atmospheric). It is to be designed on the basis of the maximum-shear theory with an allowable shear stress of $s_s = 5{,}000$ lb/sq in. (*i.e.*, with a factor of safety 3 for ordinary steel boiler plate). Find the necessary plate thickness.

232. A "penstock" is a cylindrical steel pipe in a hydraulic power plant conducting the water from the storage lake to the turbines below. Calculate the necessary wall thickness for such a pipe of 3 ft diameter for a 2,000-ft waterhead. The pipe is of ordinary boiler plate, with a yield point in tension of 30,000 lb/sq in. The factor of safety with respect to the yield point should be 2½; this figure is supposed to include the weakening effect of the riveted joints in the penstock.

233. A cylindrical oil-storage tank of 100 ft diameter is filled with oil of a specific gravity 0.85 to a height of 30 ft. If the allowable tangential stress in the steel is 10,000 lb/sq in., calculate the necessary wall thickness near the bottom of the tank and halfway up (at 15 ft oil depth). Neglect the stiffening effect of the bottom of the tank by assuming it to be just a cylinder.

234. Calculate the necessary wall thickness of a bottle-gas cylinder (for domestic cooking gas in rural districts) of 12 in. diameter, subjected to an internal pressure of 600 lb/sq in. The allowable tensile stress is 10,000 lb/sq in.

235. *a.* Derive an expression similar to Eqs. (23*a*) and (23*b*) (page 137) for the stress in a spherical vessel subjected to a constant internal pressure p.

b. Calculate the necessary wall thickness of a high-pressure spherical oil storage tank of $D = 40$ ft diameter, subjected to internal pressure $p = 100$ lb/sq in. The pressure variation due to the head of oil is to be neglected with respect to the over-all pressure. The allowable tensile stress is 15,000 lb/sq in.

236. Two steel plates of thickness t are butt-jointed with two cover plates and a single row of rivets in each main plate. Determine the rivet diameter d and the pitch p (both expressed in terms of plate thickness t) so that the three "efficiencies" of the joint all have the same value according to the A.S.M.E. Boiler Code. What is that efficiency?

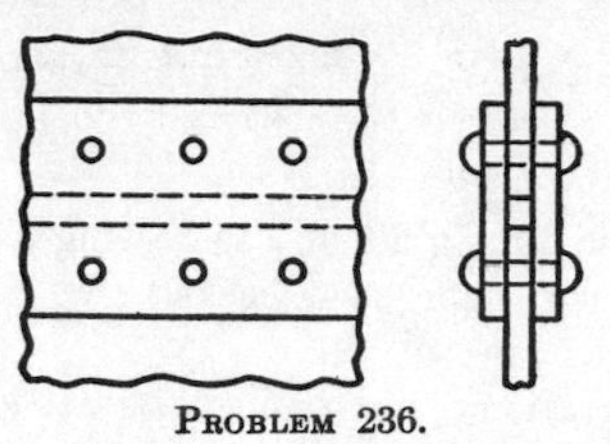

PROBLEM 236.

237. The same as Prob. 236, but now for two staggered rows of rivets in each main plate, as illustrated in Fig. 122 (page 138).

238. The same as Probs. 236 and 237, for a joint with three rows of rivets, of which the two outer rows have the same pitch p and the inner row has the pitch $p/2$. Assume that the two outer rows of rivets take half the load, so that the diminished pitch in the inner row is of no consequence on the total plate-strength efficiency.

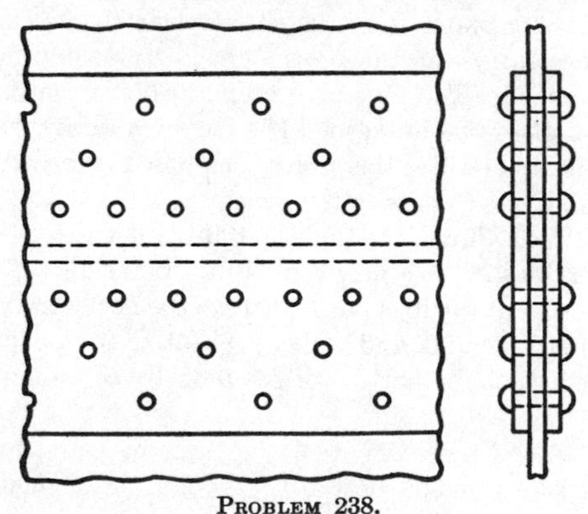

PROBLEM 238.

239. A "thin-walled" pressure vessel consists of a ¼-in. alloy-steel outer cylinder for strength and a ⅛-in. copper inner one for corrosion, put together so that for no internal pressure the two cylinders just fit without locked-up stress and without clearance.

If the maximum allowable tensile stress in the copper is just *under* the yield stress of 10,000 lb/sq in. and the maximum allowable stress in the steel is 40,000 lb/sq in. (well below its yield point), calculate the maximum allowable pressure inside the vessel. Take $E_s = 30 \times 10^6$ and $E_c = 12 \times 10^6$.

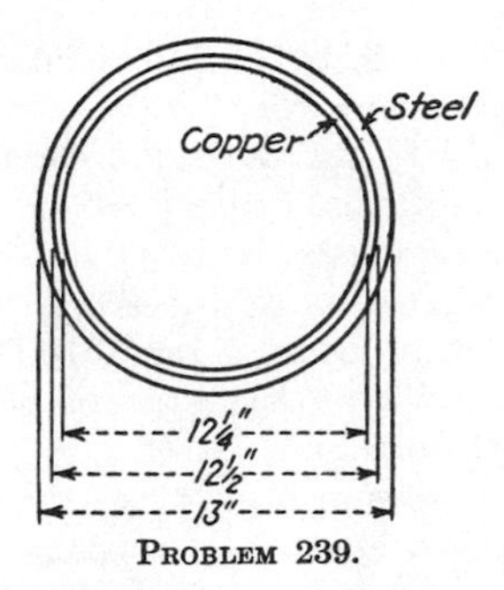

PROBLEM 239.

240. In the previous problem calculate the maximum allowable internal pressure when the copper lining is allowed to yield freely at 10,000 lb/sq in., while the stress in the steel is limited to 40,000 lb/sq in.

241. Consider the built-up cylinder of Prob. 239, without internal pressure, without locked-up stress, and without clearance between the cylinders. Calculate the tangential stresses caused by a temperature rise of 100° F, neglecting the expansion in a longitudinal direction. Take $\alpha_s = 6 \times 10^{-6}$ and $\alpha_c = 10 \times 10^{-6}$, in./in./° F.

242. The same as Prob. 241, but now the longitudinal expansion is to be considered. No longitudinal slip occurs between the steel and the copper, so that the longitudinal strain necessarily is the same for both. This causes longitudinal stresses, which, in turn, by Poisson's effect, cause tangential strains and stresses. Similarly, the tangential stresses react back on the longitudinal ones. Solve the problem for values of Poisson's ratio $\mu = 0.25$ for both materials.

243. Consider again the cylinder of Prob. 239, without pressure. This time the outside diameter of the copper cylinder was 0.006 in. greater than the inside diameter of the steel one before assembly. Calculate the tangential stresses set up by the assembly.

244. Solve the problem of the compression of a solid cylinder r by an outside radial pressure p. Do this by starting from Eqs. (*e*), (*a*), and (*b*) of article 29 (pages 141 and 143).

245. A gun with a bore diameter of 4 in. and an outside diameter of 8 in. is made of a material with a yield point in tension of 80,000 lb/sq in.

What is the maximum pressure in the bore during firing of the gun, if the maximum tangential stress is to be kept within the elastic limit?

Sketch the distribution of the tangential stress s_t across the 2-in. wall thickness.

246. Derive equations corresponding to (*f*), (*g*), and (*i*) on page 143, but for the case of external pressure, that is, $p = 0$ at $r = r_i$ and $p = p$ at $r = r_o$.

247. The construction of the gun of Prob. 245 can be improved by building it up of two cylinders of 1 in. wall thickness shrunk over each other; the internal radius still is 2 in.; the joint radius is 3 in., and the outside radius is 4 in., as before. The "radial allowance" before shrinking is $\delta = 0.005$ in. Calculate the tangential shrink stresses in the two cylinders, and plot them against r from $r = 2$ in. to $r = 4$ in.

248. When the gun of Prob. 247 is fired, the stresses due to the internal firing pressure have to be added to (superposed on) the shrink stresses. Starting from the answers to Probs. 245 and 247, find the maximum firing pressure in the gun that will keep the tangential stress at the bore below the yield stress of 80,000 lb/sq in.

249. If the internal pressure in a thick-walled cylinder is gradually increased, the tangential stress at the bore is proportional to that pressure, as long as that stress is below the yield point. The stress distribution then is as sketched in *a*. When the bore pressure is further increased, the inner layers of the cylinder will yield, giving a stress distribution as sketched in *b*, and ultimately the entire wall thickness will assume the yield stress, sketched in *c*. The situation *a* is covered by Lamé's formulas, (page 143), and the situation *c* is simple, but the intermediate state *b*) is more complicated than we care to discuss in this text.

Calculate the ratio between the bore pressure p for completely developed plasticity to that for just impending plasticity at the bore in terms of the general symbols r_i and r_o.

Plot this ratio p_{plas}/p_{elas} against r_o/r_i for the range $1 < r_o/r_i < 10$.

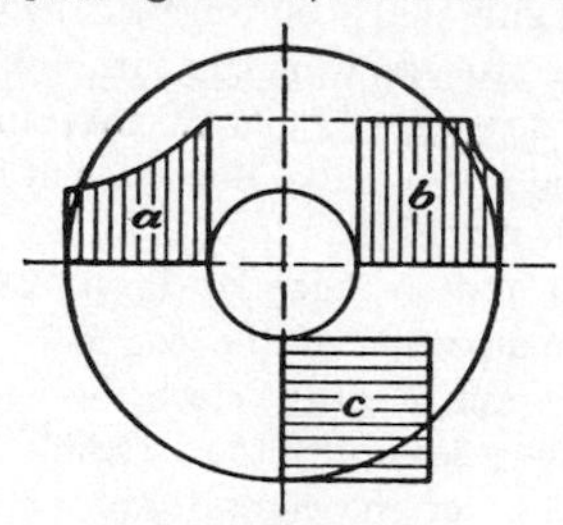

PROBLEM 249.

250. A semicircular steel spring of radius of curvature $R = 2$ in.. thickness $t = 1/16$ in., and width $b = 1/2$ in. is clamped solidly at one end A and is subjected at the other end to a single force $P = 1$ lb in the plane of the circle, directed toward A.

Calculate the maximum stress in the spring and the deflections of the end along P and perpendicular to P.

251. The spring of Prob. 250 is subjected to an end force $Q = 1$ lb only, directed tangent to the spring at its end. Calculate the maximum stress and the deflection components of the end.

252. The spring of Prob. 250 is subjected only to an end moment $M_0 = 2$ in.-lb, acting in the plane of the ring. Find the maximum stress and the components of end deflection.

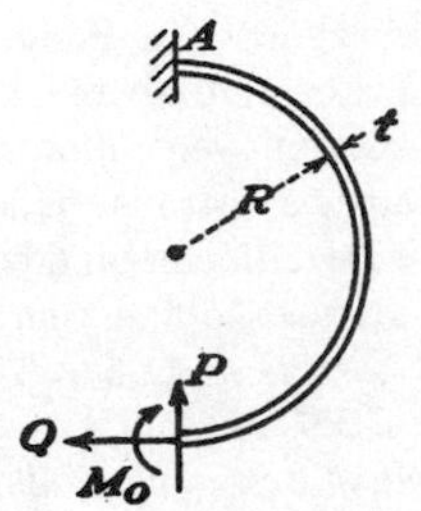

PROBLEMS 250 TO 252.

253. A semicircular spring $R = 2$ in., clamped at one end, is subjected at the free end to a force of 1 lb directed perpendicularly to the plane of the circle. This involves torsion in certain parts of the ring, and we are not in a position yet to calculate the torsion in a beam of the rectangular cross section of Probs. 250 to 252. Hence the ring of this problem has a solid circular cross section of diameter $d = \frac{1}{8}$ in. Calculate the maximum stress and deflection along the force. $E = 30 \times 10^6$; $G = 11.5 \times 10^6$.

254. The semicircular steel spring of Prob. 253 ($R = 2$ in.; $d = \frac{1}{8}$ in.) is subjected at its free end to a bending moment of 2 in.-lb in a plane perpendicular to the plane of the ring. Calculate the stresses and the vertical deflection.

255. The semicircular steel spring of Probs. 253 and 254 is subjected at its end to a twisting torque of 2 in.-lb. Find the maximum stress and the end deflection.

256. Derive some of the six Myosotis formulae for the straight cantilever beam by integration of Eqs. (26), (page 148).

257. A balcony is supported by a continuous girder of U shape 3 + 6 + 3 ft, lying in a horizontal plane as shown. The beam is loaded by a

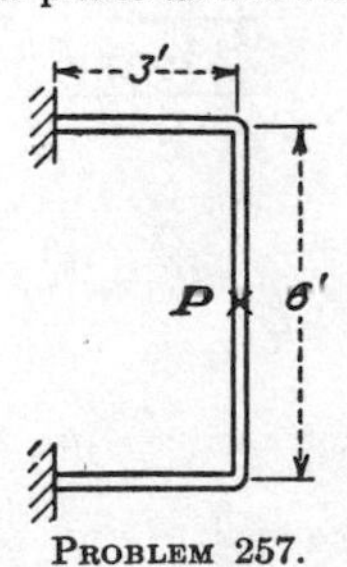

PROBLEM 257.

single vertical force P in its center. The beam is supported in a statically indeterminate manner. By imagining cuts at both built-in locations, develop an argument of symmetry, analogous to that with Fig. 130 (page 151). Reduce the number of unknowns to a single one by this procedure.

258. A balcony such as is shown in Prob. 257 usually would be constructed of a steel I section or of timbers of rectangular section. Since twist is involved in the problem, we cannot yet solve it for the practical case. Imagine instead that the beam is made of hollow steel pipe, 4 in. outside diameter 3½ in. inside diameter, and hence of ¼ in. wall thickness. Calculate the maximum stress and the center vertical deflection for a vertical load $P = 500$ lb. Let $G = 11.5 \times 10^6$ lb/sq in.

259. The same as Probs. 257 and 258, but this time the balcony beam has the form of a semicircle of 3 ft radius. The section is a 4-in.-outside-diameter pipe as before, and the load again is 500 lb.

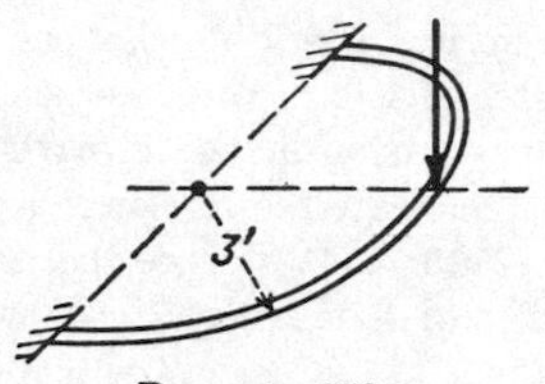

PROBLEM 259.

260. A crane hook carrying a load of 1,000 lb has a rectangular 4- by 2-in. cross section CC. Using such numbers as are available in the text and in Fig. 136 (page 159), determine the stresses at the inner and outer fibers of section CC. (Don't forget the tensile force P across the section.)

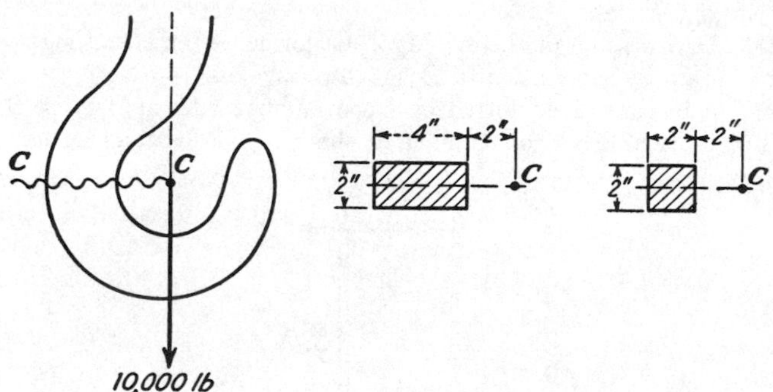

PROBLEMS 260 AND 261.

261. The dimensions of the hook of Prob. 260 are changed to 2 by 2 in., as sketched. For this case no numerical data are available in the text. Calculate the extreme stresses.

262. It is known from page 27 that a (closely coiled) common helical spring under tension has pure twist in its wire.

a. Prove that such a spring subjected to a twisting moment experiences pure bending in its wire.

b. A helical spring of coil diameter $D = 4$ in. is made up of wire of a square cross section 1 by 1 in. The spring is of steel, has $n = 20$ full turns, and is subjected to a twisting torque $M_t = 400$ ft-lb. Calculate the maximum stress in the spring, and state the location of that maximum stress.

c. Determine the "torsional stiffness" k of the spring, *i.e.*, the torque required to wind up the spring through 1 radian.

263. A highly curved beam subjected to pure bending has a symmetrical I section of height h and radius of curvature R_G. For simplicity consider the flange and web thickness to be "thin," and neglect the stiffening effect of the web altogether, so that only the two flanges are to be considered. These flanges each have a cross-sectional area A. Find the offset n between the center of gravity and the neutral axis in terms of R_G, h, and A.

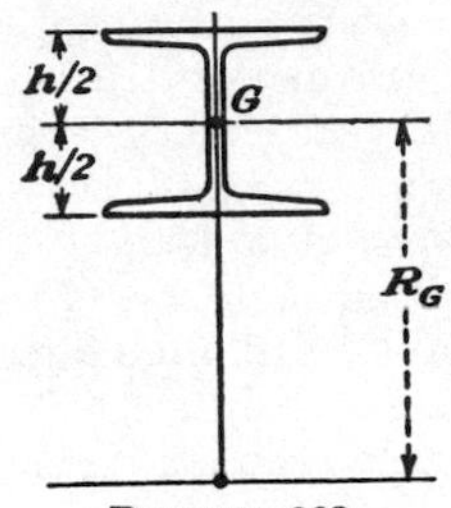

PROBLEM 263.

264. The same as Prob. 263, but now the two flanges are of different area, the outer one being smaller. Let A be the total area of the two flanges together; the outer one being αA and the inner one $(1-\alpha)A$, where α is a numerical fraction.

265. On page 164 of the text it is stated that the total energy in a beam element loaded by tensile force P and a bending moment M is equal to the sum of the two component energies. Prove this in detail by calculating the energy in a fiber dA at distance y from the neutral line and by integrating that energy over the entire cross-sectional area A.

266. Derive Eq. (32) without using energy, by the geometry of deformation only. Let the left figure $ABCD$ be a square of side a, and let it

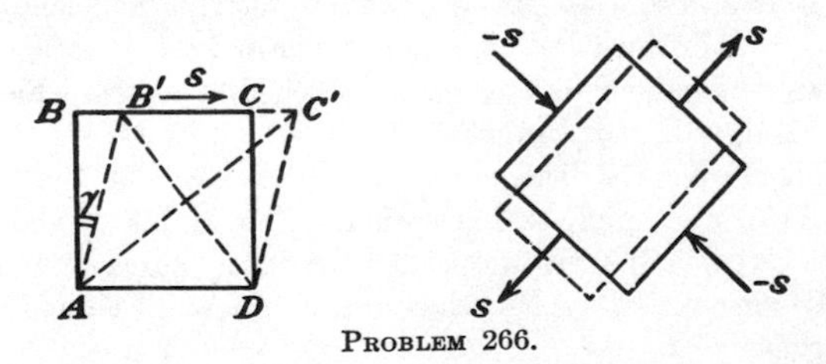

PROBLEM 266.

deform to $AB'C'D$, by an angle γ. Express the lengths AC' and $B'D$ in terms of a and γ, keeping only terms in γ to the first power and neglecting γ^2 and higher powers. Compare the strain in the AC' direction to the strain in the corresponding direction of the right figure.

267. Derive an equation for the work stored in an element $dx\ dy\ dz$, subjected to three principal stresses s_1, s_2, s_3. This result is a generalization to three dimensions of Eq. (31) (page 165).

268. To get a physical concept of the order of magnitude of the elastic energy that can be stored in a steel spring, calculate the height to which a piece of steel must be raised to contain just as much potential energy of gravitation as it can store elastically by being subjected to the yield stress in one direction only ($s_1 = s_{\text{yield}}$; $s_2 = s_3 = 0$). Do this for

a. Mild steel $s_y = 30{,}000$ lb/sq in.

b. Spring steel $s_y = 180{,}000$ lb/sq in.

269. A wooden yardstick on two simple supports at its ends is subjected to a single load P somewhere along its span. This force P is made to increase gradually until the stick breaks, which, we assume, happens when the maximum bending moment reaches a certain critical value. Where along the beam should P be located in order to break the stick with a minimum of work done by the force P?

270. A cantilever beam of length l has a rectangular cross section bh and is loaded by a vertical end load P.

a. Calculate the ratio between the stored energy of bending and the stored energy of shear.

b. Substitute numbers: $l = 2$ ft; $b = \frac{1}{2}$ in.; $h = 1$ in.; $E/G = 2\frac{1}{2}$.

271. Neglecting the energy of shear (considering bending only), calculate the "efficiency" as an elastic-energy storer of the cantilever spring of Prob. 270*b*.

272. The mainsprings in clocks and watches usually are of the "spiral" type of rectangular cross section bh, of which the larger dimension b is perpendicular to the plane of the spiral. The spring is wound by turning the central shaft and keeping the outer extremity A of the spring anchored. This is equivalent (for small wind-ups) to keeping the central shaft immovable and by pulling out the end A tangentially.

The calculation of the efficiency of such a (closely coiled) spiral is possible with the reader's present knowledge, but it is too complicated to be useful as an exercise. Instead we ask for the "efficiency" of the outer single circle of the spiral only, of radius r and cross section bh. The sketch shows the inner end of that circle attached to an arm which is to be considered rigid, so that it cannot store energy.

273. As in Prob. 272, calculate the "efficiency" of the innermost complete circle of a spiral watch spring. The radius of this circle is supposed to be negligible with respect to the outer radius r but still large enough with respect to the spring thickness h so that the circle is a "beam of small curvature."

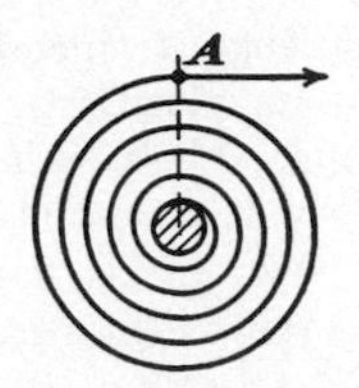

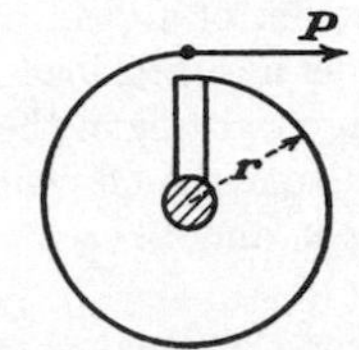

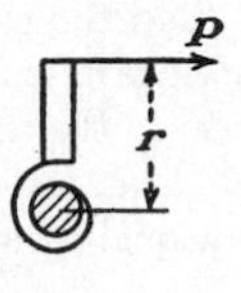

PROBLEMS 272 AND 273.

274. By Castigliano's theorem derive the end deflection and the end angle of a uniformly loaded (w) cantilever beam. Find both these results out of one calculation only, employing the procedure mentioned on page 172 of the text.

275. By means of Castigliano's theorem derive the equation for the deflection curve of a uniformly loaded cantilever beam. Express the deflection y in terms of a coordinate x, measured from the free end of the cantilever, so that $x = l$ at the built-in end.

276. *a.* Calculate the central deflection of a beam on two supports loaded by a continuous (but not uniform) loading, which increases linearly from both ends to a maximum value w_{max} in the center.

b. Substitute numbers: $l = 5$ ft; $I = 10$ in.^{4}; $E = 30 \times 10^6$ lb/sq in.; $w_{max} = 500$ lb/ft.

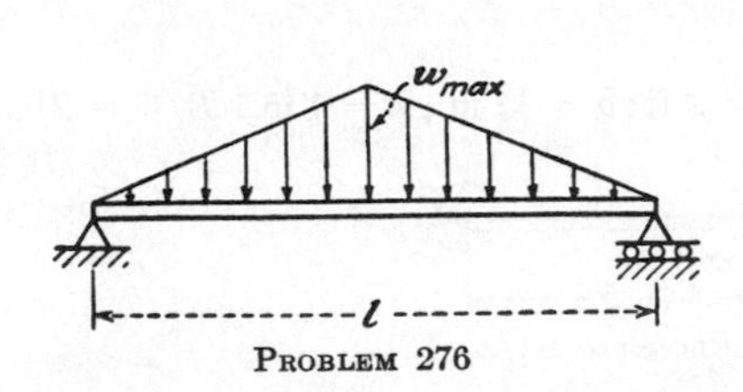

PROBLEM 276

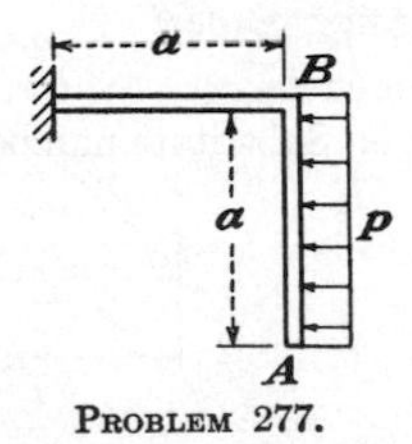

PROBLEM 277.

277. An L-shaped cantilever beam with two equal legs of length a is loaded by a uniform pressure p along one of its legs, as shown. Find the horizontal deflection at the end A and also the vertical deflection of the corner B.

278. *a.* Prove that the stored energy in a spring of stiffness k, subjected to a force F, is $F^2/2k$.

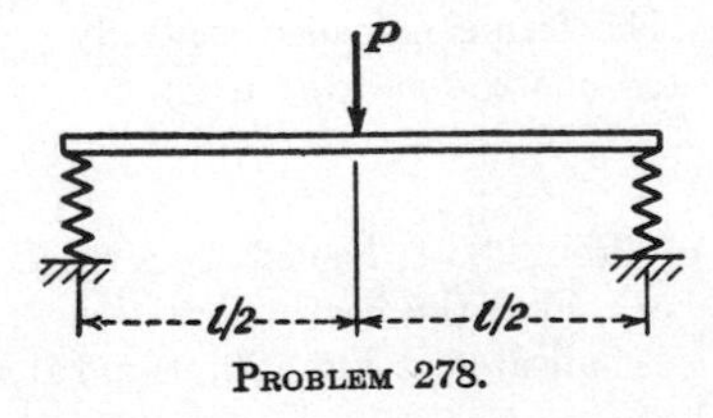

PROBLEM 278.

b. Find the central deflection of a beam EI of length l supported on end springs of stiffness k, under a central load P, by applying Castigliano's theorem to the entire system consisting of the beam and both springs.

279. Find the deflection under the off-center load P of a beam EI on two unequal spring supports k_1 and k_2.

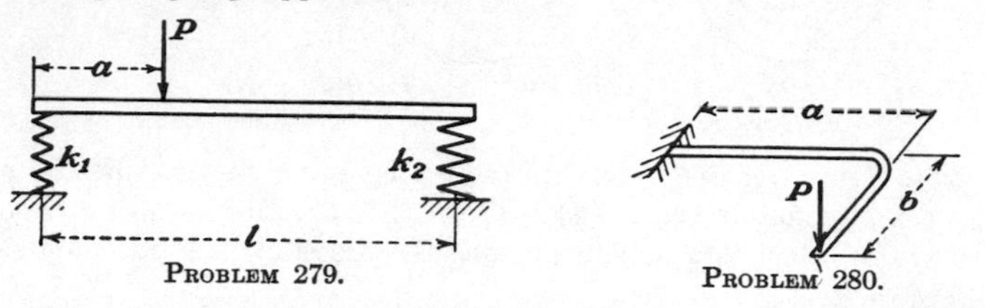

PROBLEM 279. PROBLEM 280.

280. A beam EI and GI_p is bent in an L shape in the horizontal plane. Calculate the deflection under the vertical end load P.

281. A beam of plane U shape of dimensions a, a, and $2b$ is simply supported on three points as shown. These supports are understood to be able to take an upward as well as a downward reaction force, but they do not impede the beam from small rotations at the supports. The beam is loaded by a force P at one corner; its cross section is circular.

a. Find the vertical deflection at the other corner.

b. For which ratio b/a is that deflection zero for any value of P at the first corner?

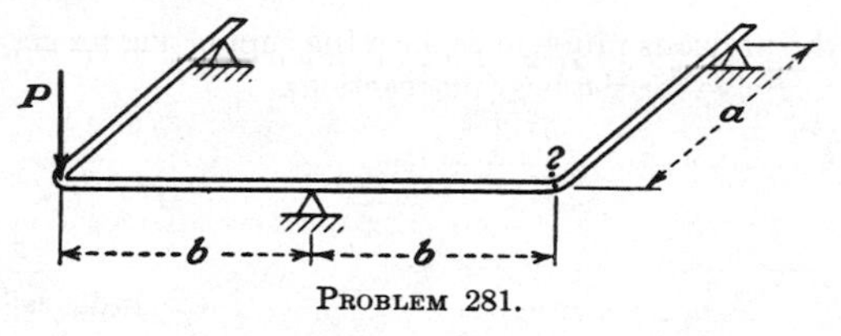

PROBLEM 281.

282. Solve the problem of Figs. 129 to 131 with Castigliano's method, starting from the result of Fig. 130.

283. A portal frame of height b, width a, and bending stiffness EI is supported by two hinges, securely anchored. (Note the absence of rollers under the hinges.) The frame is loaded centrally with a load P. Determine the bearing reactions.

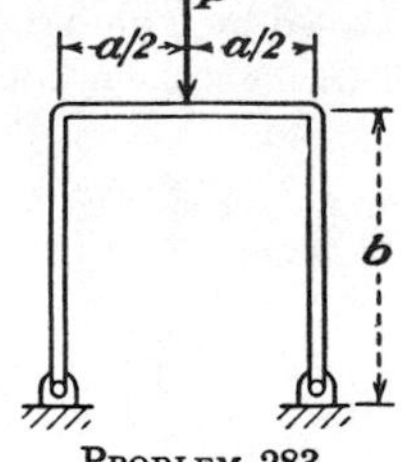

PROBLEM 283.

284. The frame of Prob. 283 is loaded by a single horizontal force P along the top member. Find the bearing reactions. (Use an argument of symmetry such as is indicated in Fig. 130, page 151.)

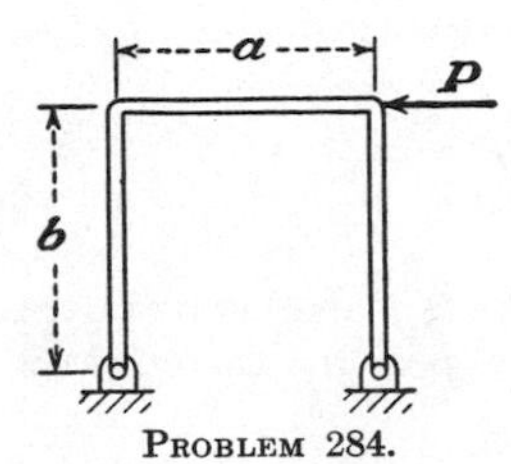

PROBLEM 284.

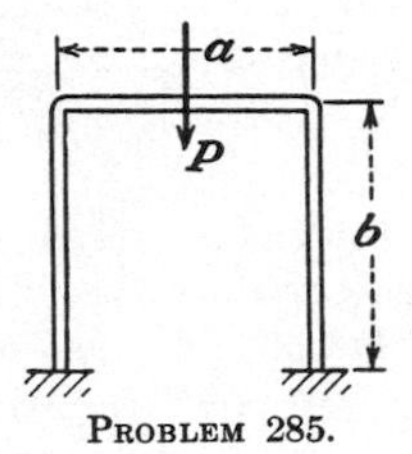

PROBLEM 285.

285. The portal frame of the previous problems is built in at both legs. Find the reactions at the supports. First reduce the problem to two unknowns by means of a symmetry analysis.

286. A semicircular arch of radius a and bending stiffness EI is supported by two hinges anchored to the ground and is loaded in the center by a force P. Find the bearing reactions.

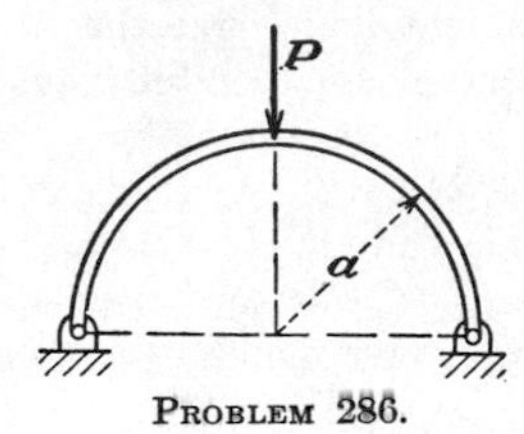

PROBLEM 286.

287. In the previous problem replace the supporting hinges by built-in connections. Solve for the bearing reactions.

288. A semicircular beam is built in at its two ends to form a horizontal balcony (see Prob. 259). The radius is a, the bending stiffness EI, and the torsional stiffness GI_p, and the load P is central and vertical. Calculate the bending and twisting moments.

289. A horizontal ring of radius a, bending stiffness EI, and torsional stiffness GI_p is supported on three points 120 deg apart. These supports are capable of taking vertical reaction forces only. The ring is loaded by three forces P in the middle between the bearings. Find the maximum bending moment and the maximum twisting torque in the ring.

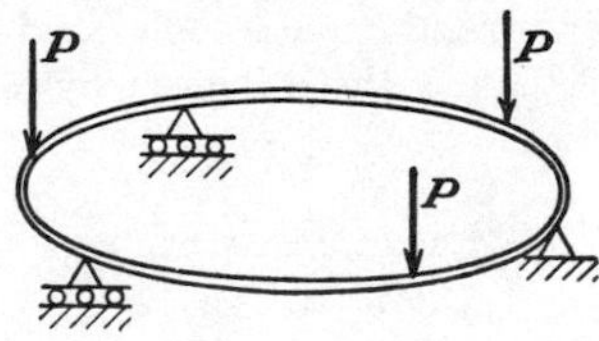

PROBLEM 289.

290. A shaft of torsional stiffness GI_p and length l is built in at both ends and is loaded by a twisting torque M_t at one-third length. Find the torque distribution along the length by Castigliano's theorem.

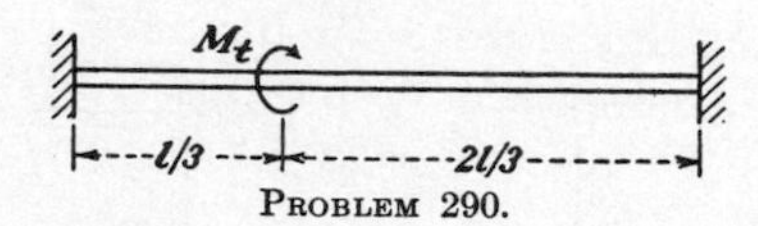

PROBLEM 290.

291. A beam of stiffness EI and length l is supported by three supports at equal height and is loaded by a force P in the center of each span. Find the three bearing reactions.

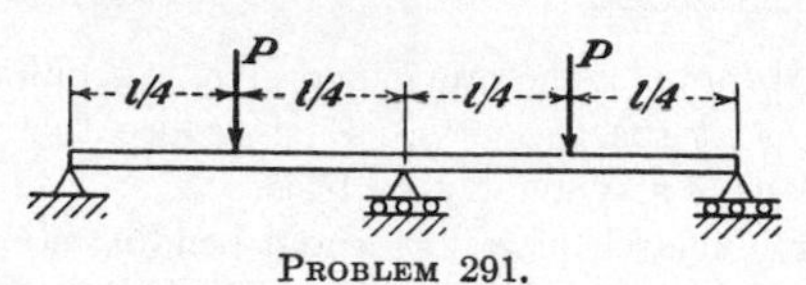

PROBLEM 291.

292. The same as Prob. 291, but now the center support is at a small distance δ below the straight line connecting the two outer supports. The beam is straight when unloaded, and it is supposed that the loads P are so large that the beam rests on the center support. Find the bearing reactions (as a function of δ) by Castigliano's theorem.

293. A cantilever EI, l has a redundant support at distance δ below the unloaded end of the beam. The beam is loaded with a force P in the center. Find the end-support reaction by Castigliano's theorem.

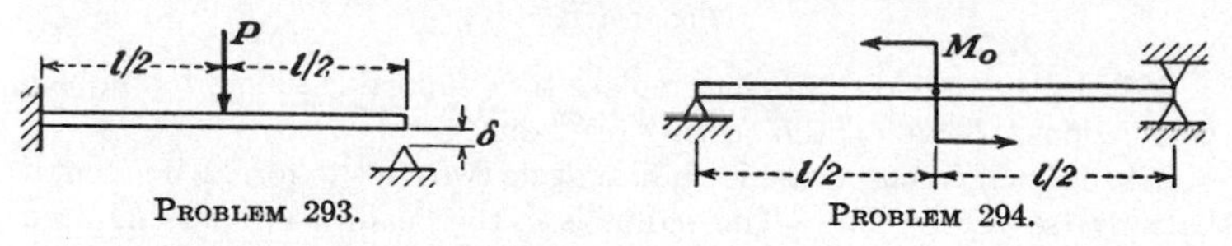

PROBLEM 293. PROBLEM 294.

294. A beam EI, l is simply supported at its ends and loaded by a bending moment M_0 at its center.

a. Calculate and plot the deflection diagram of the beam.

b. By Maxwell's theorem this diagram has another meaning. Express in a sentence what this meaning is, and check the correctness of the statement by calculation.

295. A cantilever beam EI, l is loaded by a load P in its center. Calculate the deflection diagram, and express the second meaning of this diagram. Check the correctness of the statement by calculation.

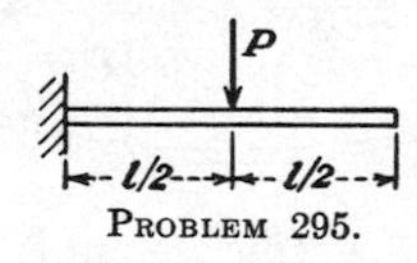

PROBLEM 295.

296. Plot the deflection diagram of a cantilever beam subjected to an end moment, and state the reciprocal significance of this diagram. Check it by calculation.

297. A beam of length l is completely stiff against bending ($EI = \infty$) along half its length, while the other half has the finite stiffness EI.

a. Calculate the angle at the right support caused by a bending moment M at the left support as shown.

b. Find the angle at the left due to a bending moment M at the right support.

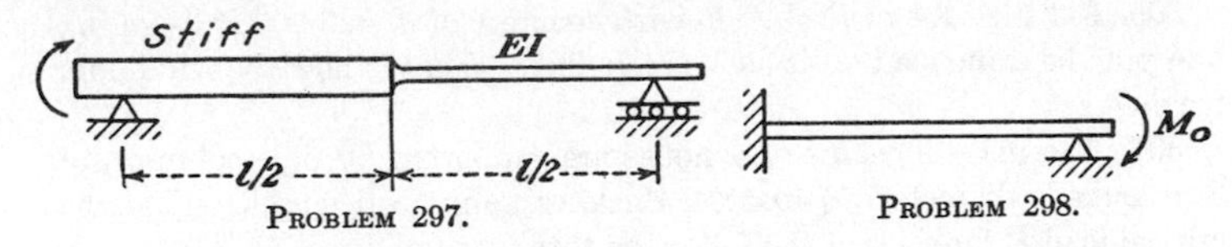

PROBLEM 297. PROBLEM 298.

298. A cantilever beam of stiffness EI and length l has an additional end support at a height such that the beam is stressless when without load. Find the deflection curve caused by a bending moment M_0 at the end. State the reciprocal property of this diagram.

299. A semicircular ring of stiffness EI and radius r is supported on an anchored hinge and on a roller hinge as shown. The load is a vertical P in the center. Find the horizontal displacement of the roller support. Which other problem have you solved thereby?

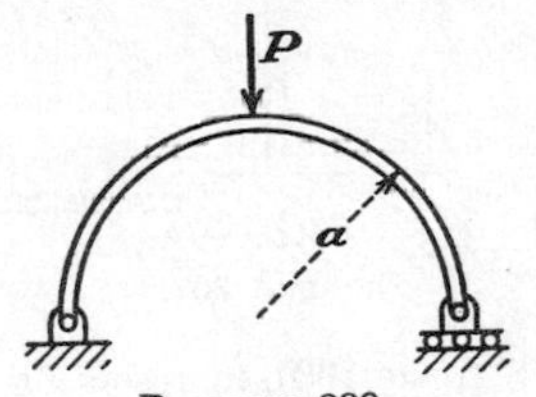

PROBLEM 299.

300. The arch of Prob. 299 is loaded at the top by a horizontal load Q (*i.e.*, a load in the plane of the arch directed parallel to the line connecting the two hinges). Find the horizontal displacement of the roller support, and state the reciprocal problem that has been solved thereby.

301. *a.* If the truss of Prob. 35 consists of steel bars of circular cross section of 1 in. diameter, find which bars, if any, buckle.

b. The same question if all bars are of rectangular cross section 2 by ½ in.

302. In the truss of Prob. 37 find the worst bar from the standpoint of buckling (assuming all bars to be of the same cross section). If the cross section is circular, the bars are of steel, $a = 6$ ft, and $P = 20{,}000$ lb, find the necessary diameter for that worst bar.

303. *a.* Calculate the buckling load of a pine 2 by 4 column of 10 ft length, when $E = 1 \times 10^6$ lb/sq in.

b. The same for a 20-ft length of 2 by 4 in. What are the compressive stresses in the two cases?

304. Calculate the ultimate compressive stress to be placed on a column of 40 ft length, being a 16-in. girder I beam with the properties described in Prob. 122.

305. What is the maximum height for a column consisting of a steel pipe of 3 in. outside diameter and ¼ in. wall thickness if the compressive stress in the column is 20,000 lb/sq in.?

306. Fill the tube of Prob. 305 with concrete of $E = 2 \times 10^6$ lb/sq in., and put the same load on it as in Prob. 305. What is now the maximum height?

307. The main I beams of a house are supported by pipe columns, of 3 in. outside diameter, ¼ in. wall thickness, and 15 ft length. Find the critical load P for each of the following three assumptions:

a. The column has hinged ends.

b. The column is built in (clamped) top and bottom.

c. The column is built in top and bottom, but the I beam can displace itself horizontally without restraint.

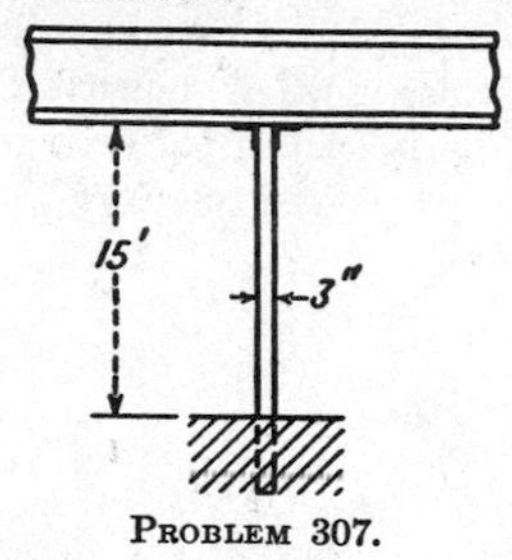

Problem 307.

308. Refer to Fig. 158 (page 190), representing between the points C and C' a column of which the ends are partially clamped. The ends can turn through a small angle y', and when they do that, they experience a bending moment $M_0 = ky'$, where k is the "stiffness" of the end connection, expressed in lb in./radian. To solve for the buckling load start from the equation of page 189, which describes the shape y in terms of two arbitrary constants C_1 and C_2. First substitute the end conditions:

$$x = 0, \qquad y = 0, \qquad \text{and} \qquad x = l, \qquad y = 0,$$

and solve for C_1 and C_2. Then differentiate the result y, so obtained, and prove that this expression for the slope y' gives equal and opposite slopes for the two ends $x = 0$ and $x = l$.

After this, substitute the third boundary condition:

$$x = 0, \qquad y' = \frac{M_0}{k},$$

and prove that the result is:

$$-\frac{EI}{kl} = \frac{1 - \cos\varphi}{\varphi \sin\varphi},$$

where $\varphi = \sqrt{Pl^2/EI}$, determining the buckling load P.

Calculate five or six points, and plot this (transcendental) equation in a diagram EI/kl versus φ for values of φ between π and 2π. [The meaning of this angle φ is the distance CC' in Fig. 158, so that $\varphi = \pi$, ($k = 0$), represents the hinged column and $\varphi = 2\pi$, ($k = \infty$), represents the fully clamped column.]

309. Apply the result of Prob. 308 to the case of Prob. 307, and calculate the buckling load of the column, assuming the top and bottom to be partially clamped with an end stiffness $k = EI/l$, where EI is the column stiffness and l its length.

310. Let the tube of Prob. 307 be built in at the bottom, but let it be completely free at the top, like a fence post. What is the maximum weight that can be placed on top of it?

311. Imagine that the girder of Prob. 307 is an 8-in. Bethlehem girder with properties as described in Prob. 105. This girder is 40 ft long and is simply supported at its ends. If the central column buckles and its ends tend to rotate, the column imposes a bending moment on the I beam. Calculate the stiffness k of the I beam (*i.e.*, the value of M/φ at the center), and find the value of the parameter EI/kl (as mentioned in Prob. 308) in connection with the column of Prob. 307.

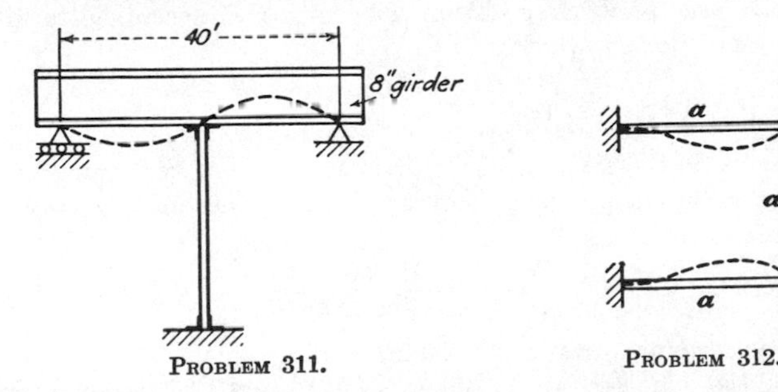

PROBLEM 311. PROBLEM 312.

312. A *U*-shaped frame with three legs of equal length a and of bending stiffness EI is built in at its two extremities as shown. The middle leg is subjected to a compressive load P. When that leg buckles, the two side legs oppose some resistance to rotation of the corners, so that the central leg is "imperfectly clamped." Calculate the angular stiffness k ($= M/\varphi$) of the two side legs, and with the help of the result of Prob. 308 find the value of the buckling load P.

313. Repeat the calculation of the previous problem, for two other cases.

a. The central leg still has the length a, but the two side legs are $a/2$, half the length of the central leg.

b. The central leg is a; the side legs $2a$.

314. Consider the column of Prob. 307, being a 3-in. pipe of ¼ in. wall thickness and 15 ft length, hinged at both ends. Calculate the eccentricity of the load that will produce yield stress (30,000 lb/sq in.)

in the column when it is subjected to a load of (a) $P = 15{,}000$ lb; (b) $P = 10{,}000$ lb. This is an application of the "secant" formula.

315. A steel column has the cross section of an 8-in. Bethlehem H girder with the properties listed in Prob. 105. In addition, the tables state that the moment of inertia about an axis through the web is $I_{min} = 33.6$ in.^{4}

Calculate the maximum allowable force on the column for the three cases:

$$l = 6 \text{ ft}; \qquad l = 12 \text{ ft}; \qquad l = 18 \text{ ft}.$$

The column should be designed by the building code of Chicago, which is:

$$s = \frac{P}{A} = 16{,}000 - \frac{70l}{k}.$$

316. Solve the previous problem, Quaker style, according to the building code of Philadelphia, which is:

$$s = \frac{P}{A} = \frac{16{,}250}{1 + (l^2/11{,}000k^2)}.$$

Plot the results of the three points so found on a diagram like Fig. 162, and compare the curve so obtained with that of the Chicago code.

317. Solve the previous problem, Pilgrim style, according to the building code of Boston, which is:

$$s = \frac{P}{A} = \frac{16{,}000}{1 + (l^2/20{,}000k^2)}.$$

Plot the results in the diagram of the previous problems, for comparison.

318. The same old problem, Gotham style, by the building code of New York, which is:

$$s = \frac{P}{A} = \frac{18{,}000}{1 + (l^2/18{,}000k^2)},$$

limited to a maximum stress of 15,000 lb/sq in.

319. Still the same 8-in. girder of lengths 6, 12, and 18 ft, respectively. This time apply the "standard" formula of the American Railway Engineering Association, which is:

$$s = \frac{P}{A} = 15{,}000 - \frac{l^2}{3k^2}.$$

Plot this relation with the rest.

320. The American Institute of Steel Construction declares the 8-in. girder column properly designed when the following formula is used:

$$s = \frac{P}{A} = 17{,}000 - 0.485\left(\frac{l}{k}\right)^2.$$

If there is any space left on your graph of Probs. 315 to 319, add this result to it.

321. The Aluminum Company of America has published a formula for the practical design of columns of 17S-T aluminum alloy (duralumin):

$$s = \frac{P}{A} = 43{,}800 - \frac{350l}{k}.$$

This formula is limited to $l/k = 83$, above which Euler's formula must be used with $E = 10 \times 10^6$ lb/sq in. This formula differs from those for structural steel (Probs. 315 to 320) in that this one for aluminum does *not* contain a safety factor but expresses the actual failure stress of the column.

Plot the above relation with a safety factor 3, and calculate the carrying capacity of a duralumin strut, being a tube of 1 in. outside diameter, $\frac{1}{16}$ in. wall thickness, and a length of (*a*) 3 ft and (*b*) 6 ft.

322. Calculate the necessary wall thickness of a duralumin tube of 2 in. outside diameter and 6 ft length, which must support a load of 8,000 lb. Use the formula of Prob. 321 with a factor of safety $2\frac{1}{2}$.

323. Timber columns almost invariably are of rectangular or square cross section bh, where we shall call h the smaller of the two sides. The Forest Products Laboratory publishes the following formula for rectangular timber columns:

$$s = \frac{P}{A} = s_c \left[1 - \frac{1}{3}\left(\frac{l}{ph}\right)^4\right],$$

where s_c = the allowable design compressive stress (no buckling, but including the factor of safety), and

$$p = \frac{\pi}{2}\sqrt{\frac{E}{5s_c}}.$$

The formula is limited to the range $l/h < p$.

a. Bring this formula into a shape that can be plotted in the diagram (Fig. 162) for the case that s_c = 1,000 lb/sq in. and $E = 10^6$ lb/sq in.

b. Calculate the maximum length of a column of 4 by 4 in. cross section to which the formula is still applicable.

c. Calculate the allowable load P for this length.

324. Find the necessary cross-sectional dimension for a square timber column of 15 ft height to carry 50,000 lb load. Use the Forest Products Laboratory formula of the previous problem with s_s = 1,000 lb/sq in. and $E = 10^6$ lb/sq in., if it happens to apply. If not, use Euler's column formula with a factor of safety 3.

325. A flat aluminum bar of $\frac{1}{2}$ in. thickness and 2 in. width carries a number of narrow saw cuts in its central portion. These cuts are $\frac{1}{32}$ in.

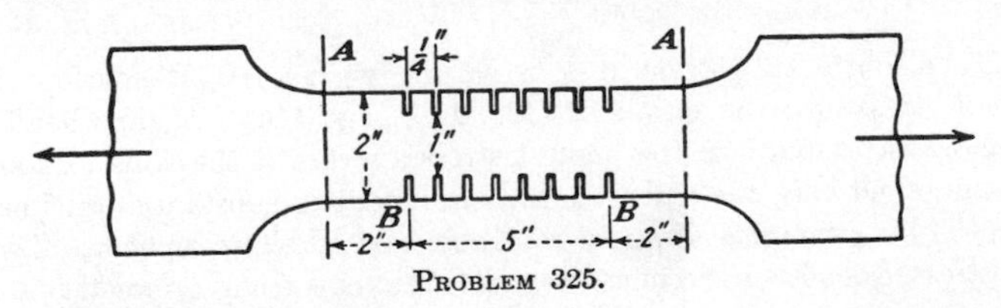

PROBLEM 325.

wide, are spaced ¼ in. apart, and are ½ in. deep, and they extend over a length of 5 in. The bar is pulled by a force of 8,000 lb, and $E = 10 \times 10^6$ lb/sq in. Calculate the elastic elongation of the section AA, being 9 in. long.

326. The bar of the previous problem is modified in that the central section BB is increased in length from 5 in. to 12 in. The bar is put in pure bending in the plane of the paper by four forces of 333 lb each, two acting downward at the points A, and two reactions upward at points B. Assuming the points B, B immovable, calculate the deflection of the bar in the center.

327. Photoelasticity applies to two-dimensional objects only. Any extension of its results to three-dimensional machine elements is necessarily of an approximate nature and involves horse sense. (If you don't know what that means, you are advised to ask one of your friends from the University of Michigan, or better still look for yourself at a fine bronze plaque in the "Engineering Arch" at Ann Arbor.)

Modify the 2- by ½-sq-in. rectangular cross section of the previous problems to a round one of 2 in. diameter. The slots, all round the circle, are ½ in. deep; the slotted section again is 12 in. long, and the shaft is subjected to pure bending with a moment of 667 in.-lb as before. Calculate the central deflection.

328. The figure shows the end of a shaft of a direct-current generator. The generator itself is to the left of the bearing A, while to the right of that bearing a cantilever extension carries the rotor of the "exciter," as is usual. Let the weight of the exciter be 100 lb, and let the moment arm a to the weak point of the shaft be 1 ft. Let further the factor of stress concentration at the fillet be 3, and let the allowable bending stress of the material be 20,000 lb/sq in. Calculate the necessary shaft diameter.

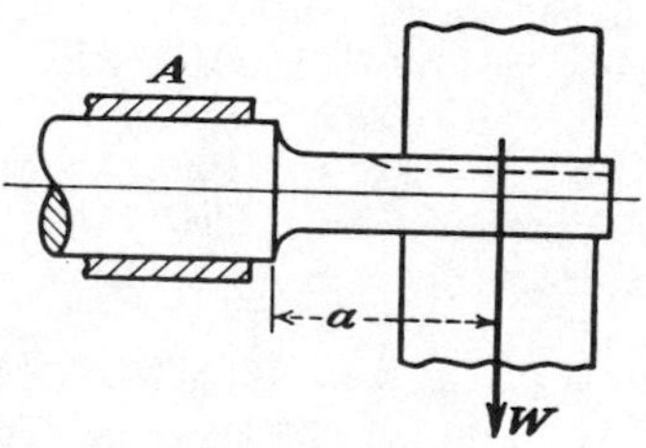

PROBLEM 328.

329. When a flat turbine disk is shrunk on a shaft, it develops its highest stress near the shrink fit (Fig. 125, page 144). Again when the assembly is rotated fast, the highest stress appears at the same location. To counteract this, these disks usually are made of substantial thickness near the hub, tapering down in thickness toward the periphery. Then the stress calculation becomes complicated, but from symmetry it is apparent that the principal directions of stress and strain are still radial and tangential. We therefore attach strain gages in these two directions and find by experiment:

$$\epsilon_t = +0.0005 \qquad \text{and} \qquad \epsilon_r = -0.0002.$$

Calculate the stresses at that point, for a steel disk with $E = 30 \times 10^6$ lb/sq in., and $\mu = 0.25$.

330. Suppose that in the previous experiment we had found:

$$\epsilon_t = +0.0005 \qquad \text{and} \qquad \epsilon_r = 0.0000.$$

Calculate the stresses.

331. *a.* Suppose that in a hydrostatic experiment (pressure acting on the surface of the test piece being equal to its compressive stress in the other two directions), we find:

$$\epsilon_1 = -0.0005 \qquad \text{and} \qquad \epsilon_2 = -0.0005.$$

Calculate the stresses if the test piece is made of aluminum with $E = 10 \times 10^6$ lb/sq in., and $\mu = 0.3$.

b. Suppose that the two equal readings of -0.0005 in two perpendicular directions were obtained on the aluminum surface without any pressure acting on that surface (*i.e.*, two-dimensional hydrostatic pressure), what would the stresses be?

332. A hollow-steel ship-propulsion shaft of 12 in. outside diameter with a bore of 6 in. inside diameter has two strain gages attached to it in a push-pull circuit as shown in Fig. 171 (page 208). The instrument shows the same reading as would be obtained with the same gages and resistors when hooked up in the circuit of Fig. 170 (with one active gage only) for a strain $\epsilon = 0.0008$. Calculate the torque in the shaft.

333. The circuit of Fig. 171 (page 208) measures only torsion and disregards bending or compression when it happens to coexist with the torsion. Similarly the "pull-pull" circuit of this problem measures only tension-compression of the shaft, disregarding whatever bending or torsion there may be. Reason this out first for yourself. The gages of this problem are the same as those of Fig. 170 (page 207), the resistors here have twice the resistance as those of Fig. 170 and the voltage applied across the bridge is twice that of Fig. 170. The reading obtained here is the same as the one that would indicate $\epsilon = -0.0001$ in Fig. 170. Find the thrust on the shaft in pounds. The shaft is the same as that of Prob. 332, that is, 12 in. outside diameter and 6 in. inside diameter.

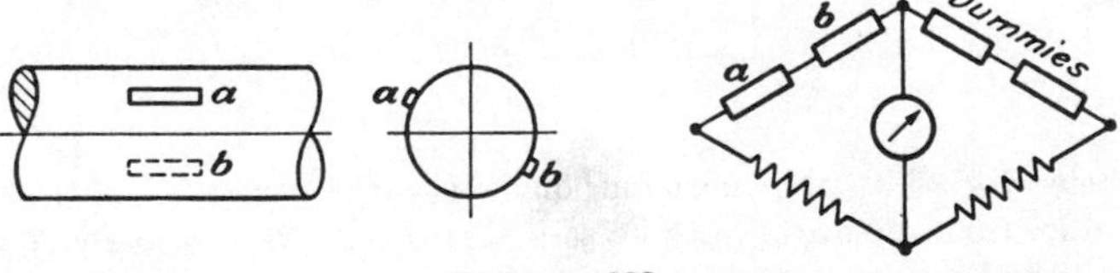

PROBLEM 333.

334. On a steel machine element the three readings of a rosette strain gage are:

$$\epsilon = +\,0.0008, \qquad \epsilon = +\,0.0008, \qquad \text{and} \qquad \epsilon = +\,0.0002.$$

Find the values of the two principal stresses and the maximum shear stress at that point.

335. In case the three rosette readings on a steel object are:

$$\epsilon = -\,0.0002, \qquad \epsilon = +\,0.0001, \qquad \text{and} \qquad \epsilon = +\,0.0004,$$

find the principal stresses.

336. The three rosette-gage readings on an aluminum object are:

$$\epsilon = -\,0.0005, \qquad \epsilon = -\,0.0001, \qquad \text{and} \qquad \epsilon = +\,0.0008.$$

Find the maximum shear stress.

337. One of the remarkable developments of the war was the FIDO project whereby gasoline pipe lines were laid across the English Channel to France. The pipe was wound on a drum of very large diameter, floating in the water with its cylindrical center line in a horizontal position. During the wind-up process new lengths of pipe were welded onto the end until the drum contained an enormous length of continuous pipe. It was then paid out into the water with the drum being towed by a ship,

a. Derive a formula relating the pipe diameter to the drum diameter and the properties of the pipe material, assuming first that no yield is to occur in the pipe.

b. What is the necessary drum diameter for 3 in. outside diameter pipe made of mild steel, if no yield is permitted? Is this a practical possibility?

338. Referring to the revolving-beam fatigue tester of Fig. 178 (page 215), the distance between the two ball bearings carrying the weight is 6 in., the distance between the outer support bearings is 9 in., and the diameter of the test piece is ¼ in. Calculate the weight W required for each 1,000 lb/sq in. stress in the test piece.

339. A rotating cantilever-fatigue-testing specimen has a length of 6 in. between the load P and the heavy section 4. The first 5 in. are of uniform

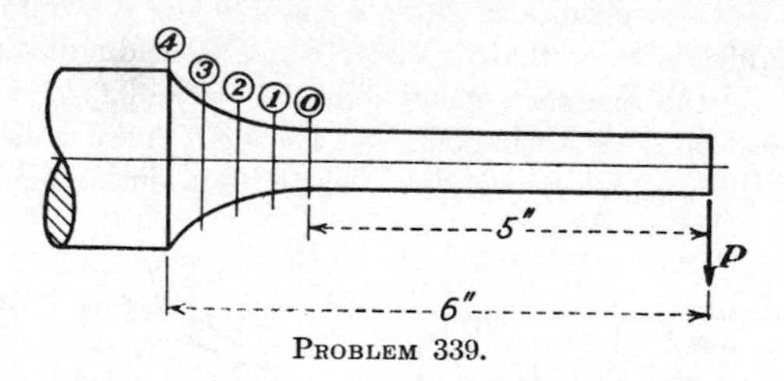

PROBLEM 339.

diameter d = ½ in. The next four diameters at distances 5¼, 5½, 5¾, and 6 in. from the load are to be so designed that between 0 and 2 the beam should be one of equal stress, the diameter 3 should be 1/32 in. greater than would correspond to equal stress, and 4 should be $\frac{1}{16}$ in. greater than would correspond to equal stress. Calculate the four diameters, and draw a graph of the contour.

340. A shaft of circular cross section is subjected to a steady bending moment of 2,000 ft-lb and simultaneously to an alternating bending moment in the same plane of 1,500 ft-lb (so that the total moment fluctuates between 3,500 ft-lb and 500 ft-lb). Calculate the necessary diameter if the factor of safety is to be $2\frac{1}{2}$. The yield stress is 30,000 lb/sq in., and the fatigue limit of the material is 25,000 lb/sq in.

341. A 16-in. girder with the properties described in Prob. 122 is laid on two simple end supports 24 ft. apart. It carries a central load varying continuously between the values $\frac{1}{2}P$ and $\frac{3}{2}P$. The yield point is 30,000 lb/sq in., and the fatigue limit is 28,000 lb/sq in. If the factor of safety is to be 3, find the value of P.

342. A steel shaft is stepped up from a diameter D to $2D$ with a bore diameter D_b. The fillets at the step in design No. 1 are generous, so that the stress-concentration factor in torsion at the fillets is 1.3, while in design No. 2 they are sharp with a stress-concentration factor 4.0

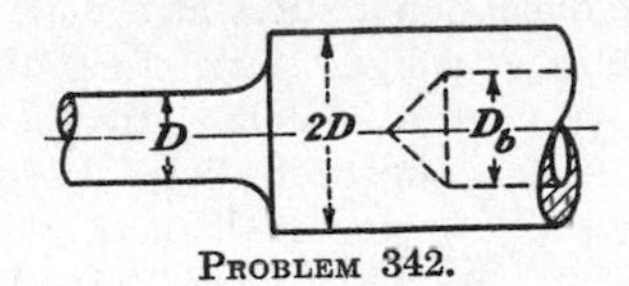

PROBLEM 342.

a. For the case that $D = 3$ in. and $D_b = 0$, calculate the maximum allowable steady-state torque, based on a torsional yield stress of 15,000 lb/sq in. with a factor of safety $2\frac{1}{2}$. (For a steady torque the stress-concentration factors are *not* considered, because a local yield is of no consequence for steady torque.)

b. For $D = 3$ in., calculate the maximum allowable alternating torque, with a factor of safety $2\frac{1}{2}$ and a torsional fatigue limit of 14,000 lb/sq in.

343. The shaft of the previous problem (where the diameters are *not* given) is subjected simultaneously to a steady torque of 1,000 ft-lb and an alternating torque of 1,500 ft-lb (so that the torque fluctuates between $-$ 500 ft-lb and $+$ 2,500 ft-lb). As before, the shear yield stress is 15,000 lb/sq in., the shear fatigue stress is 14,000 lb/sq in., and there are two fillet designs: a good one, No. 1, with a stress concentration factor 1.3 and a bad one, No. 2, with a stress concentration factor 4.0.

Find the necessary diameter D and the bore diameter D_b if the factor of safety is to be $2\frac{1}{2}$ and if the large diameter shaft is to be as strong as the thin shaft (and consequently stronger than the fillet detail).

344. When a machine with a fluctuating torque, such as a Diesel or reciprocating steam engine, is coupled to a load taking a steady torque, such as a ship's propeller or an electric generator, it is necessary to insert a soft, elastic coupling between the two. Such a coupling often has the shape of two spiders with some 12 coil springs between them. The center lines of these springs are roughly tangential to the spider periphery. Each spring is in pure compression, all these compressive forces together making up the transmitted torque. Let one such spring be of coil diameter $D = 5$ in.; $n = 10$ turns. The spring material has a yield point of 180,000 lb/sq in. and a fatigue limit of 130,000 lb/sq in.

a. If this spring is to transmit a steady force of 5,000 lb at a working stress of 90,000 lb/sq in., calculate the necessary wire diameter d.

b. With this wire diameter calculate the alternating force that, superposed on the steady force of 5,000 lb, will break the spring.

345. The crankshaft of a Diesel engine is made of forged steel with a yield point of 50,000 lb/sq in. and a fatigue limit of 45,000 lb/sq in. Both these figures are for tensile stress; for shear stress they are half as much. The shaft is designed for a steady torque with the generous factor of safety 5, based on the yield point. What alternating torque (expressed in terms of the above steady torque) must be superposed on the steady torque in order to cause failure?

346. If from a three-dimensional stress at a point with the principal stresses s_1, s_2, s_3 we subtract a hydrostatic stress of the average value $s_a = \frac{1}{3}(s_1 + s_2 + s_3)$, we find a system with the principal stresses $s_1 - s_a$, $s_2 - s_a$, and $s_3 - s_a$. Substitute these stresses into the general equation for the energy [Eq. (*a*), page 224], and work it out. Prove that the result is identical with Eq. (40), (page 224). With this you have proved that the distortion energy [Eq. (40)] and the hydrostatic energy [Eq. (*b*)] can be added algebraically to give the actual total energy as a result.

By page 164 this implies that a set of hydrostatic stresses at a point does no work on the deformation caused by a set of "distortional" stresses, and vice versa. This again means that a set of distortional stresses does not change the volume of the element.

347. A cantilever beam of rectangular section $b = 1$ in. and $h = 2$ in. is 3 ft long. The end load $P = 200$ lb.

Calculate the total energy stored in the beam and also the "hydrostatic" and "distortion" energies into which this total energy can be divided.

348. We have available a steel of which the yield point in tension is 30,000 lb/sq in.

a. Calculate the necessary diameter of a shaft to transmit 50 hp at 1,000 rpm with a factor of safety 3.0. Use the maximum-shear theory.

b. What would be the diameter of this shaft if it were designed by the maximum-distortion-energy theory, using the same factor of safety?

349. A shaft of solid circular cross section is subjected simultaneously to a torque M_t and a pure bending moment M_b. For the case that $M_t = M_b$, what percentage of the total energy is "hydrostatic" and what percentage is "distortion energy"?

350. In one of the remarkable high-pressure experiments of P. W. Bridgman a bar of circular cross section is subjected to very high pressure in its center portion, while its ends protrude from the hydraulic-pressure vessel through high-pressure packing glands. Thus there is no force in the direction of the bar, except for the gland friction force, which is comparatively small. The bar was observed to break clean through in the middle, and the two halves were forced out through the glands. This is a break across a section in which *no* tensile force acts, brought about by the effect of Poisson's ratio only.

Calculate what percentage of the total energy of the bar (in which $s_1 = s_2 = s$ and $s_3 = 0$) is "hydrostatic" and what percentage consists of "distortion energy."

ANSWERS TO PROBLEMS

1. 0.985 in.

2. (*a*) 10,050 lb/sq in. compression; (*b*) 14,070 lb/sq in. tension.

3. 10,300 ft for soft iron; 61,800 ft for piano wire.

4. $\delta = \frac{Pl}{AE}\left(1 - \frac{2x}{a} + \frac{3}{2}\frac{x^2}{a^2}\right)\cdot$ **5.** 0.01525 in. **6.** $k = 9AE/8l$.

7. (*a*) Magnification 1,536 times; (*b*) 29.8×10^6 lb/sq in.

8. (*a*) 61,100 lb/sq in.; (*b*) 45,900 lb/sq in. at top.

9. (*a*) $s_{\max} = l\gamma$ at top; (*b*) $\Delta l = \frac{\gamma x}{E}\left(l - \frac{x}{2}\right)$ at distance x from top, $\Delta l_{\max} = \frac{\gamma l^2}{2E}$ at bottom; (*c*) 18.4 ft; (*d*) 13.9 ft.

10. (*a*) $A_x = A_0 e^{\gamma x/s}$; x measured upwards from bottom; s is stress; γ is weight/cu in. (*b*) $d = 2.73$ in. (*c*) 23.7 in.

11. (*a*) $s = \frac{\gamma\omega^2}{2g}(l^2 - r^2)$; maximum stress at $r = 0$ is $\gamma\omega^2 l^2/2g$; maximum displacement at $r = l$ is $\gamma\omega^2 l^3/3gE$. (*b*) $s_{\max} = 5{,}160$ lb/sq in.; $\Delta l = 0.00138$ in.

12. ⅔ sq in.

13. (*a*) 667 lb/sq in. in 10-ft portion; 333 lb/sq in. in 20-ft portion; $\delta_C = 0.533$ in. (*b*) 1,000 lb/sq in. in 10-ft portion; zero stress in 20-ft portion; $\delta_C = 0.80$ in. (*c*) Zero stress in 10-ft portion; 1,000 lb/sq in. in 20-ft portion; $\delta_C = 1.60$ in. (*d*) 500 lb/sq in. in both ropes; $\delta = 0.60$ in.

14. (*a*) 0.02 in.; (*b*) 12 in.; (*c*) 0.02 in. **15.** 0.04 in.

16. (*a*) 0.08 in. vertically, zero horizontally; (*b*) 0.023 in. vertically, 0.067 in. horizontally.

17. (*a*), (*b*), and (*c*): 0.0267 in. for all cases; in the direction of the load, independent of α.

18. (*a*) $s = s_0 + \frac{2P}{A}\cos\alpha = s_0 + 2{,}000\cos\alpha$; (*b*) $s_0 = 2{,}000$ lb/sq in.; (*c*) $\delta = \frac{2Pl}{AE} = 0.0008$ in.; (*d*) ⅓ lb.

19. (*a*) $s = s_0 + \frac{F}{2A}\sqrt{1 + \frac{a^2}{4R^2}}$, $\delta_A = \frac{FR}{2AE}\left(1 + \frac{a^2}{4R^2}\right)^{3/2}$; (*b*) $s = 40{,}160$ lb/sq in. $\delta_A = 0.0042$ in.

20. (*a*) $s = s_0 - \frac{F}{4A\cos\alpha}\sqrt{1 + \frac{a^2}{4R^2}}$, $\delta_A = \frac{FR}{4AE\cos^2\alpha}\left(1 + \frac{a^2}{4R^2}\right)^{3/2}$; (*b*) $s = 30{,}000 + \frac{5{,}080}{\cos\alpha}$ lb/sq in., $\delta_A = \frac{0.0021}{\cos^2\alpha}$ in.

21. (*a*) $s = s_0 + \frac{4F}{A_0}\sqrt{1 + \frac{a^2}{4R^2}}\cos\alpha$; $\delta = \frac{4FR}{EA_0}\left(1 + \frac{a^2}{4R^2}\right)^{3/2}$; (*b*) 3,520 lb/sq in.

22. (*a*) $Pl/2s_0A$; (*b*) 0.12 in.

23. $P = AE(\delta/l)^3$. This is a most exceptional case. The load P is *not* proportional to the deflection δ, in spite of the fact that Hooke's law applies to all parts of the structure.

24. (*a*) $P = AE\left(\frac{\delta_0}{l}\right)^3\left[\frac{\delta}{\delta_0} - \left(\frac{\delta}{\delta_0}\right)^3\right]$. Plot $P/AE(\delta_0/l)^3$ against δ/δ_0 and see that the curve makes physical sense. (*b*) 0.53 in.

25. (*a*) 20.8 lb/sq in. above atmospheric pressure; (*b*) 16.8 lb per running inch of periphery.

26. (*a*) 3.9 in.; (*b*) 96 lb per running inch.

27. (*a*) 6,600,000 lb; (*b*) 1,320,000 lb.

28. 1,320,000 lb (the pipe yields).

29. $P = 28{,}500$ lb; $s_{\text{steel}} = 15{,}000$ lb/sq in.; $s_{\text{bronze}} = 6{,}000$ lb/sq in.

30. $s_{\text{steel}} = 4{,}230$ lb/sq in.; $s_{\text{bronze}} = -6{,}350$ lb/sq in.; elongation is 0.059 in.

31. (*a*) 0.0105 in.; (*b*) 0.0128 in.; (*c*) 0.0180 in.; (*d*) 24,500 lb.

32. 150°F. This is an important practical result; we conclude that every welded seam in steel has locked-up yield stress in it unless stress relieved by annealing afterward.

33. 0.0294 in.

34. (*a*) Bottom chords (left to right) are $-2{,}000\sqrt{3}$, $-2{,}000/\sqrt{3}$; top chord is $+4{,}000/\sqrt{3}$; diagonals (left to right) are $+4{,}000/\sqrt{3}$, $-4{,}000/\sqrt{3}$, $+4{,}000/\sqrt{3}$. (*b*) Downward 0.0208 in.; leftward 0.0055 in.

35. (*a*) Top chords (left to right) are $-5{,}000$, $-10{,}000$, $-5{,}000$; bottom chords are 0, $+5{,}000$, $+5{,}000$, 0; all four diagonals are $+5{,}000\sqrt{2}$; all four verticals are $-5{,}000$. (*b*) 0.0819 in.

36. (*a*) Areas in sq in.: top chords 1, 2, 1; bottom chords 0, 1, 1, 0; diagonals $\sqrt{2}$; uprights 1. (*b*) 0.060 in. (*c*) Sidewise collapse, because bottom chords next to supports are absent.

37. $10.83\ Pa/AE$. **38.** $27.3\ Qa/AE$. **39.** $0.396\ P$.

40. (*a*) One redundant bar; (*b*) bottom horizontals are $+P$; top horizontal is $-P$; the two outside diagonals are $-P\sqrt{2}$; all four interior bars zero.

41. $4.83Pa/AE$.

42. Bottom horizontals (left to right) are $-P/3$, zero, $+P/3$; upper girder bars (left to right) are $+P\sqrt{2}/3$, zero, $-P\sqrt{2}/3$; the two verticals are $(2\sqrt{2} - 1)\ P/3\sqrt{2}$, the left one in tension, the right one in compression; the diagonals are $P\sqrt{2}/3$, one in tension, one in compression, which?

43. $0.85Pa/AE$, under each load, in the direction of that load.

44. $5.68Pa/AE$. **45.** 0.98 in.; use 1 in.

46. $d = 1\frac{1}{2}$ in.; stress = 3,140 lb/sq in.

47. $d = 1\frac{3}{8}$ in. **48.** $p = (\pi + 1)d = 4.14d$. **49.** $\delta = 0.1$ in.

51. $d = 68.5\sqrt[3]{\dfrac{\text{hp}}{s_s \times \text{rpm.}}} = 1.47$ in.; use $1\frac{1}{2}$ in.

52. $\varphi = 640{,}000 \dfrac{\text{hp} \times l}{G(\text{rpm.})d^4}$.

53. (*a*) 185 radians = 29 full turns; (*b*) 10.9 h.p.

54. $d = 1\frac{7}{8}$ in.; $\varphi = 0.0413$ radian = 2.36 deg.

55. (*a*) $s_{s\,\max} = \dfrac{4M_t}{3\pi r^3}$; $\varphi_{\max} = \dfrac{2}{3}\dfrac{M_t l}{G\pi r^4}$.

(*b*) $s_{s\,\max} = 190$ lb/sq in.; $\varphi_{\max} = 0.0055$ deg.

56. (*a*) $s_{s\,\max} = \dfrac{2M_t}{\pi r^3}$; $\varphi_{\max} = \dfrac{2M_t l}{G\pi r^4}$.

(*b*) $s_{s\,\max} = 286$ lb/sq in.; $\varphi_{\max} = 0.0164$ deg.

57. (*a*) For $b > a$; $s_{s\,\max} = \dfrac{2bM_t}{l\pi r^3}$; $\varphi_{\max} = \dfrac{2abM_t}{\pi l r^4 G}$.

(*b*) $s_{s\,\max} = 238$ lb/sq in.; $\varphi_{\max} = 0.0068$ deg.

58. (*b*) $s_{s\,\max} = 302$ lb/sq in.; $\varphi_{\max} = 0.00027$ deg.

59. (*a*) 0.506 in.; (*b*) 0.402 in. **60.** $d = 4.8$ in. **61.** $t = 0.066$ in.

62. $D/d = 1.12$. **63.** $D/d = 1.4$. **64.** $0.71D$, $0.84D$, $0.93D$, $1.00D$.

65. $s_s = 6{,}100$ lb/sq in.; $\varphi = 2.23$ deg.

66. (*a*) $\varphi = \dfrac{32M_t}{3\pi G}\dfrac{l}{d_{\max} - d_{\min}}\left(\dfrac{1}{d_{\min}^3} - \dfrac{1}{d_{\max}^3}\right)$; (*b*) $\varphi = 1.23$ deg.

67. (*a*) $s_s = \dfrac{32M_t rG}{\pi[d_i^4 G_s + G_b(d_3^4 - d_i^4)]}$. The letters r and G without subscript apply to the fiber in question. The symbol d_i is the diameter of the joint, anywhere along the shaft. (*b*) Maximum stress in bronze at left is 6,000 lb/sq in.; maximum stress in steel at right is 7,000 lb/sq in.

68. 0.0297 radian = 1.7 deg. **69.** $I_p = a^4\sqrt{3}/48$.

70. 0.073 radian = 4.2 deg.

71. (*a*) $D = 0.76$, make it $\frac{3}{4}$ in.; $\delta = 1.56$ in.

(*b*) $D = 1.13$, make it $1\frac{1}{8}$ in.; $\delta = 3.12$ in.

72. (*a*) 128 lb; (*b*) $n = 4$ turns. For the solution of problems such as 71 or 72 Marks' "Handbook" (and other handbooks) carry numerical tables ("spring" tables). For your own practice find such a table and solve Probs. 71 and 72 by means of it.

73. (*a*) $d_{\text{outside}}/d_{\text{inside}} = \sqrt{2}$; (*b*) $s_{s\,\text{outside}}/s_{s\,\text{inside}} = 1/\sqrt{2}$.

74. $b/a = 27/8$.

75. No. 1 down by 5 units; Nos. 3 and 4 down by 2 units; No. 2 up by 1 unit.

76. Maximum shear force = − 1,100 lb at left; maximum bending moment = + 1,900 ft-lb at left.

77. Shear force is constant − 100 lb in right section, then increases linearly to − 400 lb in left section. Bending moment is 200 ft-lb at the beginning of the w load and 950 ft-lb at the wall.

78. Shear diagram is a stairway with − 625 lb in left quarter, − 125 lb in middle quarter, and + 375 lb in right half. Bending-moment diagram is a funicular with $-1{,}250$ ft-lb under left load and 1,500 ft-lb under central load.

79. Maximum moment is 3,750 ft-lb under central load P. This is not obvious. The maximum might have been in the left half of the beam. You should have shown that the bending-moment diagram does not have a horizontal tangent in that left half. For what value of P other than 1,000 lb does the tangent at P just become horizontal?

80. Shear is $+500$ lb in overhang; straight-line diagram from -650 lb at left support to $+350$ lb at right support. The bending moment at the left support is $+1500$ ft-lb; parabolic diagram in span with a maximum of -612.5 ft-lb at $3\frac{1}{2}$ ft from left.

81. $P = 1200$ lb; shear in left half is $+1200$ lb constant; in right half a linear diagram from $-2{,}400$ lb at center support to zero at right support. Bending-moment diagram has a central peak of $+7{,}200$ ft-lb; is straight at left, parabolic at right.

82. Force between planks is 125 lb. Shear is -625 lb in left 5 ft, -125 lb in central 10 ft and it varies linearly from -125 lb to $+375$ lb in right 5 ft. The bending moment at left is $+3{,}750$ ft-lb, falls to $+625$ ft-lb at the concentrated load, then to zero at the joint between planks, continues to -625 where the w load starts. In the last 5 ft at right the bending-moment diagram is a parabola with a maximum of -705 ft-lb at $3\frac{3}{4}$ ft from the right end.

83. Shear force in vertical runs linearly from zero at bottom to 500 lb at top. Shear force in horizontal is zero in right half, 300 lb in left half. Bending moment in vertical increases parabolically from zero at bottom to 1,250 ft-lb at top; it is constant 1,250 ft-lb in right half of horizontal and increases linearly to 2,150 ft-lb in left half of horizontal.

84. Perpendicular shear force is constant P everywhere. Shear force in plane is constant Q in horizontal branch, zero in vertical branch. Bending moment in plane is zero for vertical branch, runs linearly from zero to Qa in horizontal branch. Perpendicular bending moment runs from zero to Pb in vertical branch and from zero to Pa in horizontal branch. Twist is zero in vertical and constant Pb in horizontal.

85. In plane of beam at angle θ from load the tensile force is $P \sin \theta$, the shear force is $P \cos \theta$, the bending moment is $PR \sin \theta$. In the plane perpendicular to the beam all forces and moments are zero.

86. In plane of beam all zero. Perpendicular to plane of beam at angle θ from load, the tensile force is zero, the shear force is P, the bending moment is $PR \sin \theta$, the twisting moment is $PR(1 - \cos \theta)$.

87. In plane of beam all zero. Perpendicular to plane at angle θ from loads the shear force is P, the bending moment $PR \sin \theta$, and the twisting moment $PR(1 - \cos \theta)$.

88. In plane of ring at angle θ from slit, the tensile force is $pR(1 - \cos \theta)$, the shear force is $pR \sin \theta$, and the bending moment is $pR^2(1 - \cos \theta)$.

89. The tables state: $I_1 = 435.6$ in.^{4}; $I_2 = 84.9$ in.^{4}

90. The tables state: $I_1 = 128.1$ in.^{4}; $I_2 = 3.9$ in.^{4}; $x = 0.70$ in.
91. 322 lb.
92. (*a*) $w = 346$ lb/ft; (*b*) $w = 16.4$ lb/ft.
93. (*a*) $w = 692$ lb/ft; (*b*) $w = 103$ lb/ft.
94. $P = 486$ lb.

95.

h......	4″	6″	8″	10″	12″
l......	3′11″	8′10″	15′9″	24′8″	35′7″

96. 2×12 in.
97. $Z = 3.2$ in.^{3} Use 4-in. I beam
98. (*a*) $l^2 = 107{,}000\, d$; (*b*) 27 ft.
99. (*a*) $l^2 = 8k^2 s_{\text{yield}}/\gamma y_{\max}$; (*b*) $l = 133$ ft.
100. $y = 1.81$ in. from top; $I = 63.5$ in.^{4}; $Z = 12.22$ in.^{3}
102. $a = 1\frac{9}{16}$ in. **103.** Ratio 5 : 2. **104.** $c = 5hb/2l$.
105. $s_s = 1{,}940$ lb/sq in. **106.** 316 lb/in.
107. Maximum shear stress at neutral line is 2.28 times average. Horizontal shear stress is $1{,}060x$, where x is measured from end of flange.
108. (*a*) $lt = 8.4$ sq in.; (*b*) $l = 2.4$ ft; (*c*) $\frac{1}{16}$ in.
109. 13,400 lb/sq in.
110. $s_s = 900x$, where x is measured from end of flange.
111. 20.6 in.
112. (*a*) No shear in rivets for I position; (*b*) 11.3 in. rivet spacing for H position.
113. (*a*) $I_{\text{hor axis}} = 542$ in.^{4}; $I_{\text{vert axis}} = 67.2$ in.^{4}
(*b*) $w = 750$ lb/ft.
114. 19.7 in.
115. (*a*) $l = 53$ ft; (*b*) 8.6 in.
116. (*a*) $P = 890$ lb; (*b*) $d = 0.357$ in.; (*c*) $d = 0.460$ in.
117. (*a*) $P = 500$ lb; (*b*) $d = 0.206$ in.; (*c*) $d = 0.266$ in.
118. $I_1 = 66.4$ in.^{4}; $I_2 = 4.14$ in.^{4}.
119. 3.68 in.
120. (*a*) $y = 2.20$ in. from top; $I = 57.9$ in.^{4}; (*b*) $S = 1{,}600$ lb; (*c*) 3.02 in.
121. Neutral line is $1\frac{1}{2}$ in. from tension edge; maximum tensile stress is 1,000 lb/sq in.; maximum compressive stress is 3,000 lb/sq in.
122. Diamond-shaped figure with two corners on the horizontal axis at ± 1.20 in. and two corners on the vertical axis at ± 5.95 in.
123. $b = 0.63h$. **124.** $b = 0.63h$.
125. $h = w_1/4p$, independent of d. The influence of the diameter is expressed in w_1 only.
126. $\alpha = d/4h$, which for Pisa $h/d = 3.4$ gives $\alpha = 0.073$. Hence Pisa is the leaningest tower that will stand up in a breeze.
127. Regular hexagon of side length $\frac{5}{24}a$.
128. $s = 10{,}000$ lb/sq in.; $s_s = 0$.

129. $s = -5{,}000$ lb/sq in.; $s_s = 8{,}660$ lb/sq in.

130. $s = 9{,}720$; $s_s = -4{,}160$

131. $s_{max} = 12{,}070$; $\alpha = -22\frac{1}{2}$ deg.

132. $s_{max} = 10{,}500$; $\alpha = +14$ deg.

133. $s_{max} = -10{,}440$; $\alpha = +61$ deg.

134. Maximum tension is $+5{,}000\sqrt{3}$ on face $\alpha =$ zero; maximum compression is $-5{,}000\sqrt{3}$ on $\alpha = 90$ deg.

135. $s_1 = s_2 = 15{,}000$, *i.e.*, "two-dimensional hydrostatic tension." All directions are principal directions and the resulting Mohr's circle is a point.

136. $s_1 = 20{,}040$; $s_2 = -5{,}040$; $\alpha = -2.3$ deg.

137. $I_x = bh^3/12$; $I_y = hb^3/12$, $I_{xy} = 0$; $I_{AA} = \dfrac{bh}{48}(3h^2 + b^2)$.

138. $\alpha = 14.7$ deg; $I_1 = 153.4$ in.^{4}; $I_2 = 6.6$. in.^{4}

139. $\alpha = \dfrac{1}{2}\tan^{-1}\left(\dfrac{3}{2}\dfrac{ab}{b^2 - a^2}\right)$;

$$I_{1,2} = \frac{ab}{6}\left[(a^2 + b^2) \pm \frac{1}{2}\sqrt{4(a^4 + b^4) + a^2b^2}\right].$$

140. $\alpha = 45$ deg, $I_1 = 91.4$ in.^{4}, $I_2 = 55.9$ in.^{4}

141. $x = y = 1.861$ in.; $\alpha = 45$ deg; $I_1 = 55.9$ in.^{4}; $I_2 = 15.1$ in.^{4}

142. $I_x = 2tb^2\left(a + \dfrac{b}{3}\right)$; $I_y = \dfrac{2}{3}ta^2$; $I_{xy} = a^2bt$;

$$I_{1,2} = t\{[ab^2 + \tfrac{1}{3}(a^3 + b^3)] \pm \sqrt{[ab^2 - \tfrac{1}{3}(a^3 - b^3)]^2 + a^4b^2}\};$$

$$\tan 2\alpha = \frac{3a^2b}{(a^3 - b^3) - 3ab^2}.$$

143. $x = a/4$ from web; $I_x = \frac{8}{3}a^3t$; $I_y = \frac{5}{12}a^3t$; $I_{xy} = 0$; $I_{AA} = \frac{37}{24}a^3t$.

144. The core is a quadrangle with its four corners on the axes at points $y = \pm 2a/3$, $x = 0$ and at points $y = 0$, $x = 5a/12$, and $x = -5a/36$.

145. $I_{max} = \dfrac{\pi r^4}{4}$; $I_{min} = \dfrac{\pi r^4}{4}\left(1 - 4\dfrac{A}{\pi r^2}\right)$; $I_{45^\circ} = \dfrac{\pi r^4}{4}\left(1 - 2\dfrac{A}{\pi r^2}\right)$.

146. Maximum principal stress $= -4{,}820$ lb/sq in.
Maximum shear stress $= 4{,}070$ lb/sq in.

147. Maximum principal stress $= \pm 5{,}910$ lb/sq in.
Maximum shear stress $= 3{,}640$ lb/sq in.

148. $s_s = 2{,}850$.

149. At 1, $s_s = 1{,}500$; at 2, $s_s = 6{,}000$; at 3, $s_s = 3{,}200$ lb/sq in.

150. At 1, $s_s = 4{,}780$; at 2, $s_s = 6{,}050$; at 3, $s_s = 1{,}270$ lb/sq in. The principal stresses have the same values as the above maximum shear stresses.

151. (*a*) $s_{s\,max} = 1{,}200$; (*b*) $s_{s\,max} = 464$; (*c*) $s_{s\,max} = 996$ lb/sq in.

152. At 1 and 2, no shear stress; at 3, $s_{s\,max} = 7{,}250$; at 4, $s_{s\,max} = 2{,}040$ lb/sq in.

153. 1 goes with *b*, 2 goes with *a*, 3 goes with *c*.

154. $r = 0.74$ in.; use $1\frac{1}{2}$ in. diameter.

155. Maximum bending at top and bottom points of ring. Necessary dimension of cross section $a = 1.46$ in.

156. (*a*) 5.57 in.; (*b*) 5.28 in.; (*c*) 5.36 in.
157. (*a*) 9.0 in.; (*b*) 9.0 in.; (*c*) 7.1 in.
158. $y = \frac{w_0x}{12EIl}\left[\frac{x^4}{10} - \frac{l^2x^2}{3} + \frac{7l^4}{30}\right]$, with the x origin at the left end.

$\delta_{\max} = 0.00652 \frac{w_0l^4}{EI}$ at $x = 0.52l$.

159. $y = \frac{12P}{Eb\alpha^3}\left[\ln\frac{h_b}{h} - \frac{h_e}{2h} + \frac{h}{h_b}\left(1 - \frac{h_e}{2h_b}\right) + \frac{h_e}{h_b} - 1\right]$, where $h = h_e + \alpha x$ and $\alpha = \frac{h_b - h_e}{l}$; x being measured from the free end. At the free end $x = 0$. $\delta_{x=0} = \frac{12P}{Eb\alpha^3}\left[\ln\frac{h_b}{h} + 2\frac{h_e}{h_b} - \frac{1}{2}\left(\frac{h_e}{h_b}\right)^2 - \frac{3}{2}\right]$.

160. $y = \frac{6M_0}{Eb\alpha^2}\left[\frac{1}{h} + \frac{1}{h_b}\left(\frac{h}{h_b} - 2\right)\right]$, with x, α, and h as in Prob. 159.

At the free end: $\delta = \frac{6M_0}{Eb\alpha^2}\left[\frac{1}{h_e} + \frac{1}{h_b}\left(\frac{h_e}{h_b} - 2\right)\right]$.

161. $\delta_{\text{end}} = w_0l^4/2EI_0$, where I_0 pertains to the built-in end.
162. For x measured from the free end to the left:

$$y = \frac{wl}{4EI}\left(\frac{x^3}{3} - \frac{x^2l}{4} - \frac{xl^2}{2} + \frac{5}{12}l^3\right); \frac{l}{2} < x < l.$$

$$y = \frac{w}{8EI}\left(\frac{x^4}{3} - \frac{7}{6}xl^3 + \frac{41}{48}l^4\right); 0 < x < \frac{l}{2}. \quad \delta_{\text{end}} = \frac{41}{384}\frac{wl^4}{EI}.$$

163. $M_x = \frac{q_0l^2}{\pi^2}\sin\frac{\pi x}{l}$.

164. $y = \frac{q_0l^4}{\pi^4EI}\sin\frac{\pi x}{l}$. The sine-wave loading is the only one which produces a deflection curve of the same shape as the loading, a fact which is of importance in later applications (vibrations of beams).

165. $y = \frac{M_0}{6EIl}(x^3 - l^2x)$; $(y')_{x=0} = \frac{M_0l}{6EI}$; $(y')_{x=l} = \frac{M_0l}{3EI}$;

$y_{\max} = \frac{M_0l^2}{9EI\sqrt{3}}$.

166. $y_{\max} = 0.188\frac{Ml^2}{EI}$ at $0.53l$ from left end.

167. $y_{\max} = 0.016\frac{Ml^2}{EI}$ at $0.21l$ from either end.

168. 5.24×10^{-4} in. **169.** 4.35×10^{-4} in.

170. (*a*) $s_{\max} = \frac{2pl}{\pi r^3}\left(a + \frac{l}{4}\right)$;

(*b*) $\delta_O = \frac{16pl}{\pi r^4E}\left[\frac{2}{3}a^3 + \frac{1}{16}a^2l + \frac{5}{192}al^2 + \frac{5}{1{,}536}l^3\right]$

$\delta_A = \frac{16pl}{\pi r^4E}\left[\frac{2}{3}a^3 + \frac{1}{16}a^2l + \frac{1}{96}al^2\right]$.

171. 0.006 in. **172.** $X = {}^{20}\!/_{81}P$.

173. (*a*) For zero slope at $X: X = \dfrac{wl^3}{4a(2l + 3a)}$;

(*b*) for zero deflection at $X: X = \dfrac{wl^3}{8a(a + l)}$.

174. (*a*) $\delta_E = \dfrac{5wl^4}{1{,}458EI}$; (*b*) $\delta_{\max} = 0.000237 \dfrac{wl^4}{EI}$.

175. $\delta_E = \dfrac{17wl^4}{1{,}536EI}$.

176. $l = 13.6h$. With the result of Prob. 95 this gives:

Height h, in.......	4	6	8	10	12
(Strength), l, ft...	4	9	16	25	36
(Stiffness), l, ft....	4½	7	9	11	14

177. $\delta_O = -\dfrac{Pba^2}{2EI}$; $\delta_{\text{end}} = \dfrac{Pb^2}{EI}\left(a + \dfrac{b}{3}\right)$.

178. $\delta_{\max} = Pal^2/36EI\sqrt{3}$ at $x = l\sqrt{3}/6$ from ends.

179. (*a*) $\delta_P = \dfrac{Pl^3}{48EI} - \dfrac{wa^2l^2}{32EI}$, $\delta_{\text{end}} = -\dfrac{Pal^2}{16EI} + \dfrac{wa^3}{EI}\left(\dfrac{a}{8} + \dfrac{l}{6}\right)$;

(*b*) $\delta_P = 0.0495$ in.; $\delta_{\text{end}} = 0.252$ in.

180. $\delta_{\max} = 0.015 \dfrac{wl^4}{EI}$ at $0.357l$ from left.

181. $\Delta h = \dfrac{l_1}{16}\left(\dfrac{W_1l_1^2}{EI_1} + \dfrac{W_2l_2^2}{EI_2}\right)$.

182. $\dfrac{17}{960}\dfrac{Wl^3}{EI}$.

183. $k = \dfrac{8EI}{b^3}\left(1 + \dfrac{b}{a}\right)$.

184. (*a*) $k = \dfrac{12EI}{b^3}\left(1 + \dfrac{b}{a}\right)$; (*b*) $a/b = 0$; (*c*) $k = 18{,}400$ lb/in.

185. $\delta_P = \dfrac{P}{3EI}\left[l_1^3 - \dfrac{l_2}{8}(3l_1 - l_2)^2\right] = 0.81$ in.

186. (*a*) $P = \dfrac{Ebt^3}{6Rl}$; (*b*) $P = \dfrac{Ebt^3}{6(l - 2x)R}$. Notice that there is no contact pressure between the table and the flat portion of the spring.

187. $\theta_A = \dfrac{Pal}{2GI_p}$; $\delta_P = \dfrac{Pl}{2}\left[\dfrac{l^2}{24EI} + \dfrac{a^2}{GI_p}\right]$.

188. (*a*) $M = \dfrac{P}{2l^2}(3a^2l - 2l^2a - a^3)$; (*b*) $M = \dfrac{Pa^2}{2l^3}(3l^2 - 4al + a^2)$.

189. See Fig. 87, page 95.

190. $R_{\text{left}} = \dfrac{Pb^2}{l^3}(3a + b) = 156$ lb; $R_{\text{right}} = \dfrac{Pa^2}{l^3}(3b + a) = 844$ lb.

191. $R_{\text{end}} = \dfrac{55}{128}wl$.

192. $R_{\text{end}} = P\dfrac{1 + (5kl^3/768EI)}{1 + (kl^3/48EI)}$; $R_{\text{center}} = P\dfrac{(11/384)(kl^3/EI)}{1 + (kl^3/48EI)}$.

193. $\delta = wl^4/384EI$.

194. $\delta = \dfrac{wl^4}{384EI}(8\sqrt{2} - 11)$; $R_{center} = wl(2 - \sqrt{2}) = 0.59wl$;

$R_{end} = wl\left(\dfrac{1}{\sqrt{2}} - \dfrac{1}{2}\right) = 0.21wl$.

195. (*a*) $R = 20$ lb; $M_b = 40$ ft-lb;

(*b*) for the straight shaft, R is constant in direction; for a curved shaft, R rotates around the bearing, causing it to vibrate. In a straight shaft an individual fiber goes through alternating tension-compression each revolution; in a curved shaft each individual fiber has a constant stress during the rotation.

196. $R_{center} = 1\frac{1}{8}P$; $R_{end} = \frac{5}{16}P$.

197. $R_{left} = M_0/l$ down; $R_{center} = 4M_0/l$ up; $R_{right} = 3M_0/l$ down.

198. $R_{side} = \frac{7}{20}P$; $R_{middle} = \frac{23}{20}P$.

199. $\delta = \dfrac{19}{768}\dfrac{Pl^3}{EI}$; $\varphi = \dfrac{3}{64}\dfrac{Pl^2}{EI}$.

200. $l/h = 6.6\sqrt{E/G} = 10.5$.

201. $l/h = 5.9\sqrt{E/G} = 9.5$.

202. $Z = 14.3$ in.^{3}; $\Sigma EI = 1{,}444 \times 10^6$ lb in.^{2}

203. (*a*) 4,000 ft-lb; (*b*) 5,333 ft-lb.

204. (*a*) $s_{steel} = -s_{copper} = 7{,}500$ lb/sq in.

(*b*) $s_{steel} = +9{,}375$; $s_{copper} = -6{,}875$ lb/sq in.

(*c*) $s_{steel} = -s_{copper} = 1{,}240$ lb/sq in.

205. $k = 1.0 \times 10^9$ in.-lb/radian.

206. (*a*) $\Sigma AE = 1.26 \times 10^9$ lb; (*b*) $\Sigma EI_{max} = 47.5 \times 10^9$ lb in.^{2}; $\Sigma EI_{min} = 13.7 \times 10^9$ lb in.^{2}; (*c*) a diamond with corners at 4.19 in. and 1.61 in. from center.

207. $s_{steel} = -s_{bronze} = 1{,}650$ lb/sq in. This stress is uniformly distributed over the thickness of each material. At the "neutral" line of bending, *i.e.* at the joint of the two materials there is a sudden jump in the stress from $+1{,}650$ to $-1{,}650$. The bending moment is 0.74 in.-lb.

208. (*a*) 15.8 deg. (*b*) $s_A = -0.49Pl/a^2t$; $s_B = -1.28Pl/a^2t$; $s_C = +1.00Pl/a^2t$. (*c*) 53.4 deg.

209. (*a*) To the right; (*b*) 56.1 deg.

210. 87.7 deg.

211. $\delta_{hor} = 6\dfrac{h^2 - b^2}{h^3b^3}\dfrac{Pl^3}{E\pi^3}(\pi^2 - 4)$;

$$\delta_{vert} = 2\frac{h^2 + b^2}{h^3b^3}\frac{Pl^3}{E} - 12\frac{h^2 - b^2}{h^3b^3}\frac{Pl^3}{E\pi^2}.$$

212. 861% per radian or 15 per cent per degree!

213. (*a*) $\tan\alpha = \dfrac{I_{max} - I_{min}}{I_{max} + I_{min}}$. (*b*) $\alpha = -31$ deg. (*c*) The rivets have to hold a tensile force corresponding to that of a sidewise force $\dfrac{P}{2}\tan\alpha = 0.3P$ acting on the end of each angle.

214. $\dfrac{a}{2 + b/3a}$ to left of vertical web.

215. To left of G by distance $a \dfrac{4 + \pi\sqrt{2}}{2\sqrt{2} + \pi/2} = 1.93a$.

216. $a \dfrac{2b + 3a}{2b + 6a}$ to the left of the left upright.

217. Neutral line intersects upper leg at $\frac{2}{3}a$ from corner; lower leg at $\frac{2}{5}a$ from corner. Corner stress is $+3M/ta^2$; at end of vertical leg is $-9M/2ta^2$; at end of horizontal leg is $-3M/2ta^2$.

218. $s_s = 3P/4ta$; $CN = a/3$.

220. Effective height $h = 12$ in.; steel cross section $A_s = 1.60$ sq in.

221. $s_s = 20{,}500$ lb/sq in.; $s_c = 1{,}360$ lb/sq in.

222. $E_sA_s(h - x) = E_cA_cy_G \ldots (a_2)$; $M = A_ss_s(h - x + y_1) \ldots (b_2)$; $\dfrac{s_c}{s_s} = \dfrac{E_c}{E_s}\dfrac{x}{h - x} \ldots (c_2)$ where y_1 is the height of the flange.

223. $s_s = 20{,}400$ lb/sq in.; $s_c = 690$ lb/sq in.

224. $s_s = 20{,}450$ lb/sq in.; $s_c = 670$ lb/sq in.

225. (*a*) Just under the flange; (*b*) 180 per cent.

226. Residual stress of 10,000 lb/sq in. tension in the two alloy bars; 20,000 lb/sq in. compression in the weak central bar.

227. $\frac{4}{3} = 1.33$.

228. $\dfrac{4}{3}\dfrac{r_0^4 - r_0r_i^3}{r_0^4 - r_i^4}$.

229. In bronze linear distribution is from 3,870 lb/sq in. at outside to 918 lb/sq in. at the joint. In steel linear distribution is from $-3{,}173$ lb/sq in. at joint to $-15{,}000$ lb/sq in. at center.

230. 1.27. **231.** $t = 1.08$ in. **232.** $t = 1.30$ in.

233. At bottom $t = 0.66$ in.; half way up $t = 0.33$ in.

234. $t = 0.36$ in. **235.** (*a*) $s = pr/2t$; (*b*) $t = 0.8$ in.

236. $d = 1.37t$; $p = 3.73t$; 63 per cent.

237. $d = 1.37t$; $p = 6.01t$; 77.5 per cent.

238. $d = 1.37t$; outer row $p = 10.82t$, inner row $p = 5.41t$; 87.1 per cent.

239. 1,190 lb/sq in. **240.** 1,820 lb/sq in.

241. $s_{\text{steel}} = 1{,}900$ lb/sq in. tension; $s_{\text{copper}} = -3{,}800$ lb/sq in.

242. The tangential stresses are: $s_{\text{tang steel}} = 1{,}917$ lb/sq in.; $s_{\text{tang copper}} = -3{,}834$ lb/sq in. The longitudinal stresses are: $s_{\text{long steel}} = 1{,}980$ lb/sq in.; $s_{\text{long copper}} = -3{,}960$ lb/sq in.

243. $s_{\text{steel}} = 2{,}410$ lb/sq in.; $s_{\text{copper}} = -4{,}820$ lb/sq in.

244. $s_r = s_t = -p$; $u = -p_0r(1 - \mu)/E$.

245. $p = 48{,}000$ lb/sq in.; $s_{\text{tang}} = 80{,}000$ at $r = 2$ in. and $s_{\text{tang}} = 32{,}000$ at $r = 4$ in.

246. $s_r = -p\dfrac{r_0^2}{r_0^2 - r_i^2}\left(1 - \dfrac{r_i^2}{r^2}\right)$; $s_t = -p\dfrac{r_0^2}{r_0^2 - r_i^2}\left(1 + \dfrac{r_i^2}{r^2}\right)$;

$$u = -p\frac{r}{E}\frac{r_0^2}{r_0^2 - r_i^2}\left[(1 - \mu) + \frac{r_i^2}{r_0^2}(1 + \mu)\right].$$

247. The inner cylinder is in tangential compression: $-29{,}200$ lb/sq in. at $r = 2$ in., $-21{,}100$ lb/sq in. at $r = 3$ in. The outer cylinder is in tangential tension: $+28{,}900$ lb/sq in. at $r = 3$ in., $+20{,}800$ lb/sq in. at $r = 4$ in.

248. $p = 65{,}500$ lb/sq in.

249. $\dfrac{\text{Plastic pressure}}{\text{Elastic pressure}} = \dfrac{1 + (r_0/r_i)^2}{1 + (r_0/r_i)}$.

250. $s = 6{,}176$ lb/sq in. compression, and $6{,}112$ lb/sq in. tension; $\delta_{\text{up}} = 0.041$ in.; $\delta_{\text{left}} = 0.052$ in.

251. $s = 12{,}320$ lb/sq in.; $\delta_{\text{up}} = 0.052$ in.; $\delta_{\text{left}} = 0.124$ in.

252. $s = \pm 6{,}144$ lb/sq in.; $\delta_{\text{up}} = 0.052$ in.; $\delta_{\text{left}} = 0.082$ in.

253. $s_{\text{bend}} = 10{,}400$; $s_{\text{tors}} = 10{,}400$; $\delta = 0.170$ in.

254. $s_{\text{bend}} = 10{,}400$; $s_{\text{tors}} = 5{,}200$; $\delta = 0.058$ in.

255. $s_{\text{bend}} = 10{,}400$; $s_{\text{tors}} = 5{,}200$; $\delta = 0.080$ in.

256. The twisting moment at the built-in ends is the only statically indeterminate quantity.

258. $s_{\text{bend}} = 3{,}460$; $s_{\text{tors}} = 375$; $\delta = 0.042$ in.

259. $s_{\text{bend}} = 3{,}460$; $s_{\text{tors}} = 630$; $\delta = 0.038$ in.

260. $s_{\text{outer}} = -4{,}120$ lb/sq in.; $s_{\text{inner}} = 12{,}650$ lb/sq in.

261. $s_{\text{outer}} = -15{,}820$ lb/sq in.; $s_{\text{inner}} = 31{,}700$ lb/sq in.

262. (*b*) $s = 33{,}300$ lb/sq in. at inside of coil;
(*c*) $k = 9{,}940$ in lb/radian.

263. $n = h^2/4R_G$.

264. $n = \dfrac{\alpha(1-\alpha)}{1 + \dfrac{h}{R_G}(1 - 2\alpha)} \cdot \dfrac{h^2}{R_G}$.

267. $U = \dfrac{d\,\text{vol}}{2E}\,[s_1^2 + s_2^2 + s_3^2 - 2\mu\,(s_1s_2 + s_1s_3 + s_2s_3)]$.

268. (*a*) 5 ft; (*b*) 180 ft.

269. Anywhere; $M_{\max} = \dfrac{Px(l-x)}{l}$; $U = \dfrac{Px^2(l-x)^2}{6EIl} = \dfrac{M_{\max}^2\,l}{6EI}$, independent of x.

270. (*a*) $\dfrac{U_{\text{shear}}}{U_{\text{bending}}} = \dfrac{3}{10}\dfrac{E}{G}\left(\dfrac{h}{l}\right)^2$; (*b*) 0.001.

271. $\frac{1}{9}$ or 11.1 per cent. **272.** $\frac{1}{6}$ or 16.7 per cent.

273. $\frac{1}{8}$ or 12.5 per cent. **274.** Myosotis!

275. $y = \dfrac{wl^4}{8EI}\left(1 - \dfrac{4}{3}\dfrac{x}{l} + \dfrac{1}{3}\dfrac{x^4}{l^4}\right)$.

276. $\delta = \dfrac{w_{\max}\,l^4}{120EI} = 0.015$ in.

277. $\delta_{\text{hor }A} = \dfrac{5pa^4}{8EI}$; $\delta_{\text{vert }B} = \dfrac{pa^4}{4EI}$.

278. $\delta = P\left(\dfrac{l^3}{48EI} + \dfrac{1}{2k}\right)$.

279. $\delta = \dfrac{Pa^2b^2}{3EIl} + \dfrac{P}{l^2}\left(\dfrac{a^2}{k_2} + \dfrac{b^2}{k_1}\right)$, where $b = l - a$

280. $\delta = \frac{P(a^3 + b^3)}{3EI} + \frac{Pab^2}{GI_p}.$

281. (*b*) $b/a = 2.06$, for $E/G = 2\frac{1}{2}$.

283. Vertical $\frac{P}{2}$; horizontal $\frac{3Pa^2}{8b(3a + 2b)}$.

284. Vertical $\pm Pb/a$; horizontal $P/2$.

285. Vertical $\frac{P}{2}$; horizontal $\frac{3Pa^2}{8b(b + 2a)}$; moment $\frac{Pa^2}{8(b + 2a)}$.

286. Vertical $P/2$; horizontal P/π.

287. Vertical $P/2$; horizontal $P\frac{3\pi - 8}{16 - \pi^2} = 0.23P$;

moment $\frac{Pa}{2}\frac{4 + 2\pi - \pi^2}{16 - \pi^2} = 0.034Pa$.

288. $M_{\text{bend}} = Pa/2$; $M_{\text{twist}} = Pa\left(\frac{1}{2} - \frac{1}{\pi}\right)$, both at built-in ends.

289. $M_{\text{bend}} = \frac{\sqrt{3}}{6}Pa$, under the load; $M_{\text{twist}} = \frac{\sqrt{5}}{6}Pa$, between loads.

290. In short length $M_t = \frac{2}{3}M$; in long section $M_t = \frac{1}{3}M$.

291. $R_{\text{center}} = \frac{11}{8}P$; $R_{\text{side}} = \frac{5}{16}P$.

292. $R_{\text{center}} = \frac{11}{8}P - \frac{48EI\delta}{l^3}$; $R_{\text{side}} = \frac{5}{16}P + \frac{24EI\delta}{l^3}$.

293. $R = \frac{5}{16}P - \frac{3EI\delta}{l^3}$.

294. (*a*) $\delta = \frac{M_0}{12EI}(x^2 - \frac{1}{4}lx)$ where x is measured from the left end to the right. This equation is valid in the left half only. (*b*) Angle at center of beam caused by load of value M_0 at x.

295. Measure x from the free end toward the left. Then in the right half of the beam: $y = \frac{Pl^2}{48EI}(5l - 6x)$, for $x < l/2$. In the left half of the beam: $y = \frac{P}{12EI}(2x^3 - 3lx^2 + l^3)$, for $x > l/2$. The second meaning of y is the deflection in the mid-point of the beam, loaded with a force P at point x.

296. Measure x from the built-in end toward the free end. Then $y = M_0x^2/2EI$. The quantity y can also be interpreted as the angle at the free end when the beam is loaded by a vertical force M_0 at point x.

297. (*a*) $\varphi = Ml^2/12EI$; (*b*) same as (*a*) necessarily.

298. $y = \frac{M_0x^2}{4EI}\left(\frac{x}{l} - 1\right)$. This is also the angle at the right end, caused by a vertical force M_0 at point x. Count x from the built-in end towards the right.

299. $\delta = Pa^3/EI$ to the right. This is also the downward deflection of the top of the arch, caused by a force P at the right roller support, pulling horizontally to the right.

300. $\delta = \dfrac{\pi Q a^3}{4EI}$.

301. (*a*) The long top stringer bar only; (*b*) the three top stringers and the four vertical struts.

302. Any one of the upper four diagonals. $d = 1.76$ in. without factor of safety by Euler's formula; $d = 2.65$ in. with factor of safety 5 on Euler's formula.

303. (*a*) $P_{crit} = 1{,}830$ lb; $s = 229$ lb/sq in. (*b*) $P_{crit} = 457$ lb; $s = 57$ lb/sq in.

304. $s = 8{,}850$ lb/sq in.

305. 10 ft; more precisely 119 in. **306.** 122 in.

307. (*a*) 18,600 lb; (*b*) 74,400 lb; (*c*) 18,600 lb.

309. $P_{crit} = 25{,}800$ lb. **310.** $P_{crit} = 4{,}700$ lb.

311. $EI/kl = 1/12$; $k = 12EI/l$. **312.** $P_{crit} = 13.5EI/a^2$.

313. (*a*) $P_{crit} = 17.5EI/a^2$; (*b*) $P_{crit} = 19.0EI/a^2$.

314. (*a*) 0.018 in.; (*b*) 0.326 in.

315. 128,600 lb; 102,200 lb; 76,200 lb.

316. 128,600 lb; 102,000 lb; 70,800 lb.

317. 144,000 lb; 119,000 lb; 92,500 lb.

318. 145,000 lb; 131,000 lb; 100,000 lb.

319. 140,000 lb; 126,000 lb; 102,000 lb.

320. 157,500 lb; 136,500 lb; 101,500 lb.

321. (*a*) 1528 lb; (*b*) 382 lb. **322.** 0.045 in.

323. (*a*) $s = 1{,}000 - 951\left(\dfrac{l}{100k}\right)^4$; (*b*) 89 in.; (*c*) 10,700 lb.

324. $8\frac{3}{4} \times 8\frac{3}{4}$ in. **325.** 0.0112 in. **326.** 0.029 in.

327. 0.0245 in. **328.** $d = 1.22$ in.

329. $s_t = +14{,}500$ lb/sq in.; $s_r = -1{,}650$ lb/sq in.

330. $s_t = +16{,}500$ lb/sq in.; $s_r = +4{,}950$ lb/sq in.

331. (*a*) $-12{,}500$ lb/sq in.; (*b*) $-7{,}100$ lb/sq in.

332. 2.93×10^6 in.-lb. **333.** 256,000 lb.

334. $s_1 = 34{,}800$; $s_2 = 16{,}500$. The maximum shear stress is 17,400 lb/sq in. (Remember Fig. 66, page 72.)

335. $s_1 = +12{,}400$; $s_2 = -3{,}700$.

336. 17,800 lb/sq in.

337. (*a*) $d/D = s_{yield}/E$; (*b*) $D = 250$ ft. **338.** 2.30 lb.

339. 0.500; 0.508; 0.516; 0.555; 0.594 in. **340.** 3.37 in.

341. 12,800 lb. **342.** (*a*) 31,800 in.-lb; (*b*) 7,200 in.-lb.

343. $D = 3.14$ in. (good design); $D = 4.20$ in. (bad design); $D_{bore} = 1.93\,D$.

344. (*a*) 0.89 in.; (*b*) 3,600 lb.

345. $T_{alt} = 3.2T_{steady}$. Such a condition can easily occur in Diesel engines, at a "torsional critical speed," and crankshafts have often broken in this manner.

347. Distortion energy, 13.5 in.-lb; hydrostatic energy, 2.1 in.-lb; total energy, 15.6 in.-lb.

348. (*a*) 0.735 in.; (*b*) 0.705 in.

349. 5.8 per cent hydrostatic; 94.2 per cent distortive.

350. 62 per cent distortion energy and 38 per cent hydrostatic energy for $\mu = 0.3$; calculate and plot this against μ for $0.25 < \mu < 0.35$.

LIST OF FORMULAS

1. Hooke's law (page 4):

$$\frac{s}{E} = \frac{\Delta l}{l} = \epsilon \quad \text{or} \quad \Delta l = \frac{Pl}{AE}.$$

2. Elongation of bar of variable section (page 5):

$$\Delta l = \frac{P}{E}\int_0^l \frac{dx}{A}.$$

3. Hooke's law for shear (page 16):

$$s_s = G\gamma.$$

4 and 5. Torsion of round shaft (pages 19, 20):

$$s_s = \frac{M_t r}{I_p}, \quad \varphi = \frac{M_t l}{GI_p}.$$

Torsion of round shaft of several materials (page 112):

$$s_{sf} = G_f \frac{M_t r}{\Sigma GI_p}, \quad \varphi = \frac{M_t l}{\Sigma GI_p}.$$

6. Moment of inertia of solid (page 20) and hollow shaft (page 24):

$$I_p = \frac{\pi R^4}{2} = \frac{\pi D^4}{32}, \; I_p = \frac{\pi}{32}(D_o^4 - D_i^4).$$

7. St. Venant's approximation for torsion of non-circular shafts (page 26):

$$I_{p\,\text{equiv}} = \frac{A^4}{4\pi^2 I_p}.$$

8. Stress in coil spring (page 27):

$$s_s = \frac{2}{\pi}\frac{PR}{r^3} = \frac{8PD}{\pi d^3}.$$

9. Deflection of coil spring (page 28):

$$\delta = P \cdot \frac{8D^3N}{Gd^4}.$$

10. Loading, shear force, and bending moment in a beam (page 34):

$$w = \frac{dS}{dx}, \qquad S = \frac{dM}{dx}.$$

11. Stresses in bending (page 39):

$$s = \frac{My}{I}, \qquad s_{\max} = \frac{M}{Z}.$$

Beams of two materials (pages 109, 110):

$$s_f = E_f \frac{My}{\sum\limits_n E_n I_n}, \qquad Z = \frac{\sum\limits_n E_n I_n}{E_f y_{\max}}.$$

12. Beam of rectangular cross section (page 40):

$$I = \frac{bh^3}{12}, \qquad Z = \frac{bh^2}{6}.$$

13. Beam of circular cross section in bending (page 40):

$$I = \frac{\pi D^4}{64}, \qquad Z = \frac{\pi D^3}{32}.$$

14. Shear stress in bent beam (pages 43 and 110):

$$s_s = \frac{S}{bI}\int_{y_0}^{\frac{h}{2}} y\,dA, \qquad s_s = \frac{S}{b\Sigma E_n I_n}\int_{y_0}^{\text{top}} Ey\,dA.$$

15. Parallel axis theorem in moment of inertia (page 49):

$$I_O = I_G + a_G^2 A.$$

16. Neutral line for eccentric compression (page 55):

$$ae = \frac{I}{A} = k^2.$$

17. Mohr's circle for stresses (page 59):

$$\left.\begin{aligned} s_n &= \frac{s_x + s_y}{2} + \frac{s_x - s_y}{2}\cos 2\alpha. \\ s_s &= \frac{s_x - s_y}{2}\sin 2\alpha, \end{aligned}\right\}$$

Mohr's circle for moments of inertia (page 65):

$$\left.\begin{aligned} I_{x'} &= \frac{I_x + I_y}{2} + \frac{I_x - I_y}{2}\cos 2\alpha, \\ I_{x'y'} &= \frac{I_x - I_y}{2}\cos 2\alpha, \end{aligned}\right\}$$

18. Hooke's law in two dimensions (page 75):

$$\epsilon_1 = \frac{1}{E}(s_1 - \mu s_2),$$

$$\epsilon_2 = \frac{1}{E}(s_2 - \mu s_1).$$

Hooke's law in three dimensions (page 75):

$$\left.\begin{aligned} E\epsilon_1 &= s_1 - \mu(s_2 + s_3),\\ E\epsilon_2 &= s_2 - \mu(s_1 + s_3),\\ E\epsilon_3 &= s_3 - \mu(s_1 + s_2).\end{aligned}\right\}$$

19. The differential equation of flexure (page 80):

$$M = EI\frac{d^2y}{dx^2}.$$

Flexure of a beam of several materials (page 110):

$$M = \frac{d^2y}{dx^2}\sum_n E_n I_n.$$

20. The Myosotis formulas for cantilevers (page 85):

Angle	Deflection
$\frac{Ml}{EI}$	$\frac{Ml^2}{2EI}$
$\frac{Pl^2}{2EI}$	$\frac{Pl^3}{3EI}$
$\frac{wl^3}{6EI}$	$\frac{wl^4}{8EI}$

21. General beam equation (page 101):

$$w = \frac{d^2}{dx^2}\left(EI\frac{d^2y}{dx^2}\right).$$

22. Shear deflection of a beam (page 108):

$$\delta_{\text{shear}} = \frac{1}{G}\int_0^l s_s\,dx.$$

23. Stresses in a thin-walled cylinder under pressure (page 137):

$$s_t = \frac{pr}{t}, \qquad s_l = \frac{pr}{2t}.$$

24. Shrink fit of cylinder around solid shaft (page 145):

$$\frac{\delta}{r_i} = \frac{2p}{E} \cdot \frac{r_o^2}{r_o^2 - r_i^2}.$$

25. Maximum shear stress in shrink fit (page 145)

$$(s_s)_{\max} = \frac{E}{2(\delta/r_i)}.$$

26. End deflections of curved cantilever (page 148):

$$\varphi = \int_s \frac{M}{EI}\,ds, \qquad \delta_x = \int_s \frac{My}{EI}\,ds, \qquad \delta_y = -\int_s \frac{Mx}{EI}\,ds.$$

27, 28, 29. Thick curved bar (pages 156, 157):

$$\int_A \frac{y\,dA}{R - n + y} = 0, \qquad M = E\frac{d\varphi}{d\theta}nA, \qquad s_y = \frac{M}{nA}\frac{y}{R - n + y}.$$

30*a*, *b*. Energy stored in an element of volume (pages 162,163):

$$U = \frac{s^2}{2E}\cdot \text{volume}, \qquad U = \frac{s_s^2}{2G}\cdot \text{volume}.$$

30*c*, *d*, *e*. Energy stored in a beam (pages 163, 164):

$$U = \int \frac{P^2}{2EA}\,dl, \qquad U = \int \frac{M^2}{2EI}\,dl, \qquad U = \int \frac{M_t^2}{2GI_p}\,dl.$$

31. Energy in an element with two-dimensional stress (page 165):

$$dU = (s_1^2 + s_2^2 - 2\mu s_1 s_2)\frac{d\text{ vol}}{2E}.$$

32. Relation between the three elastic constants (page 32):

$$E = 2G(1 + \mu).$$

33. Castigliano's theorem (pages 169, 170):

$$\frac{\partial U}{\partial P_n} = \delta_n.$$

34. Maxwell's theorem of reciprocity (page 181):

$$\alpha_{12} = \alpha_{21}.$$

35. Buckling of column hinged at both ends (pages 186 and 195):

$$P_{\text{crit}} = \frac{\pi^2 EI}{l^2}, \qquad s = \frac{P_{\text{crit}}}{A} = \pi^2 E\left(\frac{k}{l}\right)^2.$$

35*a*. Buckling of column built-in at both ends (page 190):

$$P_{\text{crit}} = \frac{4\pi^2 EI}{l^2}.$$

36. The secant formula: buckling of eccentrically loaded beam (pages 193 and 194):

$$s_{\max} = \frac{P}{A}\left(1 + \frac{eh/2}{k^2} \sec \sqrt{\frac{Pl^2}{4EI}}\right).$$

37. Tetmajer's formula for buckling of short columns (page 196):

$$s = \frac{P}{A} = 48{,}000 - 210\,\frac{l}{k} \qquad \text{lb/sq in.}$$

38. Formulae to find the strains from rosette gage readings (page 213):

$$c = \frac{\epsilon_a + \epsilon_b + \epsilon_c}{3}, \qquad \tan 2\alpha = \sqrt{3}\,\frac{\epsilon_c - \epsilon_b}{2\epsilon_a - \epsilon_b - \epsilon_c}.$$

39. Soderberg's safety criterion for steady and alternating stress combined (page 220):

$$\frac{s_{\text{steady}}}{\text{Yield Stress}} + \frac{s_{\text{alt}}}{\text{Fatigue Limit}} = \frac{1}{\text{Factor of Safety}}.$$

40. Distortion energy in a volume element (page 224):

$$dU_{\text{distortion}} = \frac{1+\mu}{3E}\,(s_1^2 + s_2^2 + s_3^2 - s_1 s_2 - s_1 s_3 - s_2 s_3)\,d\text{ vol.}$$

INDEX

T

V

W

Y

Z

A CATALOG OF SELECTED

DOVER BOOKS

IN SCIENCE AND MATHEMATICS

A CATALOG OF SELECTED

DOVER BOOKS

IN SCIENCE AND MATHEMATICS

RELATIVITY, THERMODYNAMICS AND COSMOLOGY, Richard C. Tolman. Landmark study extends thermodynamics to special, general relativity; also applications of relativistic mechanics, thermodynamics to cosmological models. 501pp. 5⅜ × 8½. 65383-8 Pa. $12.95

APPLIED ANALYSIS, Cornelius Lanczos. Classic work on analysis and design of finite processes for approximating solution of analytical problems. Algebraic equations, matrices, harmonic analysis, quadrature methods, much more. 559pp. 5⅜ × 8½. 65656-X Pa. $13.95

SPECIAL RELATIVITY FOR PHYSICISTS, G. Stephenson and C.W. Kilmister. Concise elegant account for nonspecialists. Lorentz transformation, optical and dynamical applications, more. Bibliography. 108pp. 5⅜ × 8½. 65519-9 Pa. $4.95

INTRODUCTION TO ANALYSIS, Maxwell Rosenlicht. Unusually clear, accessible coverage of set theory, real number system, metric spaces, continuous functions, Riemann integration, multiple integrals, more. Wide range of problems. Undergraduate level. Bibliography. 254pp. 5⅜ × 8½. 65038-3 Pa. $7.95

INTRODUCTION TO QUANTUM MECHANICS With Applications to Chemistry, Linus Pauling & E. Bright Wilson, Jr. Classic undergraduate text by Nobel Prize winner applies quantum mechanics to chemical and physical problems. Numerous tables and figures enhance the text. Chapter bibliographies. Appendices. Index. 468pp. 5⅜ × 8½. 64871-0 Pa. $11.95

ASYMPTOTIC EXPANSIONS OF INTEGRALS, Norman Bleistein & Richard A. Handelsman. Best introduction to important field with applications in a variety of scientific disciplines. New preface. Problems. Diagrams. Tables. Bibliography. Index. 448pp. 5⅜ × 8½. 65082-0 Pa. $12.95

MATHEMATICS APPLIED TO CONTINUUM MECHANICS, Lee A. Segel. Analyzes models of fluid flow and solid deformation. For upper-level math, science and engineering students. 608pp. 5⅜ × 8½. 65369-2 Pa. $13.95

ELEMENTS OF REAL ANALYSIS, David A. Sprecher. Classic text covers fundamental concepts, real number system, point sets, functions of a real variable, Fourier series, much more. Over 500 exercises. 352pp. 5⅜ × 8½. 65385-4 Pa. $10.95

PHYSICAL PRINCIPLES OF THE QUANTUM THEORY, Werner Heisenberg. Nobel Laureate discusses quantum theory, uncertainty, wave mechanics, work of Dirac, Schroedinger, Compton, Wilson, Einstein, etc. 184pp. 5⅜ × 8½. 60113-7 Pa. $5.95

INTRODUCTORY REAL ANALYSIS, A.N. Kolmogorov, S.V. Fomin. Translated by Richard A. Silverman. Self-contained, evenly paced introduction to real and functional analysis. Some 350 problems. 403pp. 5⅜ × 8½. 61226-0 Pa. $9.95

PROBLEMS AND SOLUTIONS IN QUANTUM CHEMISTRY AND PHYSICS, Charles S. Johnson, Jr. and Lee G. Pedersen. Unusually varied problems, detailed solutions in coverage of quantum mechanics, wave mechanics, angular momentum, molecular spectroscopy, scattering theory, more. 280 problems plus 139 supplementary exercises. 430pp. 6½ × 9¼. 65236-X Pa. $12.95

NUMERICAL METHODS FOR SCIENTISTS AND ENGINEERS, Richard Hamming. Classic text stresses frequency approach in coverage of algorithms, polynomial approximation, Fourier approximation, exponential approximation, other topics. Revised and enlarged 2nd edition. 721pp. 5⅜ × 8½.
65241-6 Pa. $14.95

THEORETICAL SOLID STATE PHYSICS, Vol. I: Perfect Lattices in Equilibrium; Vol. II: Non-Equilibrium and Disorder, William Jones and Norman H. March. Monumental reference work covers fundamental theory of equilibrium properties of perfect crystalline solids, non-equilibrium properties, defects and disordered systems. Appendices. Problems. Preface. Diagrams. Index. Bibliography. Total of 1,301pp. 5⅜ × 8½. Two volumes. Vol. I 65015-4 Pa. $14.95
Vol. II 65016-2 Pa. $14.95

OPTIMIZATION THEORY WITH APPLICATIONS, Donald A. Pierre. Broad-spectrum approach to important topic. Classical theory of minima and maxima, calculus of variations, simplex technique and linear programming, more. Many problems, examples. 640pp. 5⅜ × 8½. 65205-X Pa. $14.95

THE CONTINUUM: A Critical Examination of the Foundation of Analysis, Hermann Weyl. Classic of 20th-century foundational research deals with the conceptual problem posed by the continuum. 156pp. 5⅜ × 8½. 67982-9 Pa. $5.95

ESSAYS ON THE THEORY OF NUMBERS, Richard Dedekind. Two classic essays by great German mathematician: on the theory of irrational numbers; and on transfinite numbers and properties of natural numbers. 115pp. 5⅜ × 8½.
21010-3 Pa. $4.95

THE FUNCTIONS OF MATHEMATICAL PHYSICS, Harry Hochstadt. Comprehensive treatment of orthogonal polynomials, hypergeometric functions, Hill's equation, much more. Bibliography. Index. 322pp. 5⅜ × 8½. 65214-9 Pa. $9.95

NUMBER THEORY AND ITS HISTORY, Oystein Ore. Unusually clear, accessible introduction covers counting, properties of numbers, prime numbers, much more. Bibliography. 380pp. 5⅜ × 8½. 65620-9 Pa. $9.95

THE VARIATIONAL PRINCIPLES OF MECHANICS, Cornelius Lanczos. Graduate level coverage of calculus of variations, equations of motion, relativistic mechanics, more. First inexpensive paperbound edition of classic treatise. Index. Bibliography. 418pp. 5⅜ × 8½. 65067-7 Pa. $11.95

MATHEMATICAL TABLES AND FORMULAS, Robert D. Carmichael and Edwin R. Smith. Logarithms, sines, tangents, trig functions, powers, roots, reciprocals, exponential and hyperbolic functions, formulas and theorems. 269pp. 5⅜ × 8½. 60111-0 Pa. $6.95

THEORETICAL PHYSICS, Georg Joos, with Ira M. Freeman. Classic overview covers essential math, mechanics, electromagnetic theory, thermodynamics, quantum mechanics, nuclear physics, other topics. First paperback edition. xxiii + 885pp. 5⅜ × 8½. 65227-0 Pa. $19.95

DE RE METALLICA, Georgius Agricola. The famous Hoover translation of greatest treatise on technological chemistry, engineering, geology, mining of early modern times (1556). All 289 original woodcuts. 638pp. 6¾ × 11.
60006-8 Pa. $18.95

SOME THEORY OF SAMPLING, William Edwards Deming. Analysis of the problems, theory and design of sampling techniques for social scientists, industrial managers and others who find statistics increasingly important in their work. 61 tables. 90 figures. xvii + 602pp. 5⅜ × 8½. 64684-X Pa. $15.95

THE VARIOUS AND INGENIOUS MACHINES OF AGOSTINO RAMELLI: A Classic Sixteenth-Century Illustrated Treatise on Technology, Agostino Ramelli. One of the most widely known and copied works on machinery in the 16th century. 194 detailed plates of water pumps, grain mills, cranes, more. 608pp. 9 × 12.
28180-9 Pa. $24.95

LINEAR PROGRAMMING AND ECONOMIC ANALYSIS, Robert Dorfman, Paul A. Samuelson and Robert M. Solow. First comprehensive treatment of linear programming in standard economic analysis. Game theory, modern welfare economics, Leontief input-output, more. 525pp. 5⅜ × 8½. 65491-5 Pa. $14.95

ELEMENTARY DECISION THEORY, Herman Chernoff and Lincoln E. Moses. Clear introduction to statistics and statistical theory covers data processing, probability and random variables, testing hypotheses, much more. Exercises. 364pp. 5⅜ × 8½. 65218-1 Pa. $9.95

THE COMPLEAT STRATEGYST: Being a Primer on the Theory of Games of Strategy, J.D. Williams. Highly entertaining classic describes, with many illustrated examples, how to select best strategies in conflict situations. Prefaces. Appendices. 268pp. 5⅜ × 8½. 25101-2 Pa. $7.95

MATHEMATICAL METHODS OF OPERATIONS RESEARCH, Thomas L. Saaty. Classic graduate-level text covers historical background, classical methods of forming models, optimization, game theory, probability, queueing theory, much more. Exercises. Bibliography. 448pp. 5⅜ × 8¼. 65703-5 Pa. $12.95

CONSTRUCTIONS AND COMBINATORIAL PROBLEMS IN DESIGN OF EXPERIMENTS, Damaraju Raghavarao. In-depth reference work examines orthogonal Latin squares, incomplete block designs, tactical configuration, partial geometry, much more. Abundant explanations, examples. 416pp. 5⅜ × 8¼.
65685-3 Pa. $10.95

THE ABSOLUTE DIFFERENTIAL CALCULUS (CALCULUS OF TENSORS), Tullio Levi-Civita. Great 20th-century mathematician's classic work on material necessary for mathematical grasp of theory of relativity. 452pp. 5⅜ × 8½.
63401-9 Pa. $9.95

VECTOR AND TENSOR ANALYSIS WITH APPLICATIONS, A.I. Borisenko and I.E. Tarapov. Concise introduction. Worked-out problems, solutions, exercises. 257pp. 5⅜ × 8¼. 63833-2 Pa. $7.95

THE FOUR-COLOR PROBLEM: Assaults and Conquest, Thomas L. Saaty and Paul G. Kainen. Engrossing, comprehensive account of the century-old combinatorial topological problem, its history and solution. Bibliographies. Index. 110 figures. 228pp. 5⅜ × 8½. 65092-8 Pa. $6.95

CATALYSIS IN CHEMISTRY AND ENZYMOLOGY, William P. Jencks. Exceptionally clear coverage of mechanisms for catalysis, forces in aqueous solution, carbonyl- and acyl-group reactions, practical kinetics, more. 864pp. 5⅜ × 8½. 65460-5 Pa. $19.95

PROBABILITY: An Introduction, Samuel Goldberg. Excellent basic text covers set theory, probability theory for finite sample spaces, binomial theorem, much more. 360 problems. Bibliographies. 322pp. 5⅜ × 8½. 65252-1 Pa. $8.95

LIGHTNING, Martin A. Uman. Revised, updated edition of classic work on the physics of lightning. Phenomena, terminology, measurement, photography, spectroscopy, thunder, more. Reviews recent research. Bibliography. Indices. 320pp. 5⅜ × 8¼. 64575-4 Pa. $8.95

PROBABILITY THEORY: A Concise Course, Y.A. Rozanov. Highly readable, self-contained introduction covers combination of events, dependent events, Bernoulli trials, etc. Translation by Richard Silverman. 148pp. 5⅜ × 8¼. 63544-9 Pa. $5.95

AN INTRODUCTION TO HAMILTONIAN OPTICS, H. A. Buchdahl. Detailed account of the Hamiltonian treatment of aberration theory in geometrical optics. Many classes of optical systems defined in terms of the symmetries they possess. Problems with detailed solutions. 1970 edition. xv + 360pp. 5⅜ × 8½. 67597-1 Pa. $10.95

STATISTICS MANUAL, Edwin L. Crow, et al. Comprehensive, practical collection of classical and modern methods prepared by U.S. Naval Ordnance Test Station. Stress on use. Basics of statistics assumed. 288pp. 5⅜ × 8½. 60599-X Pa. $6.95

DICTIONARY/OUTLINE OF BASIC STATISTICS, John E. Freund and Frank J. Williams. A clear concise dictionary of over 1,000 statistical terms and an outline of statistical formulas covering probability, nonparametric tests, much more. 208pp. 5⅜ × 8½. 66796-0 Pa. $6.95

STATISTICAL METHOD FROM THE VIEWPOINT OF QUALITY CONTROL, Walter A. Shewhart. Important text explains regulation of variables, uses of statistical control to achieve quality control in industry, agriculture, other areas. 192pp. 5⅜ × 8½. 65232-7 Pa. $7.95

THE INTERPRETATION OF GEOLOGICAL PHASE DIAGRAMS, Ernest G. Ehlers. Clear, concise text emphasizes diagrams of systems under fluid or containing pressure; also coverage of complex binary systems, hydrothermal melting, more. 288pp. 6½ × 9¼. 65389-7 Pa. $10.95

STATISTICAL ADJUSTMENT OF DATA, W. Edwards Deming. Introduction to basic concepts of statistics, curve fitting, least squares solution, conditions without parameter, conditions containing parameters. 26 exercises worked out. 271pp. 5⅜ × 8½. 64685-8 Pa. $8.95